T0321957

COMMUNICATION DISORDERS

APPLIED PSYCHOLINGUISTICS AND COMMUNICATION DISORDERS

APPLIED PSYCHOLINGUISTICS AND MENTAL HEALTH
Edited by R. W. Rieber

PSYCHOLOGY OF LANGUAGE AND THOUGHT
Essays on the Theory and History of Psycholinguistics
Edited by R. W. Rieber

COMMUNICATION DISORDERS
Edited by R. W. Rieber

COMMUNICATION DISORDERS

Edited by
R. W. Rieber
John Jay College, CUNY, and
Columbia University College of Physicians and Surgeons
New York, New York

PLENUM PRESS · NEW YORK AND LONDON

Library of Congress Cataloging in Publication Data

Main entry under title:

Communication disorders.

(Applied psycholinguistics and communication disorders)
Includes index.
1. Communicative disorders. I. Rieber, Robert W. II. Series. [DNLM: 1. Aphasia. 2. Speech disorders. 3. Hearing disorders. 4. Communication. 5. Language disorders. WM475 C734]
RC423.C64 616.85'5 80-18394
ISBN 0-306-40527-X

A Division of Plenum Publishing Corporation
227 West 17th Street, New York, N.Y. 10011

Printed in the United States of America

CONTRIBUTORS

Sheldon M. Frank • Departments of Psychiatry and Pediatrics, University of Miami, School of Medicine, Miami, Florida

Gerard L. Kupperman • Department of Communication Disorders, University of Massachusetts, Amherst, Massachusetts

Christy L. Ludlow • Communicative Disorders Program, National Institute of Neurological and Communicative Disorders and Stroke, Bethesda, Maryland

Edward D. Mysak • Department of Speech Pathology and Audiology, Teachers College, Columbia University, New York, New York

A. C. Nichols • Department of Speech Pathology, San Diego State University, San Diego, California

E. Harris Nober • Department of Communication Disorders, University of Massachusetts, Amherst, Massachusetts

Janet M. Purn • The Mount Carmel Guild, 17 Mulberry Street, Newark, New Jersey

Bruce Quarrington • York University, Downsview, Ontario, Canada

R. W. Rieber • Department of Psychology, John Jay College, CUNY, and Columbia University, College of Physicians and Surgeons, New York, New York

Carol N. Wilder • Department of Speech Pathology and Audiology, Teachers College, Columbia University, New York, New York

Harris Wintz • Division of Speech and Hearing Science, University of Missouri-Kansas City, Kansas City, Missouri

Annette R. Zaner • The Mount Carmel Guild, 17 Mulberry Street, Newark, New Jersey

CONTENTS

PART I
PERIPHERAL DISTURBANCES

CHAPTER 1

Articulation—Organic Factors 3

A.C. Nichols

CHAPTER 2

Phonatory and Resonatory Problems of Organic Origin 21

Edward D. Mysak and Carol N. Wilder

CHAPTER 3

Physiogenic Hearing Impairment in Adults 57

E. Harris Nober and Gerard L. Kupperman

CHAPTER 4

Auditory Problems in Children 101

Annette R. Zaner and Janet M. Purn

PART II
CENTRAL DISTURBANCES

CHAPTER 5

Recovery and Rehabilitation of Adult Aphasic Patients: Relevant Research Advances 149

Christy L. Ludlow

CHAPTER 6

Language Development and Language Disorders in Children and Adolescents 179

Sheldon M. Frank and R.W. Rieber

PART III
PSYCHOLOGICAL–INTERPERSONAL DISTURBANCES

CHAPTER 7

Articulation Disorders: Psychological Factors 209

A.C. Nichols

CHAPTER 8

Considerations in the Treatment of Articulation Disorders 225

Harris Winitz

CHAPTER 9

Phonatory and Resonatory Problems: Functional Voice Disorders 259

R.W. Rieber

CHAPTER 10

Pseudohypacusic Disturbance in Adults 275

E. Harris Nober

CHAPTER 11

Stuttering 299

Bruce Quarrington

AUTHOR INDEX 347

SUBJECT INDEX 353

PART I

PERIPHERAL DISTURBANCES

1

A.C. Nichols

ARTICULATION—ORGANIC FACTORS

Historically, the most common explanations by laymen and physicians for articulation disorders have involved organic factors. In the past fifty to one hundred years, a change in professional opinion has led authorities to assert that articulation disorders are psychologically based, learned phenomena (Milisen, 1954; Winitz, 1977). While it is currently acknowledged that articulation defects may have both organic and functional origins, few experimental studies involve appropriate controls for both factors. It is rare, for example, to find the tactile sensitivity of the tongue controlled in an articulation-learning study, or stimulatability controlled in a study of the articulation of the apraxic. Hence, a philosophical bias or experimental artifact based upon the investigator's theoretical position has led to a curiously dichotomized clinical approach to articulation problems.

Clearly, however, *anatomical*, *neurological*, and *audiological* anomalies may place limitations on the articulatory facility of the speaker. The present chapter deals with a review of these three factors. An effort will be made to demonstrate the descriptive (diagnostic) and predictive (therapeutic) powers of procedures based upon experimental study of the organic components involved in articulatory production.

Anatomical Factors

Lungs and Larynx. The lungs and larynx provide the physical power source for articulated speech. There is no useful study of the effects of such problems as

A.C. Nichols • Department of Speech Pathology, San Diego State University, San Diego, California 92182.

emphysema or the breathy or hoarse voice caused by vocal fold lesions upon articulation to be found in the literature. Authorities assume that intelligibility may suffer because of vocal weakness, but there are no guidelines to tell the clinician how much, and there is no empirical evidence of which phonemes may suffer. Rees (1958), however, demonstrated that the height of the vowel influences vocal harshness, that high vowels are less harsh than low vowels, and that "vowels in voiceless consonant and stop plosive environments are less harsh than vowels in voiced and fricative environments" (p. 167). The implications of these findings for intelligibility have not been studied.

The function of the laryngeal mechanism in articulation may best be seen in its absence, that is, in esophageal speech. Nichols has shown that initial consonants and clusters (1976a), vowels (1976b), and terminal consonants and clusters (1977) produced by esophageal speakers are less intelligible than those produced by normal speakers. Loss of vocal power and the noise of esophageal vocal production are important factors in accounting for the diminution of communicative effectiveness (see Nichols, 1968; Shipp, 1967). Of importance too are the less obvious factors of laryngeal–articulatory coordination. Esophageal speakers' voiceless consonants may be perceived as voiced (Christensen, Weinberg, & Alfonso, 1978; Nichols, 1976a, 1977), while the *tense* initial consonants and the *tense* vowels may be perceived as *lax* by listeners (Nichols, 1976a,b). The tenseness–laxness feature has durational implication according to Jacobson and Halle (1969). It may be inferred that durational coordination of the larynx and articulators is critical to competent articulation.

Nichols (1976a) also noted that the esophageal speaker's back consonants were often received as front consonants by listeners, while the high vowels tended to be more intelligible than the low vowels. It could be hypothesized that the concurrent involvement of lingual and pharyngeal muscles in producing the esophageal voice may interfere with the articulatory movements of the tongue. Study of this hypothesis may lead to improved therapeutic approaches.

Intelligibility practice based upon multiple-choice items derived from Nichols's studies has been shown to be effective in improving the understandability of esophageal speakers (Fitzpatrick, 1978; Nichols, 1976a,b, 1977). The demonstrated predictive (therapeutic) powers of the intelligibility training procedure should be encouraging for clients, clinicians, and researchers.

Tongue. That tongue size or strength, or interference with the freedom of movement of the tongue may be implicated in articulatory defects has often been suggested. In an early study, Fairbanks and Bebout (1950) reported that there were no tongue size, strength, or movement differences between articulatory-defective and normal-speaking adults. Dworkin (1978), however, developed equipment that demonstrated lingual strength differences between normal articulators and children with lisps. The question of intelligibility differences between such groups also remains open since this factor has not been controlled or measured in any study. Speakers who have had part or all of the tongue removed are severely handicapped in intelligibility even after therapeutic training (Skelly, Donaldson, & Fust, 1973).

The flexibility of adjustment of the tongue, its ability to compensate for its own size, strength, and symmetry problems, and for the deviations of the other articulatory structures, has been emphasized by authorities. Other articulators may also compensate, in part, for lingual defects. Glossectomy, complete or partial surgical removal of the tongue, is becoming a fairly common operation in some populations. Therapy for restoring spoken communication for the patients can be successful. An example of compensatory movement in this group is an anterior thrust of the jaw that produces clearer vowel structure, according to Skelly *et al.* (1973). They found, however, that there are limits to compensatory adjustments, e.g., none of the glossectomies in the study was able to produce the word "though" effectively; their problems of intelligibility have been noted above.

Several issues of tongue motility and tactile perception will be discussed in later sections, since they fit better into the content devoted to the neurological factors in articulation problems. It has been suggested by some investigators, notably Carlson (1972; see also, Colletti, Geffner, & Schlanger, 1976), that tongue thrust during swallowing may also involve neurological factors. This inference was based upon the relationship of tongue thrust to oral stereognostic deficits, for the latter have been shown to have strong organic (neurological) implications. For the most part, however, tongue thrust has been treated as a peripheral phenomenon, and a discussion seems best placed in the present section of this chapter.

The activity of the tongue during swallowing has been extensively studied as a factor in articulation disorders. The tongue thrust swallow (TTS) may be involved in articulatory distortions such as the lingual protrusion lisp, and may cause dental misalignments that are associated with articulatory defects, particularly sibilant distortions (Fletcher, Casteel, & Bradley, 1961; Subtelny, Mestre, & Subtelny, 1964; Ward, Malone, Jann, & Jann, 1961). It should be noted, however, that most children with TTS do not have articulation problems. Mason and Proffit (1974) have implicated complications in other anatomical factors as the basis for TTS, particularly those that have to do with the nasal pharyngeal airway, its size, and the pathologies of structures such as the tonsils and adenoids that may block it. If this pathway is obstructed, the child compensates by advancing the base of the tongue to open the airway, thus forcing the body of the tongue forward and the tip against the teeth. These authors point out that the airway will increase in size with growth, and the need for the TTS abate. Many children thus "grow out of" TTS behaviors. The articulation behaviors, however, may remain (Fletcher *et al.*, 1961), and therapy may be needed. It is of interest in this context to note that Stansell (1969) showed that TTS could be reduced by articulation therapy, but articulation could not be cleared up by teaching better swallowing patterns. Mason and Proffit (1974) suggest that speech therapy together with orthodontia may be the best therapeutic approach to this complex of problems.

If TTS is an anatomical problem, rather than a "normal" step in the development of swallowing, diagnosis based upon airway integrity as well as tongue tip observations is indicated. Another variable that has not yet been fitted into the picture is the oral stereognostic (lingual perception of shape) deficits of children with TTS behaviors reported by Carlson (1972) and Colletti *et al.* (1976). The neurological implications of such deficits will be discussed in a later section; at this point it may only be inferred that there may be more to TTS than airway obstruction.

Teeth and Jaws. Much study has been given to the influence of dental factors on articulation. The results are somewhat mixed, but enough evidence has accumulated to show that the teeth and their conformation in the jaws may play significant roles in articulatory behaviors (Fymbo, 1936; Fairbanks & Lintner, 1951; Heighton & Nichols, 1967; Snow,1961). The effects of TTS on dental malformations and the associated articulation problems have been noted in the preceding section. Moreover, it has been shown that the correction of dental and maxillary anomalies may improve articulation (Mason & Proffit, 1974; Rathbone & Snidecor, 1959).

Most authorities, however, emphasize that far more children with dental anomalies achieve good articulation than develop phonological production (e.g., Snow, 1961). In theory, this achievement is accomplished by compensating adjustments of the tongue and jaw. For example, Snidecor and Kaires (1965) pointed out that adults with "open bite" occlusion (often mentioned as the most extreme dental misalignment with the greatest potential for disturbing speech) rarely have poor articulation. It has also been shown, however, that from the early school years (Bankson & Byrne, 1962) to early adolescence (Heighton & Nichols, 1967), that children with missing incisors or "open bite" have poorer articulation than their peers. It may be concluded that the effect of dental anomalies is most commonly to slow or complicate articulation learning.

Complex, multivariable study of articulation problems has shown how subtle the interrelationships of dentition and phonology can be. Arndt, Shelton, Johnson, and Furr (1977) noted that several subgroups of their eight- and nine-year-old subjects with articulation defects were different (statistically) from their peers in x-rays and orthodontic measures of the oral structures. One group (their T3-3), made up primarily of children with /s/ defects, exhibited, among other language and reading deficits, "open bite" and somewhat narrow maxillary intercuspid distance. Another group (their T12-7) had broader than average maxillary intercuspid distance, incisor inclination, angle of the nasion-sella and mandibular planes, and pharyngeal displacement. While fully developed explanations have not yet been made of the roles played by these anatomical differences in the articulation of the groups identified, such explanations appear to be needed in the diagnostic process.

To summarize, the evidence indicates that the identification of dental configuration deviations should be part of the diagnostic approach to the remediation of articulation problems.

Palatal Anomalies. Most children with clefts of the palate, and many in the group with short or weak soft palates, have defective articulation. Some 90% of the preschool group with clefts exhibit articulation problems (Bzoch, 1965; Phillips & Harrison, 1969). Moll (1968), in surveying studies done in the 1950s and 1960s, found that correct articulation among older children with cleft palate ranged from 40% to 80%. The most common defects in such studies were found to be fricatives, with affricates, plosives, glides, and nasals following. A common substitution for plosives is the glottal stop, while various nasal emissions may distort both fricatives and plosives (Moll, 1968; Shelton, Hahn, & Morris, 1968; Van Demark, 1964, 1966).

Van Demark (1964) found that inadequate interoral pressure due to poor palatal closure is the primary factor in accounting for articulation errors; the other factor he found to be of importance was age, which of course reflects developmental functions such as learning and growth. In a further study, Van Demark (1966) related both oral pressure and age to error types and perceived nasality. Many errors were uncorrelated with other defect patterns, e.g., *substitutions–nasal* were a factor by themselves. Other error patterns clustered into complex factors accounting for pressure or developmental variances. Similarly, Hanes (1973) showed that articulation test scores comprise a factor that is independent of aural discrimination and language test scores. Such results suggest that error patterns should be assessed in therapy to identify the origins of a particular child's articulation problems.

The Iowa Pressure Test (IPT), developed by Morris, Spriestersbach, and Darley (1961) to assess the articulation of cleft palate children, reveals yet another phenomenon of diagnostic importance. That is, the majority of their 43 items consist of two-phoneme clusters. The items were particularly difficult for cleft palate children to articulate, probably because they require relatively longer maintenance of palatal closure and involve complex muscular balance in movements that may interfere with maintaining closure.

Relatively small openings between the soft palate (or velum) and the wall of the pharynx may release air pressure through the nose and reduce oral pressures thus affecting the intelligibility of speech. In one study, one-millimeter gaps (identified in x-rays) were found to produce errors (Shelton, Brooks, & Youngstrom, 1965; see also Subtelny, Koepp-Baker, & Subtelny, 1961). Correlations on the order of 0.50 to 0.60 have been reported between articulation test scores and velopharyngeal gaps (Brandt & Morris, 1965; Shelton *et al.*, 1965). Thus, it may be estimated (by r^2) that some 25% to 36% of the variances in articulation measures are attributable to the effects of the failure to close this gap.

If the velopharyngeal gap is minor, articulation can be improved with therapy (Chisum, Shelton, Arndt, & Elbert, 1969; Van Demark, 1971), but therapy will not reduce the size of the gap (Shelton, Chisum, Youngstrom, Arndt, & Elbert, 1969). Van Demark (1971) showed that potential for improvement could be demonstrated both before and after therapy by "stimulation testing" (stimulatability measures). When the gap or associated dental anomalies are large, however, dental, prosthetic, or surgical care should precede therapy. Little improvement can be expected without basic anatomical competence. Without interoral pressures adequate to the task of articulating, speech therapy may even produce maladaptive articulatory behaviors and undesirable psychological reactions. With successful anatomical management, articulation may improve rapidly. Speech therapy is often needed, however, for relearning and the development of self-concept are important variables in articulatory performance. Moreover, when anatomical competence cannot be improved, compensatory movements and postures, and light, quick, and appropriate contacts may be taught to optimize production. (For a more detailed discussion of these therapeutic considerations see Shelton *et al.*, 1968.)

Neurological Factors

The articulatory errors due to neurological dysfunction may be classified into two types: the dysarthrias and the apraxias. General theory attributes the dysarthrias to defects in the descending motor control pathways leading from the cerebral cortex to the articulators. The apraxias are attributed to lesions in the cerebral cortex that produce cognitive confusions in the speaker as to the phonetic character of the speech sounds.

Dysarthria

Congenital. The most common of the dysarthrias are congenital, the cerebral palsies. The spastic and athetoid types are by far the most frequently encountered among cerebral-palsied speakers, though other types such as the ataxic and rigidity syndromes occur. The spastic is characterized by slow, stiff movement—his reflexes are resisting his movements. The athetoid exhibits writhing, overextended movements. His reflexes do not control his movements. Ataxics are lacking in fine coordinative control. Studies of the primitive reflexive behaviors shown by older cerebral-palsied children show that many infantile feeding patterns may be seen in the behaviors of both spastics and athetoids. The mouth-opening jaw thrust was the most common behavior observed by Sheppard (1964). She found that speech proficiency is greater among children with fewer uncontrolled reflexive behaviors. A group with a mean of 2.1 such behaviors (out of 9 tested) were without speech problems, while a group with a mean of 5.7 had no speech. Children with a mean of 3.7 behaviors were severely to moderately impaired in intelligibility and had a fair prognosis in speech therapy. Those with a mean of 3.3 infantile reflexive behaviors had mild disorders and a good to excellent prognosis for therapy.

In the most extensive series of studies done on the articulation of speakers with cerebral palsy, Irwin (1955, 1956, 1961a,b) also found that overall severity of involvement has a powerful influence upon articulation scores. Differences between persons with different types of cerebral palsies may be observed, but statistical tests may have mixed results because of the individual differences in these populations. Byrne (1959), for example, found spastics to be superior to athetoids in vowel, diphthong, consonant, and blend articulation, but used a statistic that could not show these differences to be significant. Irwin (1961b) confirmed his previous findings of a superiority of spastics over athetoids in consonant and vowel articulation in using his integrated test, and by employing a more powerful statistic.

Irwin's (1961a,b) integrated test of 87 vowel and consonant items was the culmination of several years of research. It was demonstrated to have a between-alternate-forms reliability of $r = 0.98$ and part test reliabilities of somewhat lower magnitudes. The validity was confirmed by "extreme groups" methods with the general motor involvement classifications of "mild" and "severe" shown to be highly significantly different, as were groups with overall speech ratings of "very good" and "very poor." The "good" and "poor" speech groups also were very different in terms of their articulation scores.

In Irwin's research, as in the studies of several other investigators (e.g., Byrne, 1959), the initial consonants were best articulated, the medials next in order, and the final consonants a distant third in articulatory proficiency. He also found that articulatory proficiency increases with age, a finding encountered in other articulation research, which, as has been noted, reflects learning and growth factors (Irwin, 1955). Byrne (1959), studying preschool children (ages 3–5), however, did not find overall articulatory development with age. She found similarities in developmental pattern to normal children, however, e.g., bilabial consonants developed first, the tongue-tip and back-of-tongue sounds followed, and complex-tip and tongue-tip sounds were last. Using x-ray techniques, Kent and Netsell (1978) found athetoid speakers to have abnormally large ranges of jaw movements and very restricted tongue movements (particularly in the front–back dimension), instable soft palate closure, prolonged transitions, and retrusion of the lower lip. They attributed this theoretically to a motor-learning problem, as well as to the motor-coordination problem itself. Specifically, they related these problems to poor propriaception. It may be concluded that development can be long delayed in this population, but some resemblances to order of acquisition in normals may be observed.

Stimulatability is not commonly found among the cerebral palsied, but Flores and Irwin (1956) did find that their 96 children responded more correctly to spoken than to picture stimuli with medial consonant stimuli. This effect may warrant further study, for clinical experience with other types of disorders has shown that stimulatability may provide the basis for therapeutic approaches.

A study of the articulation of the cerebral-palsied speaker by Burgi and Matthews (1958) gives one of the clearest demonstrations of the relationship between articulation test measures and word and sentence intelligibility. The correlations between articulation tests were very high (r = 0.97–0.99), just as were those reported by Irwin (1961). The correlations between (PB) word intelligibility and articulation tests were nearly as high (r = 0.88–0.91), as were the correlations between sentence intelligibility and articulation measures (r = 0.87–0.89). Measures based upon the articulation of the medial phonemes in their articulation tests proved as valid (words: $r = 0.90$; sentences: $r = 0.90$). It may be inferred that some 81% of the variance in recognizing the words and understanding the sentences of cerebral-palsied speakers involves the factors measured by articulation tests.

Acquired. The acquired dysarthrias have been studied by Darley, Aronson, and Brown (1975). Their studies have validated the hypothesis that there are many dysarthrias. Moreover, they demonstrated that the dysarthric's speech pattern—a pattern due to his dysarthric type—is an indicator of the site of his brain damage. These authors have identified five types of dysarthria: *flaccid, spastic, ataxic, hypokinetic*, and *hyperkinetic*. A mix of flaccid and spastic dysarthria occurs in cases of amyotrophic lateral sclerosis.

Occasionally they found that a particular symptom will be specific to only one type of dysarthria, but more commonly a symptom characterizes several types. Of 38 variables (symptoms) studied by Darley, Aronson, and Brown (1969a,b, 1975), four relate to the topic of the present chapter: *imprecise consonants, irregular articulatory breakdown, vowels distorted,* and *intelligibility (overall)*. All dysarthrias, and almost all dysarthrics, have imprecise consonant articulation production. In spastic dysarthria (pseudobulbar palsy), ataxic dysarthria, and hyperkinetic dysarthria (dystonia and chorea), the articulation is at times adequate, but at times breaks down obviously. Vowel distortions are characteristic of all the dysarthrias, except the hypokinetic (Parkinson's disease). Though occasional dysarthrics have no intelligibilty problems, at least 80% of each of the Darley *et al.* (1975) groups were judged to have some problem in making themselves understood.

The identification of the interrelationships between the voice, timing, and articulation characteristics, their "patterns of dysarthria," is the most fruitful outcome of the Darley *et al.* (1975) studies. They showed, for example, that *imprecise consonants* are commonly associated with *hypernasality, nasal emission,* and *short phrases* in a cluster of dysarthric dysfunctions. They related this pattern to a marked reduction in force of muscular contraction in the closure of the soft palate—a problem similar to that seen in cleft palate with a similar effect on articulation. A cluster may be exhibited by more than one dysarthria. The pattern of clusters seems to explain the speech symptoms of a particular dysarthric type.

In therapy, the intercorrelations between dimensions and clusters may suggest approaches to improved production. In *chorea*, for example, there appear to be correlations between the prosodic variables of *monopitch, monoloudness,* and *reduced stress,* and the presence of *imprecise consonants*. It may thus be hypothesized that therapy targets involving improved prosodic variability may have the effect of stimulating more adequate articulation. Such approaches have been found to be effective in clinical settings (at San Diego State University), suggesting that the Darley *et al.* (1975) intercorrelations have predictive (therapeutic) power as well as descriptive (diagnostic) power.

Childhood Apraxia

Children with severe articulation problems may have low scores on neuropsychological tests as well (Frisch & Handler, 1974). Yoss and Darley (1974) noted that such children may display articulation errors that are different from their peers who do not show deficits on neurological tests. For example, the errors of children with severe neurological deficits may differ more, e.g., by two or three features, from the correct phoneme than do the errors of children who have less pronounced deficit, and who tend to make errors that may differ in only one feature from the correct sound. They concluded that there is "substantial support [for] the use of the term 'developmental apraxia of speech' as descriptive of their articulatory problem" (Yoss & Darley, 1974, p. 412).

There has been substantial evidence that sensory-motor deficits may underlie some articulation problems for many years. Dickson (1962), for example, found that tests of gross motor abilities would allow the tester to predict which children would "outgrow" articulation problems and which would not, indicating that a generalized motor deficit may delay articulatory development. Ringel, House, Burk, Dolinsky, and Scott (1970) demonstrated that children with oral form discrimination deficits have more articulation problems than those who do not, which has led to the conclusion that oral perception is involved in articulatory problems. Moreover, among the subgroups with articulation problems, children with mild, moderate, and severe problems had correlated oral sensory deficits. This correlation between oral sensory and articulatory deficits was also observed by Carlson (1972).

Other researchers such as McNutt (1977) have found similar evidence, and some have reached a conclusion as did Yoss and Darley (1974) that such deficits are organically based: "a result of a dysfunction in the left hemisphere" (Frisch & Handler, 1974, p. 442). There is conservative opinion that questions whether the evidence is sufficiently strong to warrant such conclusions. The crucial demonstration of damage to the brain in such cases has not been made. To borrow the term "apraxia" from aphasiology may also be inappropriate, for there are several differences between the behaviors of adult apraxics and "developmental apraxics," as Yoss and Darley (1974) were careful to point out. In balance, however, there seems sufficient evidence to justify the discussion of a syndrome of childhood apraxis as we have done in this chapter. The procedures for description (diagnosis) are available. The impact of the concept upon prediction (therapy) remains to be assessed.

Adult Apraxia

The apraxic has brain damage that impairs his ability to conceive, and thus to produce, the sounds of the language. Clinical studies of adult dysarthrics, aphasics, and apraxics done by Johns and Darley (1970) and others have permitted the development of procedures that differentiate them on the basis of articulatory behaviors. The apraxic, for example, tends to make substitution errors in articulating, while the dysarthric tends to make distortion errors. Speakers with language problems secondary to brain damage, but no articulatory problems, may then be classed as aphasic (Johns & Darley, 1970; Teixeira, Defran, & Nichols, 1974). Some apraxics may reduce their articulatory errors when asked to speed up their reading aloud of connected prose. Dysarthrics increase their errors when following such instructions (Johns & Darley, 1970; Teixeira *et al.*, 1974).

Apraxia has been thought of as a disorder of production, but two lines of investigation may involve input control procedures in their remediation. The first of these is the demonstration of striking oral form perception problems in the apraxic (as in the child with severe articulation problems) by Rosenbek, Wertz, and Darley (1973b) and Teixeira *et al.* (1974). In the latter study, there was virtually no population overlap between the apraxics and the other brain-damaged populations; the apraxics made many more errors. Clinical applications of these findings, that is, practice with oral stereognostics with apraxics, has led to some success in improved oral perception, but little effect upon articulation improvement.

Other studies have shown dichotic listening differences between the brain-damaged population and the normal population (e.g., Johnson, Sommers, & Weidner, 1977). Some aphasics have substantial left ear advantages, that is, when words are presented to the right and left ears simultaneously, they report more of the words presented to the left ear than they do words presented to the right ear. In contrast, normals tend to report words presented to the right ear. This finding led the author and one of his students to provide left ear input for articulatory stimulation with an apraxic client. The results were striking. He was able to echo the stimulation with no error with left ear presentation, but right ear or "both ear" presentations evoked only labored and error-ridden productions. It should be emphasized at this point that there is no conclusive evidence that other apraxics would profit from such a procedure. This remains to be tested.

Some procedures specific to the restoration of articulatory functions in severe apraxia have been developed. Rosenbek, Lemme, Ahern, Harris, and Wertz (1973a), developed an eight-step therapeutic procedure beginning with simultaneous productions by both therapist and client, and ending with spontaneous use in role playing situations. Deal and Florance (1978) reported a modification of the Rosenbek *et al.* (1973a) eight-step procedure. They required, apparently appropriately, 80% accumulated correct responses at any one step before proceeding to the next, developed a home practice program, and found that some steps could be bypassed. The program was effective with the first four clients that Deal and

Florance (1978) selected for therapy, but they agreed with Rosenbek *et al.* (1973a) that the program might not be effective for all severe apraxics.

Deaf and Hard of Hearing

There was a time when to be deaf was to be without speech, mute. Today the deaf speak, but often not very well. Many articulate so poorly as to be functionally "mute." This fact is often cited as evidence that of all the organic pathologies, a loss of hearing can be the most devastating to the acquisition of speech. Even a partial loss of hearing can affect the articulation of normal speakers. Silverman (1971b) noted that the hard of hearing may lose articulatory precision as time passes since they had good hearing, or as the hearing loss grows greater. He noted that commonly such speakers may distort the sibilants, /s/, /ʃ/, etc., or word endings.

The deaf speakers' consistency of articulation errors is noteworthy (Silverman, 1971a). Work on the phonatory acquisition of the deaf has shown that this may be attributed to the articulatory "rules" that Dodd (1976) has observed in the speech of the deaf. He also noted that "lipreading was a major input to the deaf children's phonological systems." An example was the "rule" in the speech of some children that /m/ is substituted for /b/, i.e., [baus]/[maus]. Several rules held in a second study, suggesting that they may have generality for more than one group of deaf speakers.

An early study of the articulation of the deaf by Hudgins (1934) noted distortions of vowels, nasality, and poor juncture control between phonemes. Hudgins and Numbers (1942) carried on the study, noting errors in voicing substitution and omission errors, and vowel and diphthong errors, as well as the patterns observed by Hudgins earlier. Another major factor in the intelligibility of the deaf is the rhythm of utterances according to this study.

Monsen has carried out several studies of the articulation of the deaf noting, for example, that although normal speakers tend to give different durations to vowels depending on the terminal consonant, the deaf do not (Monsen, 1974). He noted that this is one of many possible phenomena which contribute to the low intelligibility of the deaf speaker. In this study, vowel durations of deaf speakers tended to be longer than those of normal speakers, a finding also noted by Hudgins (1934). In another study of vowels, Monsen (1976) reported that vowel quality (second formant) may tend to be "neutralized" by deaf speakers, again a phenomenon mentioned by earlier observers (Hudgins & Numbers, 1942).

In a project attempting the integration of the results of several studies, Monsen (1978) prepared short "predictable" sentences for deaf speakers to read, and analyzed the listening errors made by normal receivers when the sentences were read. A multiple regression analysis of phonatory factors showed that the voice onset times of /t/ and /d/, and the vowel quality differences (second formant) of the /i/ and the /ɔ/ were correlated highly with speaker intelligibility. The overall intelligibility of the Monsen sentences was 76%, much higher than the 19–20% reported by earlier researchers (John & Howarth, 1965; Markides, 1970), no doubt because of the high predictability of the sentences he used. It must be noted then that semantic and syntactic as well as phonatory variables contributed significantly to the intelligibility scores, and the generality of the findings must, therefore, be suspect.

Potter, Kopp, and Kopp (1966, pp. 285–301) have noted that durational and transitional phenomena account for many articulation errors common to the deaf. Problems that are related to precision and timing of movement are accessible in spectrographic analysis, particularly "real-time" visual speech displays, which have shown that the critical differences in the speech of a deaf child are highly individual and require a highly individualized program of speech therapy to achieve effective communication (Kopp, 1971). The individual differences may be expected, for as has been noted earlier in this section, the deaf child has been found to structure his own phonological system, although with many visual rather than auditory perceptual biases, as does the hearing child (Dodd, 1976).

Summary

There is substantial evidence that each of the anatomical and physiological factors reviewed in this chapter may be implicated in articulation disorders. The laryngeal tone, the tongue, and the soft palate appear to be very important in articulation. Neurological factors account for many defective phenomena for they initiate articulatory movements at the highest brain levels, and transmit the commands through the nervous system to the muscles of the larynx, tongue, and soft palate to control muscular expression of the phonological concepts of speech. Hearing appears fundamental to learning articulation, particularly in terms of the time-linked phonatory factors of voice onset, phoneme duration, and articulatory juncture or linkage. It must be concluded that organic factors must be accounted for in any program of articulation study and management.

References

Arndt, W. B., Shelton, R. L., Johnson, A. F., & Furr, M. L. Identification and description of homogeneous subgroups within a sample of misarticulating children. *Journal of Speech and Hearing Research,* 1977, *20,* 263–292.

Bankson, N. W., & Byrne, M. C. The relationship between missing teeth and selected consonant sounds. *Journal of Speech and Hearing Disorders,* 1962, *27,* 341–348.

Brandt, S. D., & Morris, H. L. The linearity of the relationship between articulation errors and velo-pharyngeal incompetence. *Cleft Palate Journal,* 1965, *2,* 176–183.

Burgi, E. J., & Matthews, J. Predicting intelligibility of cerebral palsied speech. *Journal of Speech and Hearing Research,* 1958, *1,* 331–343.

Byrne, M. C. Speech and language development of athetoid and spastic children. *Journal of Speech and Hearing Disorders,* 1959, *24,* 231–240.

Bzoch, K. R. Articulatory proficiency and error patterns of preschool cleft palate and normal children. *Cleft Palate Journal,* 1965, *2,* 340–349.

Carlson, L. C. *Oral stereognostic differences between first grade children with tongue thrust and normal swallow.* Unpublished master's thesis, San Diego State University, 1972.

Chisum, L., Shelton, R. L., Arndt, W. B., & Elbert, M. The relationship between remedial speech instruction activities and articulation change. *Cleft Palate Journal,* 1969, *6,* 57–64.

Christensen, J. M., Weinberg, B., & Alfonso, P. J. Productive voice onset time characteristics of esophageal speech. *Journal of Speech and Hearing Research,* 1978, *21,* 56–62.

Colletti, E. A., Geffner, D., & Schlanger, P. Oral stereognostic ability among tongue thrusters with interdental lisp, tongue thrusters without interdental lisp, and normal children. *Perceptual Motor Skills,* 1976, *42,* 259–268.

Darley, F. L., Aronson, A. E., & Brown, J. R. Differential diagnostic patterns of dysarthria. *Journal of Speech and Hearing Research,* 1969, *12,* 246–269. (a).

Darley, F. L., Aronson, A. E., & Brown, J. P. Clusters of deviant speech dimensions in the dysarthrias. *Journal of Speech and Hearing Research,* 1969, *12,* 462–497. (b).

Darley, F. L., Aronson, A. E., & Brown, J. R. *Motor speech disorders.* Philadelphia: W. B. Saunders, 1975.

Deal, J. L., & Florance, C. L. Modification of the eight-step continuum for treatment of apraxia of speech in adults. *Journal of Speech and Hearing Disorders,* 1978, *43,* 89–95.

Dickson, S. Differences between children who spontaneously outgrow and children who retain functional articulation errors. *Journal of Speech and Hearing Research,* 1962, *5,* 263–271.

Dodd, B. The phonological systems of deaf children. *Journal of Speech and Hearing Disorders,* 1976, *41,* 185–198.

Dworkin, J. P. Protrusive lingual force and lingual diadochokinetic rates: A comparative analysis between normal and lisping speakers. *Language, Speech and Hearing Service Schedule,* 1978, *9,* 8–16.

Fairbanks, G., & Bebout, B. A study of minor organic deviations in "functional" disorders of articulation: 3. The tongue. *Journal of Speech and Hearing Disorders,* 1950, *15,* 348–352.

Fairbanks, G., & Lintner, M. V. H. A study of minor organic deviations in "functional" disorders of articulation: 4. The teeth and hard palate. *Journal of Speech and Hearing Disorders,* 1951, *16,* 273–279.

Fitzpatrick, P. *Self-administered intelligibility training with esophageal speakers.* Unpublished master's thesis, San Diego State University, 1978.

Fletcher, S., Casteel, R., & Bradley, D. Tongue-thrust swallow, speech articulation, and age. *Journal of Speech and Hearing Disorders,* 1961, *26,* 201–208.

Flores, P. M., & Irwin, O. C. Status of five front consonants in the speech of cerebral palsied children. *Journal of Speech and Hearing Disorders,* 1956, *21,* 238–244.

Frisch, B. R., & Handler, L. A neuropsychological investigation of functional disorders of speech articulation. *Journal of Speech and Hearing Research,* 1974, *17,* 432–445.

Fymbo, F. H. The relationship of malocclusion of the teeth to defects of speech. *Archives of Speech,* 1936, *1,* 204–216.

Hanes, M. M. *A study of language abilities in cleft palate children.* Unpublished master's thesis, San Diego State University, 1973.

Heighton, R. S., & Nichols, A. C. Effects of untreated "open bite" malocclusion on the articulation of children. *Journal of Communication Disorders,* 1967, *1,* 166–169.

Hudgins, C. A comparative study of the speech coordinations of deaf and normal subjects. *Journal of Genetic Psychology,* 1934, *44,* 1–48.

Hudgins, C., & Numbers, F. An investigation of the intelligibility of the speech of the deaf. *Genetic Psychology Monographs,* 1942, *25,* 289–392.

Irwin, O. C. Phonetic equipment of spastic and athetoid children. *Journal of Speech and Hearing Disorders,* 1955, *20,* 54–57.

Irwin, O. C. Short test for use with cerebral palsy children. *Journal of Speech and Hearing Disorders,* 1956, *21,* 446–449.

Irwin, O. C. An integrated articulation test for use with children with cerebral palsy. *Cerebral Palsy Review,* 1961, *22,* 8–13. (a)

Irwin, O. C. Correct status of vowels and consonants in the speech of children with cerebral palsy as measured by an integrated test. *Cerebral Palsy Review,* 1961, *22,* 21–24. (b)

Jacobson, R., & Halle, M. Tenseness and laxness, In R. Jacobson, C. G. M. Fant, and M. Halle, *Preliminaries to speech analysis.* Cambridge, Mass.: M.I.T. Press, 1969, pp. 57–61.

John, J. E. J., & Howarth, J. N. The effect of time distortions on the intelligibility of deaf children's speech. *Language and Speech,* 1965, *8,* 127–134.

Johns, D. F., & Darley, F. L. Phonemic variability in apraxia of speech. *Journal of Speech and Hearing Research,* 1970, *13,* 556–583.

Johnson, J. D., Sommers, R. K., & Weidner, W. E. Dichotic ear preference in aphasia. *Journal of Speech and Hearing Research,* 1977, *20,* 116–129.

Kent, R., & Netsell, R. Articulatory abnormalities in athetoid cerebral palsy. *Journal of Speech and Hearing Disorders,* 1978, *43,* 348–352.

Kopp, H. G. *Some functional applications of basic phonetic principles.* New York: Neyenesch Press, 1971.

Markides, A., The speech of deaf and partially hearing children with special reference to factors affecting intelligibility. *British Journal of Disorders of Communication,* 1970, *5,* 126–140.

Mason, R. M., & Proffit, W. R. The tongue thrust controversy: Background and recommendations. *Journal of Speech and Hearing Disorders,* 1974, *39,* 115–132.

McNutt, J. C. Oral sensory and motor behaviors of children with /s/ or /r/ misarticulations. *Journal of Speech and Hearing Research,* 1977, *20,* 694–703.

Milisen, R. A rationalle for articulation disorders. *Journal of Speech and Hearing Disorders,* 1954, Monograph Suppl. *4,* 5–18.

Moll, K. L. Speech characterisitics of individuals with cleft lip and palate. In D. C. Spriestersbach, and D. Sherman, (Eds.), *Cleft palate and communication.* New York: Academic Press, pp. 61–118, 1968.

Monsen, R. B. Durational aspects of vowel production in the speech of deaf children. *Journal of Speech and Hearing Research,* 1974, *17,* 386–398.

Monsen, R. B. Second formant transitions of selected consonant–vowel combinations in the [hearing impaired and] normal-hearing children. *Journal of Speech and Hearing Research,* 1976, *19,* 279–289.

Monsen, R. B. Toward measuring how well hearing-impaired children speak. *Journal of Speech and Hearing Research,* 1978, *21,* 197–219.

Morris, H. C., Spriestersbach, D. C., & Darley, F. L. An articulation test for assessing competency of velopharyngeal closure. *Journal of Speech and Hearing Research,* 1961, *4,* 48–55.

Nichols, A. C. Loudness and quality in esophageal speech. In C. Snidecort (Ed.), *Speech Rehabilitation of the Laryngectomized.* 2nd Ed. Springfield, Ill.: Charles C Thomas, 1968.

Nichols, A. C. Confusions in recognizing phonemes spoken by esophageal speakers: I. Initial consonants and clusters. *Journal of Communication Disorders,* 1976, *9,* 27–41. (a)

Nichols, A. C. Confusions in recognizing phonemes spoken by esophageal speakers: II. Vowels and diphthongs. *Journal of Communication Disorders,* 1976, *9,* 247–260. (b)

Nichols, A. C. Confusions in recognizing phonemes spoken by esophageal speakers: III. Terminal consonants and clusters. *Journal of Communication Disorders,* 1977, *10,* 285–299.

Phillips, B. J., & Harrison, R. J. Articulation patterns of preschool children. *Cleft Palate Journal,* 1969, *6,* 245–253.

Potter, R. K., Kopp, G. A., & Kopp, H. G. *Visible speech.* New York: Dover, 1966.

Rathbone, J. S., & Snidecor, J. C., Appraisal of speech defects in dental anomalies with reference to speech improvement. *Angle Orthodontist,* 1959, *29,* 54–59.

Rees, M. Some variables affecting perceived harshness. *Journal of Speech and Hearing Research,* 1958, *1,* 155–168.

Ringel, R. L., House, A. S., Burk, K. W., Dolinsky, J. P., & Scott, C. M. Some relations between orosensory discrimination and articulatory aspects of speech production. *Journal of Speech and Hearing Disorders,* 1970, *35,* 3–11.

Rosenbek, J. C., Lemme, M. L., Ahern, M. B., Harris, E. H., & Wertz, R. T. A treatment for apraxia of speech in adults. *Journal of Speech and Hearing Disorders,* 1973, *38,* 462–472. (a)

Rosenbek, J. C., Wertz, R. T., & Darley, F. L. Oral sensation and perception in apraxia of speech and aphasia. *Journal of Speech and Hearing Research,* 1973, *16,* 22–36.

Shelton, R. L., Jr., Brooks, A. R., & Youngstrom, K. A. Clinical assessment of palatopharyngeal closure. *Journal of Speech and Hearing Disorders,* 1965, *30,* 37–43.

Shelton, R. L., Jr. Hahn, E., & Morris, H. L. Diagnosis and therapy. In D. C. Spriestersbach, and D. Sherman, (Eds.), *Cleft palate and communications.* New York: Academic Press, 1968, pp. 225–268.

Shelton, R. L., Chisum, L., Youngstrom, K. A., Arndt, W. B., & Elbert, M. Effect of articulation therapy on palatopharyngeal closure, movement of the pharyngeal wall, and tongue posture. *Cleft Palate Journal,* 1969, *6,* 440–448.

Shipp, T. Frequency duration and perceptual measures in relation to judgments of laryngeal speech acceptability. *Journal of Speech and Hearing Research,* 1967, *10,* 417–427.

Silverman, S. R. The education of deaf children. In L. E. Travis (Ed.), *Handbook of speech pathology and audiology.* New York: Appleton-Century-Crofts, 1971, pp. 399–430. (a)

Silverman, S. R. Hard-of-hearing children. In L. E. Travis (Ed.), *Handbook of speech pathology and audiology.* New York: Appleton-Century-Crofts, 1971, pp. 431–438. (b)

Skelly, M., Donaldson, R. C., & Furth, R. S. *Glossectomee speech rehabilitation.* Springfield, Ill.: Charles C Thomas, 1973.

Snidecor, J. C., & Kaires, A. K. A speech corrective prosthesis for anterior ''open-bite'' malocclusion and its effect on two post-dental fricative sounds. *Journal Prosthetic Dentistry,* 1965, *15,* 779–784.

Snow, K. Articulation proficiency in relation to certain dental abnormalities. *Journal of Speech and Hearing Disorders,* 1961, *26,* 209–212.

Stansell, B. *Effects of deglutition training and speech training on dental overjet.* Unpublished doctoral dissertation. University of Southern California, 1969.

Subtelny, J. D., Koepp-Baker, H., & Subtelny, J. D. Palatal function and cleft palate speech. *Journal of Speech and Hearing Disorders,* 1961, *26,* 213–224.

Subtelny, J., Mestre, J., & Subtelny, D. Comparative study of normal and defective articulation of /s/

as related to malocclusion and deglutition. *Journal of Speech and Hearing Disorders*, 1964, *29*, 269–285.

Teixeira, L. A., Defran, R. H., & Nichols, A. C. Oral stereognostic differences between apraxics, dysarthrics, aphasics and normals. *Journal of Communication Disorders*, 1974, *7*, 213–226.

Van Demark, D. R. Misarticulation and listener judgments of the speech of individuals with cleft palate. *Cleft Palate Journal*, 1964, *1*, 232–245.

Van Demark, D. R. A factor analysis of the speech of children with cleft palate. *Cleft Palate Journal*, 1966, *3*, 159–170.

Van Demark, D. R. Articulatory changes in the therapeutic process. *Cleft Palate Journal*, 1971, *8*, 159–165.

Ward, M. M., Malone, H. D., Jann, G. R., & Jann, H. W. Articulation variations associated with visceral swallowing and malocclusion. *Journal of Speech and Hearing Disorders*, 1961, *26*, 334–341.

Winitz, H. Articulation disorders: From prescription to description. *Journal of Speech and Hearing Disorders*, 1977, *42*, 143–147.

Yoss, K. A., & Darley, F. L. Developmental apraxia of speech in children with defective articulation. *Journal of Speech and Hearing Research*, 1974, *17*, 399–416.

2

Edward D. Mysak and Carol N. Wilder

PHONATORY AND RESONATORY PROBLEMS OF ORGANIC ORIGIN

This chapter was first prepared by one of the present authors (Mysak) and published in 1966. Since that time the area of organic phonatory and resonatory problems has, like most areas of speech pathology, undergone an "information explosion." Significant advances in knowledge have occurred in the areas of anatomy and physiology of respiration and phonation, in instrumentation to study and analyze normal and abnormal respiratory and phonatory phenomena, as well as in medicosurgical and voice therapy intervention procedures. It is no longer possible for the voice researcher, physician, or clinician to keep abreast easily with all the developments in the basic and applied areas of human voicing. With this in mind, this chapter was designed to build upon the first by incorporating the most pertinent findings in the area of human voicing and its care over the last fifteen years.

The major sections of the chapter include discussions of the concept of normal and abnormal voicing and resonation, physical determinants of voice, the evolution and involution of voice, voice disorders of infraglottal, glottal, and supraglottal origin, and disorders associated with general systemic conditions.

Concept of Normal and Abnormal Phonation and Resonation

The concept of normal and abnormal phonation and resonation may be approached from several viewpoints: (a) perceptually, in terms of the way the

Edward D. Mysak and Carol N. Wilder • Department of Speech Pathology and Audiology, Teachers College, Columbia University, New York, N.Y. 10017.

voice sounds, (b) behaviorally, in terms of the way the voice is used, and (c) physically, in terms of the status of the voice-producing mechanism. Regardless of the approach, the concept of normal and abnormal voice is not simple.

Difficulties with perceptual definitions arise, at least in part, from the fact that no consensus exists regarding terms with which to describe either normal or abnormal voice. Murphy (1964) offered a list of over 60 terms that were used in the literature to describe the voice. As he noted, many of the terms (such as "warm" or "metallic") are drawn from the language of metaphor, and they largely constitute a "sensation vocabulary related principally to non-auditory experiences" (p. 26). Perkins (1977) observed that 9 current texts use 27 terms for defective voice, of which only 12 are in more than one text, and only 2 ("nasality" and "hoarseness") are used by all.

Perceptual definitions are also complicated by the fact that the concept of "normal" is related not only to the endogenous factors of age, sex, and physical build, but also to such exogenous factors as cultural preferences, socioeconomic status, and specific speaking situations. To be perceived as normal, the pitch, loudness, and quality of the voice is deemed appropriate according to most of these endogenous and exogenous factors, and it must also fulfill the linquistic requirements for inflection and breath-group length. As a consequence, a wide variety of vocal characteristics are subsumed under the heading of "normal." Moreover, certain vocal characteristics may be perceived as normal under one set of endogenous/exogenous circumstances, and abnormal under a different set.

A behavioral definition of "abnormal" voice is somewhat less difficult than a perceptual definition because vocal abuse can be defined, and in many cases, its results are apparent in laryngeal tissue change. However, "normal" vocal behavior cannot simply be defined as the absence of vocal abuse; vocal efficiency must also be considered. If maximum vocal efficiency is defined as some combined optimum of acoustic output and esthetic gratification per unit of expended energy, it is apparent that different voices that sound "normal" may vary as to their degree of vocal efficiency. Also, a set of vocal behaviors that may be efficient in one speaking situation (such as one-to-one conversation in a quiet setting) would be inefficient if carried over into a more demanding speaking situation (such as addressing a group against a background of noise).

Presumably, a physiologically based definition of abnormal and normal phonation would appear to be clear-cut, but here, too, certain ambiguities may be found. Although vocal sounds are directly related to specific patterns of vocal fold vibration (Moore, 1976), the relationship between laryngeal structures and the vibratory pattern is not fully understood. Normal voice can apparently be produced by structures in which great individual variation is possible (Sellars & Keen,

1978). Zemlin (1978) recently reported on a lengthy series of dissections of specimens with no known history of vocal pathology; he found that idiosyncratic variations in laryngeal and respiratory structures were widespread. Yet Hirano (1974, 1975) noted that even small structural abnormalities may significantly alter the vibratory pattern of the vocal folds.

Thus, when the speech pathologist attempts to classify a voice as normal or abnormal, he must consider perceptual, behavioral, and physical factors. Whereas the speech pathologist is trained to assess the perceptual and behavioral aspects of voice, information regarding the physical aspects must be obtained from the laryngologist. The laryngologist's report should be obtained prior to the voice evaluation because similar vocal symptoms may result from dissimilar causal factors. Laryngoscopic examination is essential because some vocal symptoms may be associated with certain life-threatening conditions, and others may not be responsive to voice therapies. In a few settings, the medical findings are supplemented by those of a voice scientist who can provide more detailed information about the aerodynamic, acoustic, and vibratory characteristics of the voice.

Physical Determinants of Voice

Traditionally, the conceptual framework for describing the speech mechanism has included its subdivision into respiratory, phonatory, articulatory, and sometimes, resonatory components. However, this subdivision of the "speech effector system" (Mysak, 1976) is misleading since it may suggest that any one component can function independently of the other components. Fine coordination between all of the parts of this system is required in order to produce even a single syllable. Therefore, it is more realistic to think of the speech mechanism in terms of a unified system consisting of "a large air container that is always closed in the bottom end (by the lungs) . . . with only two openings to the atmosphere . . . the nostrils and the mouth" (Netsell, 1973, p. 214). Within this container are a number of valves that can be opened, and partially or fully closed, thereby changing the overall configuration of the container. One valve, the larynx, is responsible for generating the fundamental vocal signal, and the other valves (lips, mandible, pharyngeal walls, tongue, velum) for shaping this signal into the phonetic segments of speech.

This chapter focuses on the vocal effects of aberrations in three of the components of this complex speech effector system: the respiratory mechanism, the laryngeal valve, and the velopharyngeal valve. As a result of aberrations in these system components, the voice may be impaired in terms of its pitch, loudness, quality, and/or resonance. The vocal impairment may also have linguistic significance if it adversely affects the suprasegmental aspects of speech, particularly intonation contours and stress (Lieberman, 1967, 1975).

Respiratory Component of Phonation

Through a complex interaction between the muscular forces of respiration (the diaphragm and the muscles of the abdominal and thoracic walls) and the nonmuscular forces (inherent positive and negative elastic recoil forces, gravity, abdominal hydraulic forces) the respiratory system produces a relatively constant subglottal pressure throughout most speech utterances, even though the lung volume changes continuously throughout the utterance (Lieberman, 1967). Sugglottal pressure is related closely to the loudness of the voice. For example, relatively small increases in subglottal pressure will result in significant increases in the loudness of the voice (Cavagna & Magaria, 1965; Isshiki, 1964). Although pitch changes are primarily mediated by adjustments of the laryngeal valve, an increase in subglottal pressure may also result in a slight elevation of pitch (Hixon, Klatt, & Mead, 1970).

Laryngeal Component of Phonation

The lower pitch range encompassed by the male voice is based on the anatomical fact that the adult male larynx is larger than the adult female larynx, particularly with respect to the length and mass of the vocal folds (Hollien, 1960). However, within a given individual, lengthening the vocal folds results in an elevation of pitch (Hollien & Moore, 1960). This seems paradoxical because increasing the length of a structure usually decreases its frequency of vibration. However, it has been demonstrated that an increase in vocal fold length also results in an increase of the longitudinal tension of the vocal fold and a decrease in its cross-sectional mass, both of which serve to accelerate the rate of vocal fold vibration (Hollien, 1962). The cricothyroid muscle is primarily responsible for lengthening the vocal folds.

The normal vibratory pattern of the vocal folds may also be affected by other factors, such as: (a) the degree of adduction/abduction of the arytenoid cartilages, (b) the amount of medial compression of adducted arytenoid cartilages, (c) the internal tension of the vocal folds, and (d) the degree of active shortening of the vocal folds. These factors affecting vibration are mediated by the other intrinsic laryngeal muscles, usually in combined activity. The extrinsic laryngeal muscles elevate and depress the larynx within the neck as a function of pitch (Shipp, 1975);

they may also influence the laryngeal vibratory pattern, but the type and extent of this influence is yet unclear.

Velopharyngeal Component of Phonation

The velopharyngeal valve affects the nasal resonance balance of the voice. If it is closed, all of the air generated by the respiratory system will reach the atmosphere through the mouth. If it is fully or partially open, all or part of the air will reach the atmosphere via the nasal cavity, resulting in some degree of perceived nasal resonance. Many variations in the nasal resonance of a voice are acceptable, so long as they do not impinge upon linguistic distinctions between the nasal and nonnasal phonemes of a language. Velopharyngeal valving involves synergistic action of the muscles affecting the velum and the lateral and posterior pharyngeal walls (Shprintzen, Lencione, McCall, & Skolnick, 1974).

Vocal Evolution and Involution

Specialists in human vocalization are aware of the positive changes in the male and female voice as a function of growth and development. If all goes well, pitch, loudness, quality, and time attributes of the voice gradually mature, and the result is pleasant and effective voicing. Similarly, specialists are aware of negative vocal changes that are often associated with old age. Most people are accustomed to the stereotyped portrayal of aged individuals with high-pitched, quavering voices. Until recently, however, references in the literature to the vocal changes associated with old age were infrequent, and when found, were usually based on clinical observations. Therefore, a review of data reported for vocal evolutionary and involutionary processes is in order.

Evolution and Involution of the Male Voice

It has been well established that for males, there is a perceptual entity which might be labeled "senescent speech." Indeed, Shipp and Hollien (1969) reported that naive listeners could estimate the ages of male talkers between the ages of 20 and 89 "with surprising accuracy." Further, investigations by Ptacek, Sander, Maloney, and Jackson (1966) and Ryan and Burk (1974) suggest that vocal attributes, such as quality, voice breaks, intensity, voice tremor, laryngeal tension, air loss, and mean fundamental frequency may contribute significantly to listeners' perception of speaker age. However, investigation of age-related aspects of the male voice has been confined largely to fundamental frequency.

Table I shows a progressive lowering of fundamental frequency from Fairbanks's finding (1942) of a mean of 556 Hz for a male infant cry to the mean speaking fundamental frequency of 107.1 Hz reported by Hollien and Shipp (1972) for males with a mean age of 45.5. At some time beyond that age, speaking fundamental frequency rises, with levels of 124.9 to 132.1 reported for males in their seventies, and levels of 142.6 to 146.3 reported for males in their eighties (Mysak, 1959; Hollien & Shipp, 1972). In short, the aging trend for males with respect to pitch central tendency is one of a progressive lowering of pitch level from infancy through middle age, followed by a progessive rise in old age.

In males, pitch sigma, or the standard deviation of the distribution of an individual's vocal frequencies, also varies with age. Although the pattern of change is less clear-cut than for fundamental frequency, Table I indicates a slight decrease during the period of vocal maturation, a reduction from young adult to middle age, followed by an increase in men aged 80 and over. Maximum pitch range may also be affected by advanced age. When the performance of men over 65 was compared with that of men under 40, maximum pitch range of the older

Table I. Fundamental Frequency of Males at Various Developmental Stages

Investigator and mean maturity level	Fundamental frequency (Hz)	Pitch sigma (semitones)
Fairbanks (1942)		
Infant (nine months)	556 (in hunger wail)	
Fairbanks, Wiley, & Lassman (1949b)		
Seven years	294	2.2
Eight years	297	2.0
Hollien & Malcik (1967)		
Ten years	226.4	1.82
Fourteen years	184.0	1.32
Hollien & Jackson (1973)		
Twenty years	129.4	1.6
Hollien & Shipp (1972)		
20–29 years	119.5	
30–39	112.2	
40–49	107.1	
50–59	118.4	
60–69	112.2	
70–79	132.1	
80–89	146.3	
Mysak (1959)		
47.9 years	110.3	2.9
73.3 (elder group I)	124.9	3.0
85.0 (elder group II)	142.6	3.3

group was four tones less than the range of the younger group (Ptacek, Sander, Maloney, & Jackson, 1966). In light of the higher speaking fundamental frequencies exhibited by older males, it is interesting to note that the reduction in total range was accounted for mainly by the loss of tones at the high end of the range.

The general progressive lowering of pitch levels in males from infancy to young adulthood can easily be explained in terms of laryngeal growth and development. The continued slight lowering of pitch levels in the middle-aged males cannot be ascribed to further growth and development, although it is possible that some slight thickening or lengthening of the vocal folds does take place from young to middle adulthood. If there is indeed a true difference between these two groups in fundamental frequency, it might also be conjectured that the middle-aged male exhibits less generalized tension, and hence a slightly lower pitch level than the young adult because he has, in general, already dealt with important vocational and social adjustments (Mysak, 1959). Vocal changes up to this point may be considered positive, or evolutionary.

With respect to the two older male groups whose pitch levels manifest a reversal or a return to higher pitch levels, it is likely that a combination of physiological and socioemotional factors are operating. Hollien and Shipp (1972) speculated that muscle atrophy, reducing vocal fold thickness, combined with increasing stiffness of vocal fold tissue would contribute to raising the average fundamental frequency level of the voice from middle to old age. Mysak (1959) also noted certain senescent changes in body structures that might help account for the rise in fundamental frequency, such as central nervous system atrophy, increased blood pressure, changes in the respiratory system, and various endocrinological and muscle changes. In addition, he suggested that socioemotional changes related to decreasing self-sufficiency, forced retirement, and loss of family and friends might increase tension and anxiety and, in turn, affect pitch level. The implication is that pitch changes in the older male groups are negative ones, or involutionary changes; that is, they reflect symptoms of vocal aging. The diminishing of maximum pitch range may also be considered an involutionary phenomenon.

Evolution and Involution of the Female Voice

Pitch characteristics of the female voice at various developmental stages have also been investigated, although less extensively than for the male voice. Table II summarizes the findings of these studies. A progressive lowering of mean speaking fundamental frequency level is seen from Fairbanks's finding (Fairbanks, Herbert, & Hammond, 1949a) of 273.1 and 286.5 Hz for seven- and eight-year-

Table II. Fundamental Frequency of Females at Various Developmental Stages

Investigator and mean maturity level	Fundamental frequency (Hz)	Pitch sigma (semitones)
Fairbanks, Herbert, & Hammond (1949a)		
Seven years	273.2	1.11
Eight years	286.5	1.26
Duffy (1958)		
Eleven years	258.0	1.34
Thirteen years (premenarcheal)	251.7	1.22
Thirteen years (postmenarcheal)	237.7	1.72
Fifteen years	229.5	1.33
Hollien & Paul (1969)		
15.5 years	215.7	1.53
16.5	213.9	1.48
17.5	211.5	1.67
Linke (1953)		
Young adult	199.8	1.52
McGlone & Hollien (1963)		
72.6 years (group A)	196.6	1.48
85.0 years (group B)	199.8	1.35

old girls, respectively, to the mean fundamental frequency of 199.8 Hz for young adult females reported by Linke (1953). Little if any change is appreciated from young adulthood to the aged group A of McGlone and Hollien (1963); further, no significant difference appears to exist between the two aged female populations (whose mean ages and age ranges were very similar to the groups of older men reported by Mysak, 1959, and by Hollien & Shipp, 1972). The data suggesting no significant changes in the speaking fundamental frequencies of older females is in marked contrast to the data on the older male population.

Again, in contrast with the male population of comparable age, there is no increase in pitch variability for the two oldest groups of females. However, maximum pitch range does appear to be affected in a similar way for both older women and older men. That is, women above 65 experience a reduction in pitch range when compared with women under 40, with most of the reduction accounted for by a loss of tones at the high end of the range (Ptacek, *et al.*, 1966). In the female populations cited, the general progressive lowering of pitch levels from the seven- and eight-year-old girls through young adulthood follows the male pattern and may also be attributed to laryngeal growth and development. However, unlike the males, there is virtually no difference between young adults and older females. A possible explanation for this was offered by McGlone and Hollien (1963, p. 170) who stated "since the anatomical changes in the female larynx are not as extensive at puberty as those in men, degenerative changes may not have as great an effect on women's laryngeal structures in later life. Hence the concomitant changes in pitch, if present at all, would not be as apparent as they are in men."

In light of the data presented, little evidence exists to support a concept of involutionary trends in mean pitch level or pitch sigma in advanced aged females. In spite of this, there seems to be a prevailing impression that the female voice ages earlier than the male voice. If this impression is correct, it is apparently based on vocal attributes other than fundamental frequency.

Prophylactic and Rehabilitation Implications

Since the findings on trends in female vocal aging appear at this time to be unremarkable, this section of the chapter will be limited to a discussion of the implications for vocal prophylaxis that may be drawn from scrutinizing the trends in male vocal aging. Such trends can contribute to the development of at least a preliminary program of vocal hygiene.

1. *Ear*. Since the ear is the primary monitor of the phonatory mechanism, it is essential that close attention be given to any sign of hearing difficulty, especially after middle age. Otherwise, it is probable that deficient auditory monitoring will compound the physical and socioemotional factors that may already be causing undesirable vocal changes. Accordingly, medical care should be sought as soon as possible for reversible hearing problems, and medically irreversible losses should receive amplication benefits whenever practical.

2. *Vocal mechanism*. Vocal hygiene procedures should include avoiding excessive smoking, drinking, and improper use of the voice. Respiratory deficits should also be minimized by maintaining an appropriate weight level and good general muscle tonus.

3. *Speaking*. Efforts should be made to discourage older individuals from indulging in excessive intraverbalizing. It may be observed that elders often engage in increased amounts of musing as well as self-talking at whisper or soft-voice levels. (Such self-talking is common in children from one to three years.) To counteract this regressive tendency, older individuals should be encouraged to participate in clubs, give talks, and to engage in debates and discussions.

For those males who are approaching Mysak's elder group I (see Table I), vocal exercises, such as maintaining maximum vowel sustenation at optimum pitch levels and practicing maximum articulatory diadochokinetic rates may be helpful. Additional exercises such as manipulating vocal pitch, intensity, and rate attributes should aid in maintaining positive vocal flexibility. In a center where the equipment is available, speaking exercises under amplification should also serve to maintain auditory feedback mechanisms so essential to adequate voice control.

Phonatory and Resonatory Disorders of Infraglottal Origin

Most phonatory disorders of infraglottal origin are due to problems in the generation, maintenance, and/or fine control of the breath pressure required to sustain speech. If there is inadequate subglottal air pressure, the voice may be weak or even absent; if air pressure cannot be adequately sustained, the normal breath group length may be reduced, sometimes to the point that only one or two syllables can be uttered per expiration; if air pressure cannot be controlled, normal linguistic stress patterns may be impaired. Both muscular and nonmuscular forces are involved in speech breathing in a complex interaction dependent not only on the prevailing lung volume at any given instant during speech, but also on posture (Hixon, 1973). Therefore, phonatory disorders may result from problems with the respiratory structures *per se*, and/or with the neuromotor control of those structures.

Conditions Underlying Inadequate Air Volume

The pulmonary disabilities in this category result in a decrease in vital capacity, the total supply of air available for speech, rather than in a problem with speech air pressures. Therefore, they may have a minimal effect on the voice unless the situation requires sustaining longer phrases than usually found in conversational speech, as might be the case for the professional speaker or singer.

If the skeletal framework for respiration is adversely affected by injury or structural anomalies, it may diminish the amount of air that can be moved by the respiratory system. This may occur in association with malformation of the rib cage, or with such spinal column abnormalities as lordosis, kyphosis, or scoliosis (Mysak, 1976; Skelly, Donaldson, Scheer, & Guzzardo, 1971). West and Ansberry (1968) suggested that a weak voice might be associated with an enlarged heart, which reduces the space in the thorax that can be used for expansion of the air sacs of the lungs, and therefore reduces the amount of air available for phonation. Perkins (1977) mentioned several bronchopulmonary diseases that can affect lung volume, including carcinoma, tuberculosis, and more commonly, emphysema and bronchial asthma. A reduction in air supply for phonation may also be associated with pneumothorax, or with surgical removal of all or part of the lung (Mysak, 1976; Perkins, 1977).

Conditions Underlying Inadequate Air Pressure and Air Volume

Any condition that disturbs the complex neuromuscular control of the respiratory system can be expected to influence air volumes. But of greater concern is the influence of such conditions on the fine control of air pressure, because this impinges directly upon the fundamental process of phonation.

Some conditions involve structural deviations or muscle degeneration. Infraglottic stenosis or compression of the trachea, whether from trauma or thyroid enlargement, may prevent adequate air pressures from reaching the vocal folds; indeed, traumatic stenosis may so severely impede the flow of air as to require tracheotomy in order to sustain life. Congenital absence of respiratory muscles (Ford, 1966) and anomalies of the diaphragm (Mysak, 1976) may impair respiratory control, although the exact effect of these conditions is unclear as yet. Impaired speech respiration may also be found as a symptom of muscular dystrophy (Mullendore & Stoudt, 1961). These conditions are encountered relatively rarely in clinical practice in speech pathology.

It is more common for the speech pathologist to be concerned with respiratory system control problems secondary to neurological disorders. These disorders may affect any level of nervous system organization. With respect to the lower motor neuron, Darley, Aronson, and Brown (1975) noted that reduced vital capacity and impaired control of exhalation can result from involvement of the phrenic nucleus, the phrenic nerve (supplying the diaphragm) or the spinal intercostal nerves that innervate intercostal and abdominal wall muscles. Patients with paralysis or significant paresis of the abdominal muscles demonstrate difficulty in the upright position with the rapid inspiratory movements required for speech; this difficulty is not generally found in the supine position (Hixon, Mead, & Goldman, 1976).

Respiratory symptoms may contribute to disordered speech in various disorders of the upper motor neuron, the cerebellar system, the extrapyramidal system, and multiple systems (Darley *et al.*, 1975). These include disorders frequently seen by speech pathologists, such as cerebral palsy, bulbar palsy, multiple sclerosis, and parkinsonism, among others. However, in these disorders, both the respiratory mechanism and the laryngeal valve are often involved, so that it becomes difficult to distinguish problems of respiratory control from problems of laryngeal control. However, Netsell and Hixon (1978) recently described a simple, noninvasive method for clinically estimating subglottal air pressure, which should help the clinician determine the relative contributions of the respiratory and laryngeal components of these disorders.

Management

Evaluation and treatment of voice disorders included in this section begins in the hands of appropriate physicians and surgeons who treat the underlying respiratory condition. When symptoms persist following treatment, voice therapy may be instituted with the goal of providing maximum performance given the nature of the physical involvement.

In order to minimize the effects of the underlying respiratory condition, respiratory hygiene practices should be recommended, including maintaining an appropriate weight level and good muscle tonus, and avoiding smoking and excessive use of alcohol (Mysak, 1976). Using auditory and/or visual monitoring, systematic attempts may be made to have the client progressively increase loudness and/or the amount of time he can sustain voice. If the voice is weak because of inability to generate sufficient air pressure, West and Ansberry (1968) have suggested manipulating pitch to determine whether the carrying power of the voice is facilitated at some pitch other than habitual pitch, and/or compensating for decreased loudness by attention to very precise articulation. Darley *et al.* (1975, p. 276) also suggested working on more precise articulation, but with a slightly different rationale: "Inefficient valving (of the outflowing breath stream) results in the expiration of more air per unit of speech. More precise articulatory movements mean more efficient use of air." Other suggestions from these authors include learning to produce speech at high lung volumes (so as to utilize the greater work-producing potential of elastic recoil forces at high lung volumes), and learning to rephrase speech so as to avoid speaking on residual air.

Binding of the abdominal wall has also been suggested as a means of reducing inspiratory pause time and improving expiratory control in cases of abdominal muscle involvement. "This improvement seems related to the removal of one degree of freedom in the involved chest wall and thereby enables the patient to effectively use whatever rib cage wall expiratory potential there is rather than paradoxically wasting it in driving against an abdominal wall that paradoxes outward because of lack of tautness" (Hixon *et al.*, 1976).

Phonatory Disorders of Glottal Origin

The greater portion of the research literature on organic voice disorders is concerned with disorders of glottal origin. The plan of this section of the chapter is to first discuss disorders related to laryngeal abuse, then to consider other structural and neurogenic disorders, and finally to present a brief discussion of speech rehabilitation approaches for laryngeal malignancy.

Disorders Related Primarily to Hyperfunctional Abuse of the Larynx

Autogenous laryngeal trauma, or vocal abuse, generally involves one or more forms of vocal excess, such as excessive loudness, pitches which are too low or too high for a particular laryngeal mechanism, phonating with excessive constriction of the laryngeal valve, or simply using the voice too much. The abuse may be acute or chronic. It may begin as the result of true organic disease: for example, when a person tries to "speak over a cold," the excessive vocal effort required may result in strain and possible damage to laryngeal tissues. Smoking, certain medications, and the consumption of alcohol may cause changes in laryngeal tissues, which, in turn, lead to hyperfunctional speaking efforts. Acute abuse may be associated with a particular event, such as cheering at an athletic contest or screaming on an amusement park ride. More chronic situational precipitators of laryngeal abuse are encountered by professional users of the voice (preachers, teachers, singers, actors, etc.) and by persons who must use their voices in inhospitable vocal environments (noise, chemical fumes, heavy dust, etc.). In others, laryngeal hyperfunction is related to emotion-laden situations involving anxiety, tension, anger, forcefulness, and aggression. Excessive coughing, throat clearing, and loud laughing may also be abusive. And finally, there are individuals who have become habituated to a pattern of hyperfunctional voice production in all situations. Although the exact etiology of this habitual hyperfunction cannot always be determined, Brodnitz (1967) suggests that "vocal hyperfunction belongs to the large number of psychosomatic disorders that translate inner tension into tense organic behavior" (p. 47).

As noted many years ago by Jackson and Jackson (1942), "there is a great variation in the amount of abuse the larynx of different individuals will stand; but every larynx has its limit" (p. 40). There is also a great variation in the forms of tissue reaction to vocal abuse. One form of reaction, myasthenia laryngis (also called hyperfunctional phonasthenia), has few overt signs, but involves weakness of the laryngeal muscles as a consequence of prolonged vocal abuse. Other tissue reactions are more obvious, and may be either diffuse or focal. Diffuse reactions include vasodilation, hyperemia, submucous hemorrhages, thickening of the mucosa, and chronic nonspecific laryngitis. Focal lesions which develop as a result of laryngeal hyperfunction include hematoma, vocal nodules, polyps, and contact ulcers (Moore, 1971). All of these tissue reactions alter the normal vibratory pattern of the vocal folds (Hirano, 1974, 1975; Moore, 1976; von Leden,

Moore, & Timcke, 1960) and hence affect the voice, causing a variety of vocal symptoms including hoarseness, roughness, breathiness, and/or reduced intensity. Although it has been suggested in the clinical literature that a lower fundamental frequency is associated with these laryngeal pathologies, recent research has found no significant differences between the mean speaking fundamental frequencies of a group of men with nodules, polyps, and contact ulcers and a group with normal larynges (Murry, 1978).

Vocal abuse is often self-exacerbating. A sort of vicious circle is formed wherein vocal abuse leads to tissue reaction, which, in turn, leads to an audible vocal symptom; in a conscious or unconscious effort to compensate for the vocal symptom, the speaker (or singer) increases vocal effort, leading to increased laryngeal hyperfunction (Arnold, 1962a). In some cases, the tension and anxiety caused by vocal symptoms themselves may serve to increase laryngeal tension, causing still greater abuse (Brodnitz, 1967).

Vocal nodules occur more frequently in adult females than in adult males; it has been speculated that this is because their higher fundamental frequencies result in more frequent contact of the vocal folds at the critical juncture of the anterior two-thirds of the folds (Luchsinger & Arnold, 1965). In children with voice disorders, nodules are the most common form of tissue change, and occur more frequently in boys than girls, at least until the age of 10 (Luchsinger & Arnold, 1965). The incidence of polyps in male and female adults is about equal; polyps occur only occasionally in children. Contact ulcers are found much more often in men than in women, and they are rarely found in children.

Management

The initial management decisions for these disorders are made by the physician, who will decide whether medicosurgical intervention is needed, or whether voice therapy is more appropriate. Often the tissue pathology is alleviated by elimination of vocal abuse, whether the abuse be the result of environmental, situational and/or psychological factors, or simply of habituated hyperfunctional vocal behaviors. Therefore, when a lesion related to vocal abuse is relatively small, many physicians suggest a trial period of voice therapy. When the lesions are larger, surgical removal may be recommended, followed by vocal reeducation to prevent recurrence.

Many therapeutic approaches are available for the reduction of vocal abuse, dealing with such factors as environmental and situational vocal stress, psychological aspects, relaxation of the vocal mechanism, vocal hygiene, modification of pitch and loudness, and auditory self-monitoring (Brodnitz, 1967; Greene, 1972; Holbrook, Rolnick, & Bailey, 1974; Moore, 1971; Mysak, 1966; Van Riper & Irwin, 1958; Wilson, 1972). Recent experimentation with EMG biofeedback suggests that it may be of some benefit in reducing excessive laryngeal tension during phonation (Prosek, Montgomery, Walden, & Schwartz, 1978). Selection of a specific therapy approach is conditioned by the unique combination of causal, vocal, and psychological factors presented by each client. Voice therapy is successful in a high percentage of cases (Brodnitz, 1963).

Laryngeal Trauma, Benign Lesions, and Structural Abnormalities

Voice disorders in this category are related to aberrations of the laryngeal structures that are not primarily the result of vocal hyperfunction.

The larynx may be damaged by impact or penetrating injuries, by irradiation, or by burns from the ingestion of hot or chemically injurious substances. Benign neoplasms include hemangiomas, papillomas, hyperkeratosis, and leukoplakia. Engorgement of the laryngeal tissues may result from chronic overuse of alcohol (Luchsinger & Arnold, 1965; Moore, 1971; Mysak, 1976). These disorders result in a variety of vocal symptoms because "it is the size and location of an offending mass in combination with pitch and loudness of the voice that determine its effect upon the vocal product" (Moore, 1971, p. 549). Vocal sulcus (Luchsinger & Arnold, 1965), congenital underdevelopment of the larynx, and laryngeal webs may also affect the voice. Moore (1971) noted that because a laryngeal web is apt to cause a high-pitched voice, the possibility of a web should be considered whenever a high-pitch problem is present.

Management

The management of voice disorders in this category is primarily medico-surgical. Decisions regarding the need for voice therapy are dependent on the outcome of this treatment. If the laryngeal structures are relatively normal after treatment, voice therapy is not needed unless the poor voice quality is maintained on an habitual basis, perhaps as a part of the person's vocal self-image. If laryngeal structures are permanently impaired, compensatory therapy may be undertaken "to help the patient acquire as effective voice production as possible without developing harmful vocal habits or unpleasant associated behavior" (Moore, 1971, p. 564).

Neurogenic Disorders of Laryngeal Valving

Disorders of laryngeal valving may be associated with disease or injury to the central nervous system (CNS) and/or the peripheral nervous system.

The laryngeal valve may be involved in such central nervous system disorders as pseudobulbar palsy, cerebral palsy, parkinsonism, chorea, dystonia, amyotrophic lateral sclerosis, and multiple sclerosis. These disorders affect laryngeal tonus and/or coordination, depending on the type and location of nervous system involvement. Darley *et al.* (1975) provide a thorough review of the various vocal symptoms that may be associated with CNS disorders.

Laryngeal function may also be disturbed by disease or injury to the peripheral nerves supplying the larynx: the vagus nerve, its motor nuclei, and two branches of the vagus, the superior laryngeal nerve and the recurrent laryngeal nerve. Specific symptoms vary depending on the site of lesion, but generally involve a mild to complete flaccid paralysis of one or more of the intrinsic laryngeal muscles. The superior laryngeal nerve supplies the cricothyroid muscle: impairment of this nerve affects vocal fold elongation. The recurrent laryngeal nerve supplies the other intrinsic laryngeal muscles: impairment may affect abduction–adduction, active shortening, and/or the internal tension of the vocal folds. If the vagus nerve is impaired above the level of the superior laryngeal nerve, a combination of these effects may be observed.

Because the recurrent nerve is longer and travels through a portion of the thorax, it is more susceptible to damage than the superior nerve, and because the pathway of the left recurrent nerve is longer than the right recurrent, unilateral vocal fold paralysis occurs more often on the left. Both nerves may be affected by infection, trauma (particularly during thyroid surgery), enlarged thyroid glands, and tumors. In addition, the recurrent nerve may be affected by pathology in the apex of the lung, and on the left side, by aortic aneurysm or open heart surgery.

Management

For voice problems associated with CNS disorders, the planning of therapy depends on whether the background condition is congenital or acquired and whether it is resolving, static, or progressive. In many cases, the overall communication impairment is so extensive that the voice may be considered only a minor part of the problem. If voice therapy is deemed appropriate, specific suggestions may be found in Darley *et al.* (1975) and Mysak (1976).

Arnold (1962a) stated that no amount of physiotherapy, vocal exercise, or psychotherapy will benefit individuals with paralysis of the superior laryngeal nerve: in a few cases, a form of special surgery has met with some success. The prognosis for vocal rehabilitation is much more favorable for simple recurrent nerve paralysis, where the goal of both voice therapy and surgery is to achieve better approximation of the vocal cords. Voice therapy techniques generally involve activities that promote synkinetic movements of the laryngeal musculature (Mysak, 1976). Many of these techniques were introduced by Froeschels, Kastein, and Weiss (1955) and can be subsumed under the heading of "pushing exercises" (Brodnitz, 1967). Surgical management procedures for vocal fold paralysis include intracordal injection of Teflon or other inert substances (Arnold, 1962d, 1963; Dedo, Urrea, & Lawson, 1973), surgical decompression of the recurrent laryngeal nerve (Ogura & Roper, 1961), and surgical deformation of the thyroid cartilage (Isshiki, 1975, 1977).

Spastic Dysphonia

Because of the current controversy about the etiology of spastic dysphonia, it has been placed under a separate heading. It is a voice disorder characterized by atypical hyperconstriction of the laryngeal valve, often accompanied by extraneous tension in the respiratory and supraglottal musculature. This results in a strained, constricted voice, which may even give the impression of autostrangulation. In severe cases, the ability to communicate is seriously impaired. Because of the similarities between some of its symptoms and some of the symptoms of stuttering, many authors have referred to it as "laryngeal stuttering" (Brodnitz, 1976). Shared symptoms include facial grimacing and extraneous movements of the extremities, reduction of symptoms during singing and laughing, and perhaps most importantly, variations in severity associated with degree of nervous tension and/or specific speaking situations (Brodnitz, 1976; Dedo, Izdebski, & Townsend, 1977).

Largely because of these shared symptoms, and the emotional circumstances often associated with onset, spastic dysphonia has long been thought to be of psychogenic origin (Bloch, 1965; Heaver, 1959; Luchsinger & Arnold, 1965; Moses, 1954), representing a "physical manifestation of deep-rooted emotional conflict" (Brodnitz, 1976, p. 211). However, a number of studies have suggested a possible organic basis for the disorder. Robe, Brumlick, and Moore (1960) found abnormal electroencephalograms in 10 of 14 cases. Aronson, Brown, and Pearson (1968) found little evidence of severe emotional disorders, but suggested the possibility of a neurological substrate on the basis of neurological signs found in 17 of 20 patients. Because these signs were largely extrapyramidal, they thought "spasmodic dysphonia" might be a more appropriate term than "spastic dysphonia,'' and indeed, both terms are found in current literature. McCall and colleagues (McCall, Skolnick, & Brewer, 1971; McCall, 1973; McCall, 1977) proposed that spasmodic dysphonia is due to some sort of problem with the complex neural system found in the bulbar plexus, and that there may be several subtypes of spasmodic dysphonia that can be differentiated by careful symptom analysis. Dedo (1976) hypothesized that spastic dysphonia is caused by a disturbance of proprioceptive control of the vocal cords, secondary to a neurotropic viral injury to either peripheral or central nerve fibers.

Management

Regardless of etiological orientation, both psychotherapy and voice therapy have proven to be of limited value for relieving the symptoms of spastic dysphonia (Boone, 1977; Brodnitz, 1976; Cooper, 1973). Because of the persistent nature of the disorder and the severity of the communication impairment, a more radical approach has recently been attempted. Dedo (1976) reported on a series of 34 patients who underwent deliberate unilateral resection of the recurrent laryngeal nerve, a procedure developed on the premise that the resulting unilateral vocal fold paralysis would alleviate the laryngeal hyperadduction. He stated that with this procedure plus postoperative speech therapy, approximately half of the patients returned ''close to a 'normal' but soft phonatory voice.'' He added that the rest had varying degrees of improvement, but ''all, so far, have been pleased with the improvement in ease and quality of phonation and reduction or elimination of face and neck grimaces'' (p. 451). However, Dedo cautioned that this is not yet a standard treatment for even so disabling a disease as spastic dysphonia, and suggested that it should be done only after very careful preparation and evaluation.

Laryngeal Malignancy

In cases of malignant neoplasms of the larynx, the primary goal is obviously preservation of life, and medicosurgical treatment decisions are made accordingly.

If the malignant lesion is very small, subtotal laryngectomy may be performed, enabling the patient to continue to use the larynx as a source of voice. In one sample, 25% of patients with laryngeal cancer received less than total laryngectomy (Moore, 1975). However, even limited surgical excision of the larynx can be expected to result in a significant voice disorder, requiring voice therapy as soon as healing is complete. Therapy approaches are similar to those for unilateral vocal fold paralysis, and in general, seek to foster better approximation of the glottal margins.

More commonly, malignant laryngeal neoplasms require total laryngectomy, resulting in total loss of voice. After total laryngectomy, not only is the laryngeal valve gone, but there is no longer any airpath between the lungs and the supralaryngeal vocal tract: pulmonary air passes in and out of the newly created tracheal stoma. Because speech requires that there be vibrating air in the supralaryngeal vocal tract (where it is shaped into the phonetic segments of speech), some other source must be found. The potential sources remaining after laryngectomy include: (a) air trapped in the buccal cavity or the pharynx, which is set into vibration by being forced through a constriction created by the articulators, (b) air taken into the esophagus, which is set into vibration when it passes through a constriction at the pharyngoesophageal junction, (c) sound waves transmitted from an external source directly into the vocal tract, (d) the rerouting and setting into vibration of pulmonary air by means of an external device, and (e) the rerouting and setting into vibration of pulmonary air by a surgically created passageway. Each of these potential sources has been used for alaryngeal speech in laryngectomy rehabilitation. Advantages and disadvantages inherent in each will be briefly outlined below.

Buccal or pharyngeal speech is sometimes developed spontaneously by persons dependent on alaryngeal speech. Both methods suffer from: (a) severely restricted carrying power and somewhat bizzare tone quality, (b) significantly reduced phrase length (because of the very limited quantity of air that can be trapped), and (c) restricted movements of the articulators, because they must be held so as to serve as a source of tone generation.

It has been estimated that at least 65% of laryngectomees use esophageal voice as their primary mode of communication (Horn, 1962), making it by far the most commonly used method of alaryngeal speech. To produce esophageal voice, air is taken into the esophagus (either by inhalation or injection methods) then set into vibration as it exits through a "neoglottis" created by sphincter-like activity in some portion of the pharyngoesophageal junction (Diedrich & Youngstrom, 1966). A discussion of the various techniques for teaching esophageal speech is beyond the scope of this chapter, but many excellent descriptions of therapy methods are available in the literature (Boone, 1977; Diedrich & Youngstrom, 1966; Gateley, 1971; Moore, 1971; Rigrodsky, Lerman, & Morrison, 1971; Snidecor, 1962, 1971). A number of helpful suggestions will also be found in materials written specifically for the laryngectomized patient (Lauder, 1975; Waldrop & Gould, 1956).

In a survey of esophageal speakers, 56% rated their esophageal voice as good or very good, 23% average, 11% fair, and 10% poor (Diedrich & Youngstrom, 1966). Even at its best, esophageal voice has certain drawbacks. The quality is rougher and the fundamental frequency lower than normal voice, which poses special problems for the female laryngectomee. It has a restricted intensity range, which makes communication difficult in situations where either a very loud voice or a very soft voice is required. The air volume that may be held in the esophagus is obviously smaller than the lung volumes available for speech, and some individuals have difficulty prolonging the egression of available air from the esophagus; both factors may result in the production of unusually short breath-groups. And finally, the majority of esophageal speakers reported that esophageal speech was tiring at least some of the time (Diedrich & Youngstrom, 1966). Nevertheless, when compared to other types of alaryngeal speech, its advantages outweigh its drawbacks.

For reasons that are not entirely clear, many laryngectomees do not develop a serviceable esophageal voice. After intensive study of 27 laryngectomees, Diedrich and Youngstrom (1966) concluded that "the unknown variables of personality, motivation, family environment, and aspiration level . . . may be more important than the physical variables in explaining why one third of the laryngectomee population does not learn esophageal speech" (p. 61).

The most often used alternative to esophageal speech is an artificial larynx. These may be of two basic types, pneumatic or electronic, with the latter type used more often in the United States. In the most commonly used type of electrolarynx, a vibrator is placed against the neck and the sound waves transmitted through the neck toward the oropharyngeal cavity. In other electronic models, the vibrations are transmitted to the oral cavity via a tube, or (more rarely) a vibrator attached to a prosthetic appliance within the mouth. The pneumatic artificial larynx reroutes air from the stoma through a vibrator, and thence to the mouth via a tube. Examples of the various types are pictured in Diedrich and Youngstrom (1966) and Lauder (1975).

It should not be assumed that a person can simply pick up an artificial larynx and speak intelligibly with it the first time. Therapy methods and problems related to the use of an artificial larynx are described by Diedrich and Youngstrom (1966) and Boone (1977). In the past many speech clinicians were reluctant to encourage the use of an artificial larynx because they feared it would serve as a "crutch" and reduce the patient's motivation to learn esophageal speech. However, this fear does not seem warranted on the basis of clinical experience (Boone, 1977; Diedrich & Youngstrom, 1966), and current thinking emphasizes the benefits of early postoperative introduction of the artificial larynx. A major benefit is that this serves the urgent speech needs of the laryngectomee, thereby reducing some psychosocial stress during the period required to attain reasonable proficiency in esophageal speech, which may range from 6 to 12 months postoperatively, according to Diedrich and Youngstrom (1966). Further, these authors are of the opinion that the precise articulatory movements required by the artificial larynx are not only similar to those needed for intelligible esophageal speech, but may also facilitate efficient esophageal air intake. Some of the disadvantages associated with the use of an artificial larynx as the primary method of communication include: (a) dependence on an appliance that may suffer mechanical failure, (b) one hand must be occupied during speech, (c) conspicuousness of the appliance, (d) the "mechanical" sound produced by many models, and (e) the cost of some electronic models and/or the requisite batteries.

Several surgical techniques have been developed for redirection of pulmonary air into the vocal tract. Two of these procedures involve construction of an autogenous tube for rerouting the air. In the Asai procedure, a dermal tube leads from the stoma upward along the midline of the neck, turning inward directly below the base of the tongue to enter the hypopharynx (Snidecor, Isshiki, & Kimura, 1968; Snidecor, 1971). In the other procedure, a tube that shunts the pulmonary air directly from the trachea to the esophagus is constructed (Zwitman & Calcaterra, 1973). A different type of reconstruction involves redirecting pulmonary air from the stoma through a valve-type external device that returns vibrating air to the vocal tract through a permanent fistula created in the neck (Taub and Spiro, 1972; Weinberg, Shedd, & Horii, 1978). The voice quality achieved by these procedures is very similar to that of superior esophageal speech. The major advantage is that speech is supported by pulmonary air, with the greater air volume and better pressure control that implies. Each of these procedures has inherent problems, and all require extensive surgery. Consequently, they are not yet considered applicable for most laryngectomees.

Finally, much practical help can be obtained from the International Association of Laryngectomees (777 Third Avenue, New York, N.Y. 10017), an autonomous agency supported by the American Cancer Society, with constituent local chapters located throughout the world. This association also publishes a directory of qualified instructors of alaryngeal voice, with geographical listings.

Phonatory and Resonatory Disorders of Supraglottal Origin

The most commonly encountered disorder of velopharyngeal valving is associated with congenital cleft palate. The specialty of cleft palate habilitation has grown rapidly. As an indication of this growth, there is in the United States a multidisciplinary group of professionals who meet under the name of the American Cleft Palate Association, and a complete journal, *The Cleft Palate Journal*, is devoted to this disorder. Many books have been written on the subject. An important one from the standpoint of the specialist in oral communicative disorders is the one by Spriestersbach and Sherman (1968). Another, edited by Grabb, Rosenstein, and Bzoch (1971) covers the areas of classification, incidence, embryology, etiology, psychology, anatomy and physiology, and management (surgical, orthodontic, prosthodontic, otologic, dental, and speech and hearing). While such a comprehensive review is not possible here, some discussion of the cleft palate speech syndrome and its management by the speech and hearing clinician is appropriate.

Cleft Palate Speech Syndrome

One of the authors (Mysak, 1961a) reviewed the following speech symptoms of individuals with cleft palate: abnormal phonation and resonation, interference with speech sound development, speech-associated audible and visible compensatory reactions, and problems in auditory acuity and imagery. All, or various combinations of the above symptoms, together with delayed language and/or linguistic variations may contribute to the oral communicative disorder associated with cleft palate. In keeping with the specific purpose of this chapter, the discussion will be limited to the phonatory and resonatory aspects of the syndrome.

Mysak (1961a) described the phonatory and resonatory aspects in the following way: "To most listeners the outstanding feature of cleft palate speech is the nasal sounding tone produced by the affected individual. Frequently associated structural deviations in the oronasal-pharyngeal resonators add to the disturbance of the normal resonance of the voice." He also described phonatory involvements related to vocal fold abuse resulting from the production of compensatory glottal plosives and fricatives, or from the laryngeal mucosa inflammation stemming from chronic respiratory system disease. These impressions were confirmed by Bzoch's (1971) report of 1000 consecutive longitudinal case studies of individuals with cleft palate. In this clinical sample, hypernasal resonance imbalance was exhibited in 431 cases. As Bzoch noted, "in the light of the advancement in early surgical and prosthetic treatment for the correction of velopharyngeal insufficiency and the supposed relationship between hypernasality and velopharyngeal competency for speech, the problem of nasality remains a remarkably significant area of deviant speech needing improved management, as indicated by this high frequency in a series within excellent cleft palate treatment centers" (p. 721). A weak, aspirate voice was found in 313 cases, and was said to be a functional disorder, which may have developed because it serves to mask, to some extent, both the resonance imbalance and the nasal emission of air. In addition, 150 of the cases were found to have "hoarseness characterized by roughness of laryngeal vibration" (p. 730). Of these, 36 were found to have small nodules, but the rest had normal larynges. The source of the hoarseness remained unspecified in the cases without nodules.

Traditionally, the nasalized airflow is simply attributed to the fact that the oral and nasal chambers are in communication, and because of that, air escapes into the nasal cavities. Mysak (1961a) hypothesized that the resonation factor in the cleft palate speech syndrome arises not only from the lack of separation of oral and nasal chambers, but also from the development of abnormal airflow patterns for voicing. He suggested that this might occur when the infant constricts or occludes the airstream with his lips or tongue while attempting to produce pressure sounds during babbling; in the cleft palate infant, this would result in a rerouting of voiced or voiceless airflows from the momentarily occluded oral chamber to the always open nasal chamber. In time, this rerouting might contribute to the establishment of an abnormal or nasalized pneumatic pattern; and therefore, instead of developing oralized airflow patterns for voice production, the child develops and is habituated to nasalized airflow patterns for phonation. This abnormal airflow pattern may persist even after what appears to be successful surgical management. Conversely, the discovery of an occasional patient who speaks almost normally despite an open cleft palate suggests that near-normal oropneumodynamics may not be completely dependent on an intact velopharyngeal mechanism.

Management of the Cleft Palate Resonatory-Phonatory Complex

The management of the cleft palate speech syndrome may be divided into three parts: secondary preventive measures, causal, and symptom therapies.

Secondary Preventive Measures. It may be hypothesized that if normal oropneumodynamics can be encouraged during the first months of life, individuals with palatal clefts may develop more normal phonatory patterns irrespective, to a certain degree, of the excellence of eventual surgical or prosthetic habilitation procedures. Therefore, as soon after birth as feasible, it would be helpful to employ techniques that would contribute to the formation of more normal vegetative and signal airflow patterns. For example, some form of infant oral prosthesis might not only assist in effecting more normal sucking and swallowing activities, but might also encourage oral airflow patterns. It would be ideal, of course, if a satisfactory oral prosthesis, complete with pharyngeal bulb, could be constructed and used regularly up to the time of palatal surgery. Unfortunately, the rapidity of the child's growth during this period makes this impracticable.

Alternatively, through the joint efforts of the speech pathologist and the parents, the infant may be encouraged in activities that will focus attention on the anterior part of the vocal tract, and on a relatively wide-open mouth posture during vocalization. Since airflow will follow the path of least resistance, this should lead to greater oralization of the airstream, and may also discourage the development of glottal stop substitutions. Westlake and Rutherford (1966), Philips (1971), and Hahn (1971) offer a number of helpful suggestions along these lines for use with cleft palate infants and children during the speech acquisition period.

Causal Therapy. This portion of the habilitation procedure includes work by the surgeon, prosthodontist, orthodontist, otolaryngologist, pediatrician, psychologist, audiologist, and so on: that is, it is expected that these specialists will do all that can be done to facilitate normal voice production. This will include, as needed, lip and palatal closure, and any secondary procedures required to improve velopharyngeal valving; the fitting of an appropriate speech appliance, if that is the procedure of choice; attention to occlusal and dental problems, even during infancy; and active management of hearing problems, particularly middle-ear involvements during the speech-learning period. More specifically related to voicing behavior would be the employment of techniques designed to stimulate velar activity and compensatory activity of the lateral and posterior pharyngeal walls (see Mysak, 1976, p. 189).

Symptom Therapy. If the resonatory-phonatory symptoms of the cleft palate speech syndrome persist after all secondary preventive and causal therapies have been applied, symptom therapy is appropriate. The selection of symptom therapy, however, should be based upon objective assessment of the competency of the velopharyngeal valving mechanism. Fortunately, recent advances in multiview videofluoroscopy (Shprintzen *et al.*, 1974; Skolnick, Shprintzen, McCall, & Rakoff, 1975) and nasopharyngoscopy (Miyasaki, Matsuya, & Yamaoku, 1975) have made more accurate assessment of velopharyngeal competence possible, particularly during running speech.

If there is evidence that velopharyngeal closure is at least marginally adequate, several symptom therapies may be considered. Mysak (1976, pp. 267–269) suggested techniques aimed at readjusting the individual's manner of auditory scanning and monitoring. This approach may be particularly appropriate for the child who did not have the potential for adequate velopharyngeal closure during the speech-learning period. If such a child has consequently listened only to nasalized airflow patterns ever since he developed auditory self-awareness, he will eventually equate his hypernasal voice production with the vocalization he hears in his environment. When this occurs, the child may not make use of all his potential for normal voice production after surgical or prosthetic management has made velopharyngeal closure possible. If, however, the child is led to develop a "therapeutic error signal" in response to his own nasal speech, self-adjusting speech corrective activity may be triggered. Related to this goal, Mysak has also recommended use of the "naso-tympano-velar (NTV) reflex technique." To apply this technique, the clinician first creates a device composed of a nasal olive fitted to the child's naris (unilateral or bilateral) with a length of plastic tubing leading to the entrance of the child's external auditory meatus. (If only one side is used, it is preferable to lead the tubing to the right ear.) Speaking with the device in place has causal as well as symptom therapy implications. Not only does it allow for direct feedback of error resonance to the auditory sensor, but the nasally emitted air striking the tympanum may also cause reflexive contraction of velopharyngeal musculature.

Another approach which may help the child become more aware of hypernasality relies on a small instrument, the Voice Intensity Controller (VIC) (Holbrook *et al.*, 1974), to heighten the error pattern. A small contact microphone is placed against the child's nose and he wears a small earphone; the VIC is adjusted so that a signal tone is delivered to the earphone in response to nasal resonance. An electroacoustic instrument for quantifying nasality, the TONAR II, has been used to provide feedback for a behavior modification program for reduction of nasality (Fletcher, 1970). Shprintzen, McCall, & Skolnick (1975) described a behavior modification program based on research demonstrating similarities between the pattern of velopharyngeal closure used for blowing and that used for speech. In this program, a device called a "scape-scope" provides a visual cue for monitoring nasalized airflow. Visual feedback of velopharyngeal closure during symptom therapy has also been provided by endoscopic monitoring (Myasaki *et al.*, 1975; Shelton, Beaumont, Trier, & Fuir, 1978).

Two important considerations should be kept in mind when considering the symptom therapies described above. First, it is obvious that these techniques should be used only when there is evidence that velopharyngeal valving is at least marginally adequate. Otherwise, such techniques could prove frustrating and counterproductive for the client. Second, these techniques are focused primarily on the hypernasality component of the cleft palate speech syndrome. Hopefully, if an oralized pneumatic matrix is established for speech, it may also alleviate such other symptoms as nasal emission of air, nasopharyngeal snorting, and certain articulatory problems. If it does not, techniques that are directed more specifically at these other symptoms may be required.

If surgical and/or prosthetic habilitation procedures fail to provide a competent mechanism for velopharyngeal valving, certain compensatory techniques may sometimes be useful in reducing perceived hypernasality. Techniques suggested by Mysak include exaggerated and slowed articulation patterns, locating pitch and/or loudness levels that lessen perceived nasality, and adoption of light-contact articulatory patterns. "When these compensatory techniques prove useful, the change may be attributed to encouraging oralization of the air stream, modification of the vowel spectra, and minimizing the rerouting of the speech air stream, respectively" (Mysak, 1976, p. 190).

Other Disorders of Velopharyngeal Valving

Other types of organically based nasal resonance problems may be associated with either structural or neurological impairments.

Structurally Based. Submucous cleft palate often results in hypernasality, but some children with this problem develop normal resonance balance because a relatively large adenoid pad aids velopharyngeal closure. Although temporary hypernasality (4–6 weeks) is common in normal children following adenoidectomy (Hanley & Manning, 1958), it may persist following adenoidectomy in a child with an undetected submucous cleft palate (see film study, Mysak, 1961b). While examiners should remain alert to the fairly obvious oral signs of a submucous cleft palate, a less readily apparent condition, called an occult submucous cleft, has recently been described (Croft, Shprintzen, Daniller, & Lewin, 1978; Kaplan, 1975). Although there is hypernasality in these cases, they do not exhibit the classical physical signs of submucous cleft palate, such as bifid uvula, bony notch at the posterior midline of the hard palate, or midline furrow of the velum. Because oral examination does not suggest any structural abnormality, these cases are frequently misdiagnosed (Croft *et al.*, 1978). However, diagnostic surgical exploration has revealed an anatomic abnormality that is common in classic submucous cleft palate, namely, insertion of the palatal muscles onto the hard palate rather than onto the midline raphe of the velum (Kaplan, 1975). More recently, nasopharyngoscopic examination revealed the additional structural abnormality of the absence of the uvular muscle and a small central gap in the velopharyngeal sphincter in 20 cases where a submucous cleft palate had not been previously suspected as the cause of hypernasality (Croft *et al.*, 1978).

Velopharyngeal incompetence may also result from loss of palatal tissue due to infection, trauma, or malignancy. Congenital palatal insufficiency is suspected when there is no obvious structural cause for hypernasality. In such cases, radiographic examination may reveal structural deviations that contribute to inadequate velopharyngeal valving, such as occipitalization of the atlas (resulting in unusual antero-posterior pharyngeal depth), or widespread pterygoid plates (Calnan, 1956; Fletcher, 1960; Osborne, Pruzansky, & Koepp-Baker, 1971; Randall, Bakes, & Kennedy, 1960). Anomalous muscle origins and insertions may also contribute to a velopharyngeal deficit. Because there are no obvious signs upon oral examination, very careful evaluative procedures, involving radiography, nasopharyngoscopy, and detailed clinical examination are required to distinguish these structurally based conditions from functional hypernasality.

If it is determined that hypernasality is based upon a submucous cleft palate or any other structural abnormality, management considerations are almost identical to those for overt cleft palate. That is, causal therapy should be instituted, followed by appropriate symptom therapy, if needed.

Neurologically Based. Both peripheral and central neuropathologies may cause varying degrees of velopharyngeal dysfunction. The many neuromotor disorders with which hypernasality may be associated include bulbar palsy, pseudobulbar palsy, myasthenia gravis, amyotrophic lateral sclerosis, multiple sclerosis, and chorea. Darley *et al*. (1975) discuss these problems in considerable detail. Speech prognosis is, of course, dependent upon the prognosis for the underlying neurological disorder: in some cases, compensatory therapies may serve to improve intelligibility and/or to delay deterioration.

Other Disorders of Supraglottal Origin

Obstructions or other deviations of the nares or nasal cavities may lead to hyponasal resonance imbalance. Enlarged adenoids are commonly thought to be culprits in this respect. Among other causes for hyponasality are "swollen and irritated mucous membranes, hypertrophied turbinates, misshapen septum, elevated palatal arch, broken or deformed nose, and various growths such as nasal 'tonsils,' septal spurs, synechiae polyps, syphilitic gummas " (West & Ansberry, 1968, p. 110) and "the thin nose and constricted nares of the allergic facies and the adenoid facies" (Bloomer, 1971, p. 751). Lingual tonsils are not frequently referred to as possible contributors to resonatory problem; however, Brodnitz (1967) has indicated that enlargement of the lingual tonsils may have a "muffling effect" on oral resonance.

West and Ansberry (1968, p. 224) described interference with phonation due to various kinds of obstructions in the supraglottic passageway. The direct result of such an obstruction is a reduction in vocal intensity and possibly in pitch variation. West and Ansberry (1968, p. 226) indicated that in the case of a "greatly constricted epiglottis the air pressure during phonation increases above the glottis to a point that makes vibration of the vocal bands impossible." Other conditions that interfere with the supraglottic passageway include a small faucial arch (congenital or due to scarring after oral surgery) and enlarged tonsils. Goates and Bosma (1958) studied individuals with speech disability resulting from paralysis and contracture of neck muscles and cervical lordosis and scoliosis associated with bulbar–spinal poliomyelitis. Voice problems, for example, an aspirate voice, or "a harsh, relatively low pitch, inconstant and mostly uncontrolled tone" appear to be related to "lateral displacement, compression on and immobilization of the larynx; distortion, displacement and compression with occlusion of the pharynx." It was shown that "when the lordosis was relieved, patency of the pharynx was restored, with mobilization of the mandible and release from compression of the larynx." Such procedures allowed for satisfactory speech therapy and consequent voice improvement. Here again treatment for this type of voice problem is first the job of the physician and surgeon. However, following the removal of an obstruction or the widening of a narrow vocal outlet, a period of vocal rehabilitation may be necessary because of the possibility that affected individuals have made irregular compensatory adjustments of the phonatory mechanism and have become habituated to these adjustments.

Phonatory Disorders Associated with General Systemic Conditions

Certain vocal symptoms are related to more general, systemic conditions of the organism, rather than to specific lesions, structural abnormalities, or neurological impairment of the respiratory system, the laryngeal valve, or the velopharyngeal valve.

Disorders Related to Endocrine Function

The endocrine system directly affects the larynx, both in its general growth, and in its development as a secondary sex characteristic. Examples of general growth disturbance include the laryngeal hypoplasias associated with such disorders as congenital hypothyroidism or idiopathic pituitary dwarfism. Because of the small size of the larynx, the voice is high-pitched (Mysak, 1976). Conversely, in acromegaly overproduction of the growth hormone by the pituitary results in hypertrophy of the laryngeal structures and a consequent lowering and roughening of the voice. Luchsinger and Arnold (1965) noted that the vocal symptoms of this disorder were particularly striking when they occurred in females. The fundamental frequency characteristics of patients with acromegaly have recently been documented by Weinberg, Dexter, & Horii (1975). Laryngeal growth may also be stimulated by hyperplasia or tumors of the adrenal cortex. The growth may result in premature vocal mutation in the male, or in virilization of the voice in the female (Luchsinger & Arnold, 1965). Even if the underlying causal factors can be brought under medical control, hormone-stimulated laryngeal growth is not reversible.

The persistence of a high-pitched voice in the adult male is one of the cardinal symptoms of eunuchoidism, whether the condition is primary (due to hypofunction of the testis) or secondary (due to insufficient production of the gonadotropic hormone by the pituitary gland). Although androgen treatment is often effective in these cases, voice change occurs more slowly than in normal vocal mutation, often taking 6 months to 2 years (Hirose, Sawashima, Ushijima, & Kumamoto, 1977). Luchsinger and Arnold also mentioned delayed, prolonged, or incomplete vocal mutation due to "slight delays in physical maturation on a constitutional basis" (1965, p. 194).

In the female, excessive secretion of the androgenic hormone can result in what has been termed "perverse mutation" (Luchsinger & Arnold, 1965, p. 200): the female larynx will grow to resemble the male larynx in its dimensions. Certain hormone-like drugs have also proved virilizing to female laryngeal structures (Damste, 1967). These changes in laryngeal structure result in a permanent lowering of the pitch range of the voice.

Ovarian function may also be related to changes in the female voice. Because edematous conditions are common in the two or three days prior to the onset of the menses, several authors suggest that edema of the vocal folds results in hoarseness and/or a slight lowering of pitch during the premenstrual period (Brodnitz, 1971; Luchsinger & Arnold, 1965; Moore, 1971; Smith-Frable, 1962). However, many of their clinical observations appear to have been made on women whose careers place unusual demands on the voice. It is unclear whether premenstrual vocal changes are commonly experienced by the average woman. Although Whitehead, Kohler, & Schluter (1974) reported a significant increase in vowel spectral noise levels during premenstruation, Silverman and Zimmer (1978) found no increase in hoarseness nor any lowering of fundamental frequency of vowels produced at comfortable pitch and loudness levels. However, if the demands of vocal performance required that the voice be pushed closer to its physiological limits with respect to pitch range, loudness, or endurance, premenstrual vocal changes might be more apparent. Indeed, female operatic singers very often avoid scheduling heavy performances during the premenstrual period. It has also been reported that women who use their voices intensively may experience a low in vocal efficiency when taking various kinds of ovulation inhibitors (Wendler, 1972).

In certain endocrine disorders, dysphonias may result from muscle involvement. Hypothyroidism in adults or myxedema may be associated with vocal fatigue, elevated pitch, and a tremulous quality. If an enlarged thyroid presses on the trachea or larynx, there may be pitch and quality changes (Mysak, 1976). Dysphonia has also been reported in association with hyperparathyroidism (Simpson, 1954).

Allergy, Gout, Arthritis

Brodnitz (1957) described perennial allergies such as sensitivity to dust or to certain foods as being responsible for impairing the voice. ''Unchecked allergies may interfere with breath control through asthmatic tightening of the bronchi.'' Colds that affect ears, sinuses, tonsils, bronchi, can also be mentioned here. Fatigue or illness may be causative of weak or breathy voices due to their effect on the muscles of respiration and phonation. Hoarseness resulting from gout affecting the larynx was reported by Okada (1964). A patient was described who had a history of generalized gout for 30 years with hoarseness during much of the time. Autopsy revealed a gouty tophus embedded between the true and false vocal cords, which would have interfered with normal function of the cords. Lofgren and Montgomery (1962) reported that 26% of 100 patients with rheumatoid arthritis were found to have involvement of the cricoarytenoid joint. Hoarseness from rheumatoid arthritis of the cricoarytenoid joint was reported by Bienenstock, Ehrleck, and Freyburg (1963).

Management

Vocal management of the above voice conditions should follow, when necessary, the work of the physician responsible for treating the general problem. Requests for vocal rehabilitation may be made if (a) the condition is chronic and certain compensatory vocal techniques may be helpful, and (b) the condition is resolved but faulty phonatory patterns were habituated by the patient.

Conclusions

A chapter devoted to such a dynamic area of study as human voicing and its care cannot be ''concluded.'' Undoubtedly, as this chapter is being written, or by the time it is published, new information and findings will be reported that may refute or support many of the concepts presented herein. Hence, the authors decided to end the chapter by inviting the reader to join in and take part in the challenging and rewarding exploration of the many areas of human voicing that remain obscure: such as various aspects of vocal aging, control systems in voicing, the neurophysiology of voicing, vocal changes related to specific diseases of the respiratory, glandular, cardiovascular, and nervous systems, and studies of the comparative effectiveness of various forms and combinations of vocal rehabilitation. We are also hopeful that as the chapter was read a great many more areas of fruitful research became apparent.

References

Arnold, G. E. Vocal nodules and polyps: laryngeal tissue reaction to hyperkinetic dysphonia. *Journal of Speech and Hearing Disorders,* 1962, *27,* 205–217. (a)

Arnold, G. E. Vocal rehabilitation of paralytic dysphonia, VII, Paralysis of the superior laryngeal nerve. *Archives of Otolaryngology,* 1962, *75,* 549–570. (b)

Arnold, G. E. Vocal rehabilitation of paralytic dysphonia, VIII, Phoniatric methods of vocal compensation. *Archives of Otolaryngology,* 1962, *76,* 76–83. (c)

Arnold G. E. Vocal rehabilitation of paralytic dysphonia, IX, Techniques of intracordal injection. *Archives of Otolaryngology,* 1962, *76,* 358. (d)

Arnold, G. E. Vocal rehabilitation of paralytic dysphonia, X, Functional results of intracordal injection. *Archives of Otolaryngology,* 1963, *78,* 179–186.

Aronson, A., Brown, J., & Pearson, J. Spastic Dysphonia, I and II. *Journal of Speech and Hearing Disorders,* 1968, *33,* 203–231.

Bienenstock, H., Ehrleck, G., & Freyburg, R. Rheumatoid arthritis of the cricoartenoid joint: a clinicopathologic study. *Arthritis and Rheumatism,* 1963, *6,* 48–63.

Bloch, P. Neuro-psychiatric aspects of spastic dysphonia. *Folia Phoniatrica,* 1965, *17,* 301–364.

Bloomer, H. H. Speech defects associated with dental malocclusions and related abnormalities. In L. E. Travis (Ed.), *Handbook of speech pathology and audiology.* New York: Appleton-Century-Crofts, 1971.

Boone, D. R. *The voice and its disorders.* Englewood Cliffs, N.J.: Prentice-Hall, 1977.

Brodnitz, F. The singing teacher and the laryngologist. *The Bulletin,* February 1957.

Brodnitz, F. Goals, results, and limitations of vocal rehabilitation. *Acta Otolaryngology,* 1963, *77,* 148–156.

Brodnitz, F. *Vocal rehabilitation.* Rochester, Minn.: Whiting Press, 1967.

Brodnitz, F. Hormones and the human voice. *Bulletin of New York Academy of Medicine,* 1971, *47,* 183–191.

Brodnitz, F. Spastic dysphonia. *Annals of Otology, Rhinology, & Laryngology,* 1976, *85,* 210–214.

Bzoch, R. Categorical aspects of cleft palate speech. In W. C. Grabb, S. W. Rosenstein, & K. R. Bzoch (Eds.), *Cleft lip and palate.* Boston: Little, Brown, 1971, pp. 713–733.

Calnan, J. Diagnosis, prognosis, and treatment of "palatopharyngeal incompetence," with special reference to radiographic investigations. *British Journal of Plastic Surgery,* 1956, *8,* 265–273.

Cavagna, G. A., & Magaria, R. An analysis of the mechanics of phonation. *Journal of Applied Physiology,* 1965, *20,* 301–307.

Cooper, M. *Modern techniques of vocal rehabilitation.* Springfield, Ill.: Charles C Thomas, 1973.

Croft, C., Shprintzen, R., Daniller, A., & Lewin, M. The occult submucous cleft palate and the musculus uvulae. *Cleft Palate Journal,* 1978, *15,* 150–154.

Damste, P. H. Voice change in adult women caused by virilizing agents. *Journal of Speech and Hearing Disorders,* 1967, *32,* 126–132.

Darley, F. L., Aronson, A. E., & Brown, J. R. *Motor speech disorders*. Philadelphia: W.B. Saunders, 1975.

Dedo, H. Recurrent laryngeal nerve section for spastic dysphonia. *Annals of Otology, Rhinology, & Laryngology*, 1976, *85*, 451–459.

Dedo, H., Urrea, R., & Lawson, L. Intracordal injection of Teflon in the treatment of 135 patients with dysphonia. *Annals of Otology, Rhinology, & Laryngology*, 1973, *82*, 661–667.

Dedo, H., Izdebski, K., Townsend, J. Recurrent laryngeal nerve histopathology in spastic dysphonia. *Annals of Otology, Rhinology, & Laryngology*, 1977, *86*, 806–812.

Diedrich, W. M., & Youngstrom, K. A. *Alaryngeal speech*. Springfield, Ill.: Charles C Thomas, 1966.

Duffy, R. *The vocal pitch characteristics of eleven, thirteen, and fifteen-year old female speakers*. Unpublished doctoral dissertation, University of Iowa, 1958.

Fairbanks, G. An acoustical study of the pitch of infant hunger wails. *Child Development*, 1942, *13*, 227–232.

Fairbanks, G., Herbert, E., & Hammond, J. An acoustical study of vocal pitch in seven- and eight-year old girls. *Child Development*, 1949, *20*, 71–78. (a)

Fairbanks, G., Wiley, J., & Lassman, F. An acoustic study of vocal pitch in seven- and eight-year old boys. *Child Development*, 1949, *20*, 63–69. (b)

Fletcher, S. Hypernasal voice as an indication of regional growth and development disturbances. *Logos*, 1960, *3*, 3–12.

Fletcher, S. Theory and instrumentation for quantitative measurement of nasality. *Cleft Palate Journal*, 1970, *7*, 610–621.

Ford, F. R. *Diseases of the nervous system in infancy, childhood, and adolescence*. Springfield, Ill.: Charles C Thomas, 1966.

Froeschels, E., Kastein, S., & Weiss, D. A method of therapy for paralytic conditions of the mechanisms of phonation, respiration, and glutination. *Journal of Speech and Hearing Disorders*, 1955, *20*, 365–370.

Gately, G. A technique for teaching the laryngectomized to trap air for the production of esophageal speech. *Journal of Speech and Hearing Disorders*, 1971, *36*, 484–485.

Goates, W., & Bosma, J. Disability of speech resulting from malpositioned cervical spine following poliomyelitis. *Journal of Speech and Hearing Disorders*, 1958, *23*, 283–293.

Grabb, W. C., Rosenstein, S. W., & Bzoch, K. R. (Eds.), *Cleft lip and palate*. Boston: Little, Brown, 1971.

Greene, M. *The voice and its disorders*. Philadelphia: J.B. Lippincott, 1972.

Hahn, E. Directed home training programs for cleft palate infants. In W. C. Grabb, S. W. Rosenstein, & K. R. Bzoch (Eds.), *Cleft lip and palate*. Boston: Little, Brown, 1971, pp. 830–834.

Hanley, T., & Manning, C. Voice quality after adenotonsillectomy. *Journal of Speech and Hearing Disorders*, 1958, *23*, 257–262.

Heaver, L. Spastic dysphonia: psychiatric considerations. *Logos*, 1959, 2, 15–24.

Hirano, M. Morphological structure of the vocal cord as a vibrator and its variations. *Folia Phoniatrica*, 1974, *26*, 89–94.

Hirano, M. Phonosurgery: basic and clinical investigation. *Proceedings of the 76th Annual Convention of ORL Society of Japan*, Nara, 1975.

Hirose, H., Sawashima, M., Ushijima, T., & Kumamoto, Y. Eunuchoidism: voice pitch abnormality as an autonomous syndrome. *Folia Phoniatrica*, 1977, *29*, 261–269.

Hixon, T. J. Respiratory function in speech. In F. D. Minifie, T. J. Hixon, & F. Williams (Eds.), *Normal aspects of speech, hearing, and language*, Englewood Cliffs, N.J.: Prentice-Hall, 1973.

Hixon, T., Klatt, D., & Mead, J. *Influence of forced transglottal pressure changes on vocal fundamental frequency*. Paper presented to the Acoustical Society of America, Houston, Texas, 1970.

Hixon, T. J., Mead, J., & Goldman, M. D. Dynamics of the chest wall during speech production: function of the thorax, rib cage, diaphragm, and abdomen. *Journal of Speech and Hearing Research,* 1976, *19,* 297–356.

Holbrook, A., Rolnick, M. I., & Bailey, C.W. Treatment of vocal abuse disorders using a vocal intensity controller. *Journal of Speech and Hearing Disorders,* 1974, *39,* 298–303.

Hollien, H. Vocal pitch variation related to changes in vocal fold length. *Journal of Speech and Hearing Research,* 1960, *3,* 150–156.

Hollien, H. Vocal fold thickness and fundamental frequency of phonation. *Journal of Speech and Hearing Research,* 1962, *5,* 237–243.

Hollien, H., & Jackson, B. Normative data on the speaking fundamental frequency characteristics of young adult males. *Journal of Phonetics,* 1973, *1,* 117–120.

Hollien, H., & Malcik, E. Evaluation of cross-sectional studies of adolescent voice change in males. *Speech Monographs,* 1967, *34,* 80–84.

Hollien, H., & Moore, G. F. Measurement of the vocal folds during changes in pitch. *Journal of Speech and Hearing Research,* 1960, *3,* 157–165.

Hollien, H., & Paul, P. A second evaluation of the speaking fundamental frequency characteristics of post-adolescent girls. *Language and Speech,* 1969, *12,* 119–124.

Hollien, H., & Shipp, T. Speaking fundamental frequency and chronological age in males. *Journal of Speech and Hearing Research,* 1972, *15,* 155–159.

Horn, D. *Laryngectomee survey report.* Presented at the Eleventh Annual Meeting, International Association of Laryngectomees, Memphis, Tenn., 1962.

Isshiki, N. Regulatory mechanism of voice intensity variation. *Journal of Speech and Hearing Research,* 1964, *7,* 17–29.

Isshiki, N. *Clinical experience with thyroplasty for dysphonia.* Presented at the Second International Symposium on Plastic and Reconstructive Surgery of the Head and Neck, Chicago, 1975.

Isshiki, N. Functional surgery of the larynx. In V. Lawrence (Ed.), *Transcripts of the Sixth Symposium: Care of the Professional Voice,* The Voice Foundation, 320 Park Avenue, New York, 1977.

Jackson, C., & Jackson, C. L. *Diseases and injuries of the larynx.* New York: Macmillan, 1942.

Kaplan, E. N. The occult submucous cleft palate. *Cleft Palate Journal,* 1975, *12,* 357–368.

Lauder, E. *Self-help for the laryngectomee,* 11115 Whisper Hollow, San Antonio, Texas, 78230, 1975.

Lieberman, P. *Intonation, perception, and language.* Cambridge, Mass.: M.I.T Press, 1967.

Lieberman, P. *On the origins of language.* New York: Macmillan, 1975.

Linke, C. *A study of pitch characteristics of female voices and their relationship to vocal effectiveness.* Unpublished doctoral dissertation, University of Iowa, 1953.

Lofgren, R., & Montgomery, W. Incidence of laryngeal involvement in rheumatoid arthritis. *New England Journal of Medicine,* 1962, *267,* 193–195.

Luchsinger, R., & Arnold, G. E. *Voice–Speech–Language.* Belmont, Calif.: Wadsworth, 1965.

McCall, G. Acoustic impedance measurement in the study of patients with spasmodic dysphonia. *Journal of Speech and Hearing Disorders,* 1973, *38,* 250–255.

McCall, G. Studies in spastic dysphonia: projected research concerned with central neurologic pathologies in laryngeal dysfunction. In V. Lawrence (Ed.), *Transcripts of the Sixth Symposium: Care of the Professional Voice,* The Voice Foundation, 320 Park Avenue, New York, 1977.

McCall, G., Skolnick, M. L., & Brewer, D. A preliminary report of some atypical movement patterns in the tongue, palate, hypopharynx, and larynx of patients with spasmodic dysphonia. *Journal of Speech and Hearing Disorders,* 1971, *36,* 466–470.

McGlone, R., & Hollien, H. Vocal pitch characteristics of aged women. *Journal of Speech and Hearing Research,* 1963, *6,* 164–170.

Miyasaki, T., Matsuya, T., & Yamaoku, M. Fiberscopic methods for assessment of velopharyngeal closure during various activities. *Cleft Palate Journal,* 1975, *12,* 107–114.
Moore, G. P. Voice disorders organically based. In L. E. Travis (Ed.), *Handbook of speech pathology and audiology.* New York: Appleton-Century-Crofts, 1971.
Moore, G. P. Voice problems following limited surgical excision. *Laryngoscope,* 1975, *85,* 619–625.
Moore, G. P. Observations on laryngeal disease, laryngeal behavior, and voice. *Annals of Otology, Rhinology, & Laryngology,* 1976, *85,* 553–565.
Moses, P. J. *The voice of neurosis.* New York: Grune and Stratton, 1954.
Mullendore, J. M., & Stoudt, R. J. Speech patterns of muscular distrophic individuals. *Journal of Speech and Hearing Disorders,* 1961, *26,* 252–257.
Murphy, A. *Functional voice disorders.* Englewood Cliffs, N.J.: Prentice-Hall, 1964.
Murry, T. Speaking fundamental frequency characteristics associated with voice pathologies. *Journal of Speech and Hearing Disorders,* 1978, *43,* 374–379.
Mysak, E. D. Pitch and duration characteristics of aging males. *Journal of Speech and Hearing Research,* 1959, *2,* 46–54.
Mysak, E. D. Pneumodynamics as a factor in cleft palate speech. *Plastic and Reconstructive Surgery,* 1961, *28,* 588–591. (a)
Mysak, E. D. *The unmasked palatal cleft.* Newington, Conn.: Newington Children's Hospital, 1961. (Film) (b)
Mysak, E. D. *Speech pathology and feedback theory.* Springfield, Ill.: Charles C Thomas, 1966.
Mysak, E. D. *Pathologies of speech systems.* Baltimore: Williams and Wilkins, 1976.
Netsell, R. Speech physiology. In F. D. Minifie, T. J. Hixon, & F. Williams (Eds.), *Normal aspects of speech, hearing, and language.* Englewood Cliffs, N.J.: Prentice-Hall, 1973.
Netsell, R., & Hixon, T. A non-invasive method for clinically estimating subglottal air pressure. *Journal of Speech and Hearing Disorders,* 1978, *43,* 326–330.
Ogura, J. H., & Roper, O. Surgical decompression of the recurrent laryngeal nerve in idiopathic unilateral vocal cord paresis. *Annals of Otology, Rhinology, & Laryngology,* 1961, *70,* 451–462.
Okada, T. Hoarseness due to gouty tophus in vocal cords. *Archives of Otolaryngology,* 1964, *79,* 407–411.
Osborne, G. S., Pruzansky, S., & Koepp-Baker, H. Upper cervical spine anomalies and osseus nasopharyngeal depth. *Journal of Speech and Hearing Research,* 1971, *14,* 14–22.
Perkins, W. H. *Speech pathology: an applied behavioral science.* St. Louis: C. V. Mosby, 1977.
Philips, B. J. W. Stimulating language and speech development in cleft palate infants. In W. C. Grabb, S. W. Rosenstein, & K. R. Bzoch (Eds.), *Cleft lip and palate.* Boston: Little, Brown, 1971, pp. 835–842.
Prosek, R., Montgomery, A., Walden, B., & Schwartz, D. EMG biofeedback in the treatment of hyperfunctional voice disorders. *Journal of Speech and Hearing Disorders,* 1978, *43,* 282–294.
Ptacek, P., Sander, E., Maloney, N., & Jackson, C. Phonatory and related changes with advanced age. *Journal of Speech and Hearing Research,* 1966, *9,* 353–360.
Randall, R., Bakes, F. P., & Kennedy, C. Cleft palate-type speech in the absence of cleft palate. *Plastic and Reconstructive Surgery,* 1960, *25,* 484–494.
Rigrodsky, S., Lerman, J., & Morrison, E. *Therapy for the laryngectomized patient.* New York: Teachers College Press, 1971.
Robe, E., Brumlick, J., & Moore, P. A study of spastic dysphonia: neurological and electroencephalographic abnormalities. *Laryngoscope,* 1960, *79,* 219–245.
Ryan, W., & Burk, K. Perceptual and acoustic correlates of aging in the speech of the males. *Journal of Communication Disorders,* 1974, *7,* 181–192.
Sellars, I., & Keen, E. The anatomy and movements of the cricoarytenoid joint. *Laryngoscope,* 1978, *88,* 667–674.
Shelton, R., Beaumont, K., Trier, W., & Fuir, M. Videoendoscopic feedback in training in velopharyngeal closure. *Cleft Palate Journal,* 1978, *15,* 6–12.

Shipp, T. Vertical and laryngeal position during continuous and discrete vocal frequency change. *Journal of Speech and Hearing Research,* 1975, *18,* 707–718.

Shipp, T., & Hollien, H. Perception of the aging male voice. *Journal of Speech and Hearing Research,* 1969, *12,* 703–710.

Shprintzen, R. J., Lencione, R. M., McCall, G., & Skolnick, M. L. A three dimensional analysis of velopharyngeal closure during speech and nonspeech activities in normals. *Cleft Palate Journal,* 1974, *11,* 412–428.

Shprintzen, R. J., McCall, G., & Skolnick, M. L. A new therapeutic technique for the treatment of velopharyngeal incompetence. *Journal of Speech and Hearing Disorders,* 1975, *40,* 69–83.

Silverman, E., & Zimmer, C. Effects of the menstrual cycle on voice quality. *Archives of Otolaryngology,* 1978, *104,* 7–10.

Simpson, J. Aphonia and deafness in hyperparathyroidism. *British Medical Journal,* 1954, *4869,* 494–499.

Skelly, M., Donaldson, R., Scheer, G., & Guzzardo, M. Dysphonias associated with spinal bracing in scoliosis. *Journal of Speech and Hearing Disorders,* 1971, *36,* 368–376.

Skolnick, M., Shprintzen, R., McCall, G., & Rakoff, S. Patterns of velopharyngeal closure in subjects with repaired cleft palates and normal speech: A multi-view videofluoroscopic analysis. *Cleft Palate Journal,* 1975, *12,* 369–376.

Smith-Frable, M. Hoarseness, a symptom of premenstrual tension. *Archives of Otolaryngology,* 1962, *75,* 66–68.

Snidecor, J. *Speech rehabilitation of the laryngectomized.* Springfield, Ill.: Charles C Thomas, 1962.

Snidecor, J. Speech without a larynx. In L. E. Travis (Ed.), *Handbook of speech pathology and audiology.* New York: Appleton-Century-Crofts, 1971.

Snidecor, J., Isshiki, N., & Kimura, T. Speech after laryngoplasty (the Asai operation). *Proceedings des Seances du XIV Congres de L.I.A.P.,* Paris, 1968.

Spriesterbach, D. C., & Sherman, D. (Eds.) *Cleft palate and communication.* New York: Academic Press, 1968.

Taub, S., & Spiro, R. Vocal rehabilitation of laryngectomees: preliminary report on a new technique. *American Journal of Surgery,* 1972, *124,* 87–90.

Van Riper, C., & Irwin, J. *Voice and articulation.* Englewood Cliffs, N.J.: Prentice-Hall, 1958.

von Leden, H., Moore, P., & Timcke, R. Laryngeal vibrations: measurements of the glottic wave, III, The pathological larynx. *Archives of Otolaryngology,* 1960, *71.*

Waldrop, W. F., & Gould, M. A. *Your new voice.* New York: American Cancer Society, 1956.

Weinberg, B., Dexter, R., & Horii, Y. Selected speech and fundamental frequency characteristics of patients with acromegaly. *Journal of Speech and Hearing Disorders,* 1975, *40,* 253–259.

Weinberg, B., Shedd, D., & Horii, Y. Reed-fistula speech following pharyngolaryngectomy. *Journal of Speech and Hearing Disorders,* 1978, *43,* 401–413.

Wendler, J. Cyclically dependent variations in the efficiency of the voice and its influencing by ovulation inhibitors. *Folia Phoniatrica,* 1972, *24,* 259–277.

West, R. W., & Ansberry, M. *The rehabilitation of speech.* New York: Harper and Row, 1968.

Westlake, H., & Rutherford, D. *Cleft palate.* Englewood Cliffs, N.J.: Prentice-Hall, 1966.

Whitehead, R. Kohler, R., & Schluter, S. *The effect of the female menstrual cycle of vowel spectra.* Paper presented at the annual meeting of the American Speech and Hearing Association, Las Vegas, November 1974.

Wilson, D. K. *Voice problems of children.* Baltimore: Williams and Wilkins, 1972.

Zemlin, W. A two-hour time lapse dissection of the respiratory system, the larynx, and articulatory system through the use of slides. Presented at the Seventh Symposium: Care of the Professional Voice, The Juilliard School, New York, 1978.

Zwitman, D., & Calcaterra, T. Phonation using the tracheo-esophageal shunt after total laryngectomy. *Journal of Speech and Hearing Disorders,* 1973, *38,* 369–373.

3

E. Harris Nober and Gerard L. Kupperman

PHYSIOGENIC HEARING IMPAIRMENT IN ADULTS

Modern Trends in Audiology

Expansion and development in the field of audiology during the last decade have been vast. The availability of improved diagnostic and remedial procedures has broadened the range of benefits and resources now employed for the auditorily impaired. Many of the advances in audiology are spinoffs from allied fields. From medicine, otologic surgeons developed middle ear reconstruction techniques that often restore hearing to normal acuity. Medical treatment using improved antibiotics and preventative otologic procedures have also resulted in notable contributions. Innovations in engineering and electronics are responsible for smaller, more efficient, and more flexible amplification devices worn as individual or group hearing aids. Computer technology has become the basis for several audiological procedures. The audiologist's hearing test battery contains new diagnostic tests, many based on modern technological advances. Hence, in the past decade, significant improvements have occurred in hearing assessment, prevention, aural rehabilitation, and medical and surgical restoration. Furthermore, there have been noteworthy advances in educational management and federal legislation mandating rehabilitative services. This chapter will attempt to inform the reader of the current status of audiometry.

Diagnostic audiometry has evolved to a relatively sophisticated component of

E. Harris Nober and Gerard L. Kupperman • Department of Communication Disorders, University of Massachusetts, Amherst, Massachusetts 01003.

audiology over the years. While many of the techniques are an outgrowth of the basic pure tone and speech tests, modern procedures employing electronic measurements have opened up new dimensions in diagnosis, treatment, and research. The foremost goal in diagnostic audiometry is to locate the site of lesion and to assist the otologist in determining the nature and etiology of the lesion. Most new tests still employ pure tones, speech, and noise as the basic stimuli but with improved methods for delivery. A good deal of effort has been expended to refine tests that differentiate between peripheral and central disorders. This differentiation is crucial for rehabilitation since the two systems functionally behave quite differently.

Terminology

Peripheral Auditory Disorders. Lesions located within the outer and middle ear conductive mechanism, the cochlea and the eighth nerve, are considered peripheral disorders. In other words, all lesions that are distal to the synapse between the auditory nerve and the dorsal and ventral cochlea nuclei are considered peripheral (J. Jerger, 1973).

The most common terms used today to describe the types of peripheral auditory pathology are *conductive* and *sensorineural.* A hearing loss associated with a *conductive impairment* of the outer or middle ear mechanism results from a blockage or reduction of sound transmission into the inner ear. Any pathogenic or structural aberration to the external auditory meatus, the ossicular chain, the oval and round windows, the tympanic membrane (eardrum), and other structures may result in a conductive hearing loss or a reduction in auditory sensitivity.

Sensorineural Impairment is a broadly defined term which designates disorders of the cochlea (sensory), the eighth cranial-auditory nerve (neural) or both. Sensorineural impairments include a wide range of physiological structures and functions that are quite complex. Still another term associated with peripheral pathology is *mixed loss,* which refers to the presence of both a conductive and a sensorineural impairment in the same ear.

Central Auditory Impairment occurs when the lesion is proximal to the boundary at the first synapse between the first and second order neurons of the afferent auditory pathway in the dorsal and ventral cochlear nuclei (J. Jerger, 1973). Hearing sensitivity to pure tones is usually not affected but speech discrimination and language perception are affected. Thus, central auditory impairment does not necessarily imply hearing "loss" to pure tone stimuli.

Hearing Loss is a reduction in auditory sensitivity to pure tones and speech that exceeds the upper limit of the normal hearing range. The normal range extends from −10 to 25 dB with the latter the upper limit for normal hearing sensitivity. Clearly, while hearing impairment technically begins at 26 dB (averaging the pure tone thresholds at 500, 1000, and 2000 Hz), the deleterious effects on communication for a child who has not yet developed language will differ vastly from that for the adult with fully developed language. There is a trend to use terms such as *hearing level* or *hearing threshold level* but "hearing loss" is preferred to designate type of impairment (i.e., conductive hearing loss). These terms were established to avoid confusion in clinical reports that used to read "there is a 20-dB hearing loss" (referring to hearing level) "but the hearing is normal" (referring to upper hearing range). Reports should now indicate there is a 20-dB threshold shift or a 20-dB hearing threshold level, but thresholds are still within the range of normalcy.

Deafness is a generic term that can refer to the extent of hearing loss such as severe or complete loss of hearing (Davis & Silverman, 1978) or can refer to a type of loss, i.e., sensorieneural deafness. Deafness technically starts at 93 dB, and at this level, the loss of hearing is sufficient to make communication for ordinary life purposes very difficult, even with the use of a hearing aid or some other means of amplification.

Hard of Hearing refers to a partial loss of hearing where the impairment affects normal communication.

Hearing Impaired or Auditorily Impaired are terms being used in educational settings as general designations or as noncategorical terms to indicate any problem in audition regardless of severity.

Dysacusis is a medical term used to describe any auditory complication that may or may not be associated with a loss of sensitivity. It usually refers to a dysfunction of the cochlea, the auditory nerve, the auditory pathways, the temporal cortex, or any combinations. The various types of dysacuses are: discrimination loss, phonemic regression, recruitment, diplacusis, auditory agnosia, and receptive aphasia.

Audiometer Calibration for Pure Tones

Prior to January 1, 1965, the American Standards Association (ASA) reference threshold levels for audiometers (ASA, 1951) were officially used for the calibration of all audiometers in the United States. The American values were based on the United States National Health Survey of 1937. But hearing threshold studies conducted in England in the early 1950s yielded values that differed significantly from the American threshold values. Since the English thresholds were obtained with improved instrumentation, better psychometric controls, and more favorable acoustic conditions, other European countries readily adopted the improved British standard. As the years progressed, the International Standards Organization (ISO) sponsored about 15 separate studies to explore the difference between the American and British standards. The agreements were universal—the ASA standard of 1951 was found too lenient. Several more years passed and finally all parties agreed to a universal and uniform calibration system. On January 1, 1965, the American Academy of Ophthalmology and Otolaryngology, the American Otological Association, and the American Speech and Hearing Association collectively adopted the ISO reference levels (ISO, 1964) for audiometers.

The acceptance of the ISO standard in the U.S. was remarkably swift. The ISO (1964) calibration for pure tones was an average of 11 dB above the ASA (1951) scale at 500, 1000, and 2000 Hz. To convert from ASA (1951) to ISO (1964) the following number of decibels were added to each frequency:

Hz	125	250	1000	1500	2000	3000	4000	6000	8000
dB difference	9	15	10	10	8.5	8.5	6.0	9.6	11.5

In 1969, the American National Standards Institute (ANSI) published the currently used "Specifications for Audiometers" known as "ANSI-S3.6-1969" or the ANSI-1969 Standard. Although the pure tone reference values were identical to the ISO (1964) standard, there were more stringent requirements for other aspects of calibration. For example, when speech is used as the stimulus instead of pure tones, the sound pressure level (SPL) reference will depend on whether a loudspeaker or an earphone delivers the sound. If the earphone being used is the TDH-39 model, then the 0-dB reference level is 20 dB SPL. If, however, words are presented through a loudspeaker, then the reference level changes to 15 dB SPL. The difference compensates for the perceptual loudness gain associated with sound-field audition vs. earphone audition.

Air and Bone Conductive Testing

Air Conduction. The purpose of pure tone air conduction threshold audiometry is to determine hearing threshold sensitivity by testing the entire auditory processing mechanism as a functional unit. Individual threshold values are compared to a standard statistical norm in the hearing range from 125 Hz to 8000 Hz. Consequently, pure tone thresholds show a relationship between an individual and the population norm.

Auditory thresholds are measured by a variety of techniques. The most common procedure involves an established psychophysical technique known as the "method of limits." The limits procedure was orginally adopted for determining auditory threshold by Hughson and Westlake (1944), modified by Carhart and J. Jerger (1959) and officially endorsed by the American Speech and Hearing Association in 1978. In the limits technique, auditory threshold is determined by increasing intensity in 5 dB increments until the subject responds. Then, intensity is reduced and the process repeated. The lowest intensity level at which one-half of the tones are detected (with a minimum of three responses) is designated auditory threshold. The American Speech and Hearing Associations (1978) official adoption of the limits procedure was based on the simple instructions, the ease with which thresholds can be determined, and excellent test–retest reliability obtained. Clearly, pure tone air conduction audiometry is still the core of the clinical hearing evaluation and is the first test employed in the clinical battery (Martin & Forbis, 1978).

Bone Conduction. The purpose of pure tone bone conduction audiometry is to assess the hearing threshold sensitivity of the sensorineural mechanism while theoretically excluding the conductive mechanism (Tonndorf, 1966). The conductive mechanism is bypassed functionally by directing tones on to the mastoid bone as an alternate route to the cochlea. The threshold difference between the air and bone conduction values (the air–bone gap) indicates the degree of impairment attributed to the sensorineural mechanism. Hence, bone conduction is a measure of cochlea integrity, while air conduction is a measure of outer, middle ear and cochlea integrity.

Bone conduction audiometry as a site of lesion test has always been a source of voluminous research. One area that has received extensive investigation involves techniques to eliminate the participation of the ear not being tested called the *contralateral* ear. The ear under test is called the *ipsilateral* ear. The best solution to exclude the contralateral ear during a test has been to use masking (Goetzinger, Proud, & Embrez, 1962; Hood, 1960; Palva & Palva, 1962; Studebaker, 1964). The masking stimulus is usually a form of noise. Although various kinds of noise, such as white, saw tooth, and complex noise, have been tried through the years, the preferred masking noise is white noise divided into critical frequency bands. Each critical frequency band has as its center frequency the test tone (Egan & Hake, 1950; Fletcher, 1940; Greenwood, 1961a,b).

Critical band (narrow band) masking helps to minimize the spurious artifacts caused by *remote* and *central* masking phenomena (Ward, 1963). *Remote masking* refers to the threshold shift that occurs at frequencies that are away (remote) from the masking stimulus. *Central masking* is a neurological phenomenon in which the contralateral threshold is increased by a masking stimulus supposedly too weak to cross transcranially to the contralateral side. In addition to the difficulties incurred by remote and central masking, there is a low frequency attenuation caused by the middle ear reflex (contractions of the stapedius and tensor tympani muscles) that can amount to 20 dB (Ward, 1963). While many masking procedures have been employed, the Hood (1960) technique has achieved wide acceptance since it uses critical band masking.

a. *Rainville Test.* The first major improvisation to circumvent the multitudinous complexities of masking was the Rainville test. Rainville (1959) determined an absolute bone-conduction threshold by masking a pulsed air-conduction threshold with a white noise given first through a bone-conduction oscillator on the ipsilateral mastoid and then masked by a white noise delivered through the same air-conduction receiver that delivered the tone stimulus. Rainville's technique is questionable once the difference between the two ears exceeds 40 dB; a difference between 40–60 dB has marginal value and beyond 60 dB the Rainville technique appears valueless.

b. *Modified Rainville Test.* Lightfoot (1960) modified the Rainville technique by placing the masking bone oscillator on the forehead to mask a pulsed air-conduction tone that is presented to the ear through a receiver. The amount of bone-conduction noise required to mask the air-conduction threshold of normal subjects was predetermined and the difference between this amount and the amount required for the hearing-impaired subjects indicated the sensorineural loss.

c. *Sensorineural Acuity Level Test.* J. Jerger and Tillman (1960) introduced the sensorineural acuity level test. It is very similar to the modified Rainville. They obtained the amount of noise needed to shift the air-conduction thresholds on normal subjects 20 dB at 250 Hz, 45 dB at 500 Hz, and 50 dB at 1000, 2000, and 4000 Hz, respectively. The threshold shifts of defective ears are subsequently compared with the established normal shift values. The difference indicates the sensorineural loss.

The sensorineural acuity level test has been used successfully for testing conductive impairments (Lynn & Pirkey, 1962) and for predicting stapedectomy results (Michael, 1963). However, the validity is dubious at the low frequencies, i.e., 250 and 500 Hz in conductive and mixed-loss impairments (Dirks & Malquist, 1969; Goldstein, Hayes, & Peterson, 1962). The thresholds are actually poorer than those obtained by conventional methods. Tillman (1963) reevaluated his sensorineural acuity level test and agreed that the rest was not a suitable substitute for conventional bone-conduction testing in all instances. Other problems started to emerge and eventually a series of studies by J. Jerger and S. Jerger (1965), Tillman (1967), and Tillman and Gretis (1969) seriously challenged the value of the technique in various populations.

Speech Audiometry

Since the very foundation of audiology is steeped in the speech communication process, it is natural that speech recognition testing has become part of basic audiometry. Pure tone audiometry fulfills an enormous assessment function, but speech audiometry brings the hearing evaluation process a step closer to daily application. If any generalization is permissible, it is that pure tone audiometry is basically diagnostic for peripheral hearing assessment, while speech audiometry assesses perceptual processing of the speech stimuli at both the peripheral and central levels.

While it can be argued that modern speech audiometry had its inception in the 1940s at the Harvard Psycho-Acoustic Laboratory, the foundation was actually laid several years earlier by Fletcher and Steinberg at the Bell Telephone Laboratory, where the "articulation function" was conceived (Fletcher & Steinberg, 1929). This function established a quantified relationship between speech perception and speech intensity. It is noteworthy that the first audiometer was the Western Electric 4A developed in 1927 and introduced by Fletcher in 1929.

Like pure tone audiometry, which includes threshold and suprathreshold testing, speech audiometry employs parallel classifications. The speech threshold measure was designated the "speech reception threshold" by Hughson and Thompson (1942); it is the decibel intensity level at which 50% of the words (usually bisyllabic spondee words) are correctly identified by the listener. The speech reception threshold can be equated to the pure tone audiogram by determining the "pure tone average" of the threshold values for 500, 1000, and 2000 Hz. Hence, the pure tone average can be used as an estimate of the speech reception threshold when it is not possible to obtain a speech reception threshold directly. The relationship between the pure tone average and the speech reception threshold diminishes where there is a dysacusis or other processing problem. Lack of pure tone average/speech reception threshold (PTA/SRT) agreement is suggestive of complications such as central disorders, functional auditory dysfunction (Carhart & Porter, 1971; Davis & Silverman, 1978).

The suprathreshold speech test in popular use today is an auditory discrimination test that uses "phonetically balanced" word lists. Like the spondee words, phonetically balanced word lists were developed at the Harvard Psycho-Acoustic Laboratory (Egan, 1948) and revised at the Central Institute for the Deaf (Hirsh, Davis, Silverman, Reynolds, Eldert, & Benson, 1952). The objective of auditory word discrimination testing is to determine the maximum discrimination score of monosyllabic words delivered at a comfortable listening level, usually 40–60 dB above the speech reception threshold in normal or conductively involved ears. A comfortable listening level may be less for sensorineural ears with recruitment since the operating range of hearing is reduced. The latter is the decibel span between the threshold of discomfort and the speech reception threshold. Most phonetically balanced word lists contain 50 words, so each correctly identified word is weighted 2%, with 100% as the perfect score. Any score between 90–100% is considered within normal limits. Numerous other tests have been devised, but the Central Institute for the Deaf W-1 spondee lists and the Central Institute for the Deaf W-22 lists are commercially available on recorded discs and are most commonly used (Martin & Forbis, 1978).

Audiologists observed that recordings of phonetically balanced words by Rush Hughes provided word discrimination scores that differed from scores obtained with the Ira Hirsh W-22 discs for subjects with retrocochlear and central lesions. The critical element for the difference was that the Hughes recordings had an unusual amount of background noise. As a result of this observation, several procedures systematically adding noise to the word discrimination lists were developed (J. Jerger & S. Jerger, 1975a; Olsen, Noffsinger, & Kurdziel, 1975).

A commonly used phonetically balanced word discrimination procedure to differentiate between cochlear and retrocochlear lesions is the performance intensity–phonetically balanced (PI–PB) function (J. Jerger & S. Jerger, 1971). This measures the performance intensity function (articulation function) for phonetically balanced words. In the normal or conductive ear, there is a sharp ascending growth of word discrimination scores from 0 to 100% as intensity is increased. At maximum discrimination, the score reaches a plateau and generally remains stable. In instances of eighth nerve disorder, however, the word discrimination score decreases abruptly after it peaks with additional increases in intensity. This decline in performance, called the "rollover effect," can be quite dramatic and pronounced.

It is noteworthy that speech audiometry has traditionally occupied a major role in audiometry for differential diagnosis (J. Jerger, 1964), aural rehabilitation (High, Fairbanks, & Glorig, 1964), and in hearing aid selection (Olsen & Carhart, 1967).

Special Site of Lesion Tests

One major attribute of diagnostic audiology is the information it can provide the otologist in determining the nature, extent, and locus of pathology in the auditory system. Audiometric tests combined with other pertinent information, i.e., patient's description of symptoms, medical history, and otological, neurological, and radiological reports, can collectively provide adequate data to identify the impairment to the outer or middle ear, the cochlea, the auditory nerve and its pathways, or to the temporal cortex (J. Jerger, 1962). Identifying the site of the pathology with certainty requires an integrated battery of tests. The results of any one test or symptom must be confirmed, as no one symptom or test is universally valid. Recruitment, for example, is absent in some cases with cochlear pathology because there may be an overlay of retrocochlear pathology. Also, reliable bone-conduction thresholds do not always yield a valid estimate of cochlear sensitivity. In many instances of stapes fixation, the invalid bone-conduction thresholds are symptomatic of otosclerosis (Carhart, 1950). Nober (1964, 1967, 1970) demonstrated consistent pseudoauditory bone-conduction responses with deaf children and demonstrated that these responses paralleled the tactile thresholds elicited from nonauditory areas of the body. Hence, it is apparent there is need for a comprehensive auditory clinical test battery to identify the site of lesion (J. Jerger & S. Jerger, 1975a; Johnson, 1965, 1966, 1968; Tillman, 1969).

Bekesy Audiometry. Carhart (1958) discussed in elaborate detail the kinds of functional and diagnostic information that can be gleaned from the classicial pure tone air- and bone-conduction audiogram. Additional diagnostic information is provided when the air-conduction thresholds are ascertained by employing a different psychophysical technique known as the *method of adjustment*. With this technique the intensity dial is controlled by the subject rather than by the examiner. This procedure was developed by Bekesy (1947) and has acquired the name Bekesy audiometry.

J. Jerger (1960) observed that there was considerable diagnostic significance about the site of lesion when Bekesy threshold traces were obtained under two methods of tone presentation: (1) a continuous tracing and (2) an interrupted tracing. The relationship between these two methods of obtaining thresholds led Jerger to categorize the Bekesy tracings into four types. Later he described a fifth (J. Jerger & Herer, 1961). Types I–IV reflect some type of organic pathology while Type V is symptomatic of a functional hearing impairment. (See Nober, chapter 10.) Specifically, Type I shows the continuous and interrupted tracings intertwined; in this category are normal èars, ears with conductive impairments and occasionally a cochlear impairment. Type II shows overlapping of the continuous and interrupted tracings up to 1000 Hz but then the continuous tracing drops 10–15 dB below the interrupted and runs parallel to it for the remainder of the range; cochlear impairments fall into this group. In Type III, the intensive separation between the two tracings starts immediately as the continuous tracing runs below the interrupted and may even terminate at 1000 Hz. This type occurs with eight-nerve tumors. Type IV depicts the continuous tracing parallel to but significantly lower than the interrupted; eighth-nerve tumors commonly produce this pattern too.

Variations of the Bekesy procedure are constantly being devised. Palva, Karja, & Palva (1970) and J. Jerger, S. Jerger, and Mauldin (1972a) observed patterns characteristic of retrocochlear pathology when thresholds were measured first from low-to-high frequency (forward) and then from high-to-low frequency (backward). Thresholds obtained from continuous backward tracing were lower than those obtained from forward tracing. This discrepancy between forward and backward tracings may be a more sensitive means of identifying retrocochlear pathology than conventional Bekesy audiometry. Karja and Palva (1970) attributed the forward and backward threshold discrepancy to abnormal adaptations.

J. Jerger and S. Jerger (1974b) described another modification of threshold Bekesy measures, in which subjects adjust the level of the tone to keep it within a range of comfortable loudness. These suprathreshold measurements are based on the concept that abnormal auditory adaptation is first manifested at suprathreshold levels and is found closer to threshold as the causative factors worsen. J. Jerger and S. Jerger (1974a,b) and colleagues (J. Jerger *et al.*, 1972a) consider both modifications of the conventional threshold Bekesy procedure to be useful diagnostic tests.

Tone Decay Test. Another test used to differentiate between cochlear and retrocochlear impairments is the tone decay test. The original research on the test is based on the work of Kobrak, Lindsay, and Pearlman (1941), who observed that a continuous tone presented at thresholds for a period of time may become inaudible with certain types of auditory pathologies. This phenomenon forms the basis of the continuous and interrupted tracing differences of the Type III and IV Bekesy tracings described above. Continuous tone presentation precipitates "perstimulatory fatigue" in some pathological conditions. In the Carhart (1957) version of the tone decay test, a pure tone is presented and sustained for 60 seconds at threshold unless the tone becomes inaudible. Then the tone is increased 5 dB and presented again until audible for a full 60 seconds. Green (1963) modified the Carhart test by presenting the tone for a total of 60 seconds and increasing the intensity when necessary to maintain threshold audibility during the 60-second period. Conductive losses and normal ears should not show any auditory adaptation. Occasionally a sensorineural ear will give negative results or show small amounts of adaptation in the high frequencies but rarely in the low frequencies. Eighth nerve lesions show extreme adaptation (Carhart, 1958; Green, 1963; Hood, 1955; Sanders, Josey, & Glasscock, 1974). Whereas the original tone decay test and its modifications are performed at or near pure tone threshold, J. Jerger and S. Jerger (1975b) proposed the "suprathreshold adaptation test." A pure tone is presented at 110 dB SPL and must be perceived for 60 seconds. Individuals with retrocochlear lesions are unable to perceive the tone for the full test period while those with cochlear lesions can. The test is useful in identifying retrocochlear lesions while reducing the false positive rate.

Short-Increment Sensitivity Index Test. Another air-conduction test used for differential diagnosis is the short-increment sensitivity index test (J. Jerger, Sheed, & Harford, 1959). The short-increment sensitivity index tests the ability to detect one-decibel incremental pulses that are added to a pure tone presented 20 dB above the individual's threshold for a given frequency. The final score is the percentage of correctly identified increments of 20 presentations. Since there are 20 pulses given at random intervals, each correct identification is weighed 5%. Most audiologists consider this test an indirect test of recruitment but J. Jerger (1962) contends that the short-increment sensitivity index is not a test of recruitment; rather it is a test of the ability to hear small intensity increments, which is associated with recruitment. High short-increment sensitivity index scores (60% or more) are associated with cochlear lesions while low short-increment sensitivity index scores (0–20%) are associated with conductive losses, normal hearing, or retrocochlear lesions. The test derives its popularity from the ease with which it can be administered.

Alternate-Binaural Loudness-Balance Test. Many audiologists consider the alternate-binaural loudness-balance test the only direct measure of loudness recruitment. Recruitment is an abnormal or nonlinear increase in perceived loudness just above detection threshold. It is measured relative to linear increases in intensity (Fowler, 1936, 1963, 1965). Thus, the loudness and intensity parameters do not parallel each other when recruitment is present. The alternate-binaural loudness-balance procedure requires one normal ear since the subject matches the loudness growth of a normal ear to the loudness growth of the abnormal ear. In the instance of binaural hearing impairment, a "monaural loudness-balance test" is given by comparing the loudness growth of a frequency heard with normal sensitivity to the loudness growth of a frequency where the hearing sensitivity is abnormal.

There are several ways to administer the alternate-binaural loudness-balance test. Generally, the tone is presented 20 dB above the threshold of the normal ear at a chosen frequency. As the audiometer alternates the tone from ear to ear, the subject adjusts the tone intensity in the abnormal ear until he judges the loudness equal to the normal ear. The alternate-binaural loudness-balance procedure is repeated in successive 20-dB increments (i.e., 20 dB, 40 dB, 60 dB, 80 dB) until the discomfort level is reached. The values of the two ears are plotted on an audiogram and connected by lines for each 20-dB level. The graphic result is called a "laddergram." Actually, the first threshold responses represent the bottom rung of the audiometric ladder and the discomfort level is the top rung. The arithmetic difference between the top rung (the threshold of discomfort) and the bottom rung (detection of threshold) is extremely significant. If the rungs are parallel at each 20-dB level (or the first and last rung) recruitment is probably not present; if the rungs are not parallel, recruitment may be present. As an example, suppose the normal and abnormal ears vary by 25 dB at detection threshold. If the 25-dB difference between the two ears prevails at a higher level (such as 80 dB vs. 105 dB), loudness growth was equal for the two ears. Alternately, if the loudness difference dissipates and both ears perceive equal loudness at 80 dB, then recruitment is probably present. On occasions, hyperrecruitment occurs, in which the defective ear perceives loudness greater than the normal ear to the higher intensity levels. Recruitment is pathognomonic to cochlear pathology.

Acoustic Impedance. The term acoustic impedance refers to the opposition to the flow of energy. Clinical measurement of acoustic impedance of the middle ear is now commonplace. Early research was initiated by Webster (1919) and continued by Metz (1946) and Zwislocki (1957a,b). Intensive study during the last two decades has made acoustic impedance measurements an integral part of audiological evaluation.

Impedance audiometry has a wide range of applicability to both children and adults. One advantage of the procedure is that it does not require active participation by the patient. Thus, it can assist the audiologist in the evaluation of very young children. Some of the more noteworthy applications include the evaluation of Eustachian tube patency (Harper, 1961; Metz, 1953), middle ear fluid (Liden, Peterson, & Bjorkman, 1970a,b; Thomsen, 1961; Wilber, Goodhill, & Houge, 1970), eardrum scar tissue (Lilly, 1970), eardrum perforations (Terkildsen, 1964), middle ear cavity pressure (Alberti & Kristeneen, 1970), otitis media (Brooks, 1970), otosclerosis (Anderson & Barr, 1971), ossicular chain discontinuity (Djupesland, 1969; Elbrond, 1970), abnormal auditory adaptation (Anderson, Barr, & Wedenberg, 1970a), acoustic tumors (Anderson, Barr, & Wedenberg, 1970b), differential diagnosis (Robertson, Peterson, & Lamb, 1968), loudness recruitment (Beedle & Harford, 1970), facial nerve damage (Liden, Peterson, & Harford, 1970c), brainstem lesions (Griesen and Rasmussen, 1970), and intraaural muscle contraction (J. Jerger, 1970).

Two types of impedance measurements may be made. Static (absolute) impedance measurements determine the opposition to the flow of energy while the middle ear is at rest. Dynamic impedance refers to changes in impedance that occur when the eardrum is under varying amounts of air pressure. Dynamic impedance measurements are more useful clinically. A chart called a tympanogram plots changes in compliance of the eardrum as air pressure is varied from

+200 mm of water pressure to −400 mm of water pressure. The point of maximum compliance corresponds to the amount of air pressure behind the eardrum. Different auditory aberrations show characteristic patterns that have been labeled types A, B, C (J. Jerger, 1970).

Type A pattern is characterized by a normal size compliance peak between 0 and −100 mm of water pressure (J. Jerger, S. Jerger, & Mauldin, 1972b). *Type B* pattern is essentially flat, with little or no compliance peak present within the operating range of the tympanometer. This pattern indicates excessive stiffness of the ossicular chain and is associated with fluid present in the middle ear. *Type C* pattern is associated with a lack of sufficient air pressure in the middle ear space often due to Eustachian tube dysfunction. The peak may be normal or reduced in size, but is shifted in the negative direction and may occur at or beyond −100 mm of water pressure.

The usefulness of tympanometry in the diagnosis of middle ear disease is well documented. J. Jerger, Anthony, S. Jerger, and Mauldin (1974a), Berry, Andrus, Bluestone, and Cantekin (1975), Paradise, Smith, and Bluestone (1976), Shurin, Pelton, and Finkelstein (1977), and Neel, Keating, and McDonald (1977) are among many who have demonstrated that tympanometry can assist in improved medical management of middle ear effusion. Because tympanometry is a painless and quick procedure that does not require active participation by the patient, it has been used to determine the middle ear studies of neonates within the first hours of life (Keith, 1973, 1975).

Tympanometry can supplement pure tone screening of children in the public schools. Fulminating (and sometimes undetected) middle ear problems, not yet severe enough to cause hearing loss, can be identified by tympanometry. Tympanometry can serve as an early detection measure of middle ear impairment and consequently as a catalyst for early medical intervention. Renvall, Liden, Jungert, and Nilsson (1973), Brooks (1974, 1977), Cooper, Gates, Owen, and Dickison (1975), and Walton (1975) employed large-scale screening procedures and reported that results were most accurate when pure tone and impedance data were combined.

One major aspect of impedance audiometry is acoustic reflex measurement. The acoustic reflex is a bilateral simultaneous contraction of the stapedius muscle elicited by loud sound. Muscular contraction increases middle ear impedance. The presence or absence of the reflex, and the stimulus sensation level necessary to elicit it, are of considerable importance.

Acoustic reflex assessment provides crucial information for differential diagnosis of cochlear vs. retrocochlear disorders (J. Jerger, 1975). J. Jerger, Burney, Mauldin, and Crump (1974b) were among the first to report that the acoustic reflex occurred at reduced sensation levels in ears with cochlear impairment. Decay of the reflex in ears with retrocochlear lesions was described by Anderson, Barr, and Wedenberg (1969). Reflex levels in cochlear-impaired ears occur below 60-dB sensation level. The reflex for normal ears occurs approximately 70 to 100 dB above threshold. Mild or moderate sensorineural-impaired ears that do not show reduced acoustic reflex levels or demonstrate reflex decay may have retrocochlear pathology.

The procedure to obtain the acoustic reflex is to stimulate one ear with a pure tone and measure changes in impedance of the opposite or contralateral ear; this procedure is also called the "crossed" reflex. Instrumentation is now available to stimulate and record the reflex on the same or ipsilateral ear. The ipsilateral or "uncrossed" reflexes are useful in differentiating eighth-nerve tumors from brain stem tumors (S. Jerger & Jerger, 1975; S. Jerger, Neely, & J. Jerger, 1975; S. Jerger & J. Jerger, 1977).

Because contraction of the stapedius muscle is controlled by the seventh cranial (facial) nerve, the integrity of this nerve can be assessed by measuring the reflex. Alford, J. Jerger, Coats, Peterson, and Weber (1973) have demonstrated that reflex testing combined with the Schirmer Test of Lacrimation, electrogustometry, and submaxillary gland secretion studies can zero in on the locus of a facial nerve lesion.

Several researchers have gone beyond the use of pure tones as the reflex stimulus and used filtered bands of noise to estimate the extent and configuration of hearing thresholds. Niemeyer and Sesterhenn (1974) were able to determine the average amount of hearing loss for 73% of the 223 sensorineural ears tested to within ±10 dB by comparing the difference between the reflex thresholds for pure tones and for white noise. J. Jerger *et al.* (1974b) extended these observations to 1043 ears with sensorineural loss and was successful in predicting thresholds in all but 4% of the cases. Van Wagoner and Goodwine (1977) stressed the usefulness of filtered noise for estimating auditory threshold in pseudohypacusis, particularly in medical–legal cases.

Electrophysiological Audiometry. In electrophysiological audiometry, the response is a covert physiological reaction to an induced sound stimulus that is recorded by some electronic process. The recording process determines the nature of the test procedure. Electrophysiological audiometry is essentially used for assessing children and other difficult-to-test persons. An exception to this statement is electrodermal audiometry, which is used basically for malingerers and people with functional auditory impairments.

a. *Electrodermal response audiometry*. When applied to adults, this procedure is used principally to identify functional auditory impairments. In the 1950s when the procedure was being tested for reliability and validity, it was used on special populations such as the brain injured or the mentally retarded, but the results were variable and inconsistent with these populations. At the outset, the technique was called the *psychogalvanic skin response test,* which was subsequently abbreviated to the *galvanic skin response test,* and later changed to *electrodermal response*. The technique worked better on adults and older children. In many instances thresholds obtained on these people using electrodermal audiometry were within 10 dB of standard audiometric techniques (Nober, 1958). In electrodermal audiometry, the response is a reduction in skin resistance to an auditory stimulus, usually given as a conditioned tonal stimulus; hence, the recorded value is a conditioned response. The reduction in skin resistance is crucial since the recording circuitry uses a Wheatstone bridge arrangement and a reduction in skin resistance allows a greater flow of current through the Wheatstone bridge; changes in current flow are recorded by a stylus onto a chart. The electrodermal audiometry is mediated by the autonomic nervous system in contrast to most of the other audiometric data that are mediated via the central nervous system. Details of this procedure are outlined by Nober in Chapter 10.

b. *Electroencephalographic audiometry*. Electroencephalographic audiometry has been reviewed in detail by Nober in Chapter 10. This technique is the electronic recording of ongoing brain wave activity in which patterns change due to auditory stimulation. The neurological impulses are picked up by electrodes that are located on the scalp and subsequently transmitted to an electroencephalograph that records the brain wave activity. Since the test can last over an hour, an enormous quantity of paper accumulates along with voluminous analogued data from different areas of the brain. Data analysis by visual inspection is at best extremely difficult, if not impossible. For this reason, an electronic-averaging computer was added to the technique and this revolutionized the procedure. The computer mathematically eliminates the random cortical activity that is not time locked to the tone stimulus, and adds all cortical responses that are time locked to the tone stimulus. Thus, the computer memory stores dozens of stimulus–response time-locked data and develops a cumulative response called the average evoked response or the averaged encephalographic response. Since all other cortical activities are not time locked to the tone stimulus, they are essentially random and the mathematical addition of random activity equals zero or self-elimination.

Three time segments of averaged evoked responses may be identified. Responses that occur in the first 9 msec after stimulation are designated the "early component"; this component represents the brain stem response (Jewitt, Romano, & Williston, 1970). The "mid-component" occurs up to about 50 msec (Goldstein & Rodman, 1967; Geisler, Frishkopf, & Rosenblith, 1958) and the "late component'' occurs up to about 500 msec (Davis, 1965, 1968). Since the averaged evoked response amplitude is related to the stimulus sensation level, each response segment has been used successfully to estimate auditory threshold (Davis, 1971; Kupperman & Mendel, 1974; Rapin, Schimmel, Tourk, Krasnegor, & Pollak, 1966; Ruhm, Walker, & Flanigin, 1967).

Tumors on the eighth nerve may compress the nerve fibers, thereby decreasing the speed at which the neural activity is conducted. Reduced neural activity electronically translates to a reduction of the response amplitude. Comparisons of responses of normal and abnormal ears stimulated by sounds at equal sensation level show that the response latency in the abnormal ear will exceed that of the normal ear; amplitude differences will also occur. Shimizu (1968) detected eighth-nerve tumors by recording latency and amplitude of the middle components. Selters and Brackmann (1977) made similar measurements using the early components.

c. *Electrocardiographic audiometry*. The electrocardiographic technique, designed to measure pure tone hearing thresholds, is based on the acceleration and deceleration of heart rate. It has been used almost exclusively with neonates and retarded children. Like electrodermal audiometry, the response is mediated through the autonomic nervous system, but unlike electrodermal audiometry, thresholds are not within 10 dB of actual threshold even for adults and older children. Electrocardiographic responses seem to appear at about 35 dB above threshold.

Electronystagmography. Nystagmus is spontaneous abnormal eye movements consisting of a slow-movement component in one direction and a corresponding rapid-movement component in the opposite direction. The slow (drift) component reaches a maximum point and then the vestibulooccular reflex contracts to reverse the drift rapidly. This repetitious, rhythmic pattern can be rotational, horizontal, or vertical in direction. It is not a disease *per se* but rather a symptom of brain pathology. Nystagmus signifies an imbalance of bilateral neural synchronization.

Tests for induced nystagmus are precipitated by caloric (temperature) manipulations of water used to irrigate the ear canal. These tests have been the otologists' major diagnostic procedures for decades because they significantly reflect vestibular function, i.e., the integrity of the body orientation and equilibrium under diagnostically controlled conditions. The electronystagmographic response is an electronic recording of the ''corneoretinal potential'' (caused by positional change of cornea relative to retina), which is printed as a permanent record on a graph. Clearly, electronystagmography is not a hearing test, but it provides the otologist with valuable diagnostic information.

Central Auditory Processing Tests. Generally speaking, the auditory processing area proximal to the cochlear nuclei constitutes the central auditory system, while the area distal to the cochlear nuclei designates the peripheral auditory system (middle ear, cochlea, and eighth nerve). Most of the audiometric tests for the middle ear and cochlea are distinctively ipsilateral peripheral tests for sensitivity, distortion, and auditory adaptation (J. Jerger, 1973). On the other hand, tests of eighth-nerve integrity overlap with tests of central auditory processing. One notable response difference of the central auditory system is the contralateral effect of a lesion proximal to the brainstem crossover demarcation. Hence, the crucial information needed to determine the site of lesion is whether the defective ear is ipsilateral or contralateral to the lesion.

The transmission of sound along the auditory processing system follows an orderly systematic progression as the sound ascends from the peripheral auditory system to the central auditory system. For example, as the auditory pathways ascend from the cochlear nuclei to the temporal cortex, the usual parameters of intensity and frequency become progressively less meaningful. Pure tone thresholds may be normal when the lesion is in the cortical auditory system; consequently, pure tone tests are useless for identifying central dysacuses (Bocca & Calearo, 1963). Clearly, tones can be perceived on a subcortical level. As early as 1943, Kryter and Ades were able to demonstrate that a cat's conditioned threshold to a tone was unchanged after bilateral ablation of its auditory cortex. Even fine frequency discriminations have been elicited after bilateral removal of the auditory cortex (Meyer & Woolsey, 1952; Butler, Diamond, & Neff, 1957).

The closer the lesion is in proximity to the cochlea, the more direct is the relationship between the degree of discrimination loss and the degree of pure tone sensitivity loss. In the area where the eighth nerve passes through the lower brain stem the auditory pathways are "bottlenecked" (J. Jerger & Harford, 1960; Schuknecht, 1958) into a narrow constriction. Air-conduction thresholds can be slightly above normal but there may be a drastic reduction in discrimination. In order to test this "bottleneck" area, J. Jerger and S. Jerger (1971) advocated multiple measurement of monosyllabic speech discrimination ability as a function of sensation level to determine whether scores actually worsen or "rollover" as the stimulus level is increased. A rollover pattern is present in eighth-nerve lesions but not usually in higher order lesions.

In order to distinguish between cortical and subcortical lesions, the auditory testing material must precipitate a critical listening condition. This is accomplished by reducing the redundancy of the speech material and requiring cortical integration and summation (Bocca & Calearo, 1963). Tests designed for this purpose are called central auditory tests. Central auditory tests fall into two categories: (1) supraliminal pure tone tests and (2) supraliminal speech tests. Pure tone tests involve spatial localization or dichotic listening. In one type of spatial localization test the phase adjustment between the ears is varied, while in another type the intensity between the two ears is varied (Jerger & Harford, 1960). In the first type, normal-hearing subjects localize the sound to the ear that leads in phase (Bocca & Calearo, 1963). Matzker (1959) varied his phase between 0.2 and 0.6 msec in the two ears and found crude errors in subjects with central auditory impairment. J. Jerger and Harford (1960) employed a simultaneous binaural median-plane-localization test, which is actually a modification of the alternate-binaural loudness-balance test. Here the phase between the two ears is identical but the intensity is varied in one ear until the sound is localized to the center of the head. Cases with unilateral lesions of the auditory pathways localize the sound to the side contralateral to the lesion (Sanchez-Longo, Forster, & Auth, 1957). Davis and Goodman (1966) identified derecruitment with this test. Dichotic pure tone tests are rarely used today.

Central auditory speech tests are either monaural or binaural. The binaural tests require central summation (Bocca & Calearo, 1963; Smith & Pesnick, 1972) as each ear is given only a restricted or distorted portion of the total message. Independently, each part carries inadequate information for appropriate recognition of the material. The signals fed independently to the two ears will complement each other only when they are integrated centrally, i.e., at the cortical level. The cortically integrated material is completely intelligible to a normal-hearing subject. In cases of a unilateral temporal cortex lesion, the ear that is contralateral to the side with the lesion yields the poorer response. Presbycusic subjects have great difficulty with these tests (Bocca, 1958; Bocca & Calearo, 1963).

Speech tests based on a word or a sentence message can be alternated between the two ears to make them more critical for identifying subcortical lesions (Bocca & Calearo, 1963). Bocca (1955) used a subliminal unfiltered message in one ear and low-pass filtered supraliminal messages in the other ear. Normal subjects attained excellent discrimination scores but subjects with cortical impairment had poorer results. J. Jerger and Harford (1960), Calearo (1960), and Lynn and Gilroy (1972) substantiated these findings with modified versions of this test.

Monaural speech integration tests can involve *distorted* speech material. Phonetically balanced words that are distorted with low-pass filters (500 Hz) have been used (Bocca, Calearo, & Cassinari, 1954; J. Jerger & Harford, 1960). One phonetically balanced recording that is inherently poor in quality was used successfully by Goldstein, Goodman, and King (1956). Some of the tests involve monosyllabic words mixed with a masking noise. The test goal for both the filtered and masked speech stimuli is to make the listening condition critical, precipitating a breakdown in the listener's word discrimination. Many investigators have used this procedure in various forms (Heilman, Hamner, & Wilder, 1973; J. Jerger & S. Jerger, 1974a; Morales-Garcia & Poole, 1972; Noffsinger, Kurdziel, & Applebaum, 1975; Olsen *et al.*, 1975). Abnormal breakdown in speech discrimination occurs in patients with brain stem involvement. In more recent binaural versions of the masking effect of speech mixed with noise, Olsen and Noffsinger (1976) and Olsen, Noffsinger, & Carhart (1976) employed phase differences between the ears. Again subjects with brain stem lesions were differentiated from normal subjects.

Another group of tests in current use is called the *competing message tests*. Various speech signals have been used, i.e., monosyllabic words, bisyllabic spondee words, synthetic sentences, and nonsense syllables. In all instances, the basic principle is the same—the primary message is presented to one ear while the other ear simultaneously receives a competing message.

J. Jerger, Speaks, and Trammell (1968) and J. Jerger and S. Jerger (1975a) developed the *Synthetic Sentence Identification Test* with both ipsilateral and contralateral competing messages at varying signal-to-noise ratios. Test results differentiate cortical from brain stem lesions (S. Jerger & J. Jerger, 1975).

A current competing message test receiving considerable attention among audiologists is the *Staggered Spondaic Word Test* (Katz, 1962, 1968). Two bisyllabic spondee words are presented, one to each ear. The timing is such that the first syllable of the second spondee overlaps with the second syllable for the first spondee. Subjects have to repeat all four syllables in order of occurrence. Results pertaining to the competing and noncompeting conditions differentiate cortical (particularly Heschl's gyrus) from upper and lower brain stem lesions (Katz, 1970; Lynn & Gilroy, 1972, 1975).

Dichotic nonsense syllable presentation is a procedure used by Berlin and McNeill (1976). Subjects receive pairs of voiceless and voiced nonsense syllables at the two ears simultaneously or within a fraction of simultaneity. The listener selects the syllables heard from a list. This technique has met with limited success (Sparks, Goodglass, & Nickel, 1970; Speaks, 1975).

Disorders and Treatments of the Conductive Mechanism

Disorders

Otitis Media. There are several types of otitis media and a number of etiologic agents. Perhaps the major cause is malfunction of the Eustachian tube. It is through this tube that the air pressure in the middle ear is equalized and drainage is provided (Jackson & Jackson, 1959; Juhn, Huft, & Paparella, 1973).

The tube may malfunction due to adenoid hypertrophy, nasopharyngeal tumors, or anatomical abnormality of the tensor veli palatine muscle, which results in excessively negative middle ear pressure (Payne & Paparella, 1976). Any obstruction of the Eustachian tube or a reduction of its patency impairs its normal functioning and the middle ear becomes a closed chamber. Once the tube closes, the trapped air within the tympanum is absorbed leaving a partial vacuum. This vacuum precipitates an extravasation of sterile fluid from the surrounding mucosa into the middle ear cavity. If secondary infection sets in, it becomes a suppurative otitis; if infection does not occur, it is a nonsuppurative otitis. In the former, the accumulated fluid or exudate becomes infected and is transformed to pus (purulent). The nonsuppurative effusion is nonpurulent. If enough fluid accumulates, the internal pressure causes rupture of the tympanic membrane. Hearing loss may or may not occur depending on the size and location of the rupture. Otitis media has significantly higher incidence in children than in adults. Eagles, Wishik, and Doerfler (1967) reported in a survey of elementary school children that hearing loss due to otitis media occurred in about 33% of the children.

a. *Suppurative otitis media*. This disorder has a high incidence in children under ten years because of frequent respiratory diseases and hypertrophied lymphoid tissue in the nasopharynx (Sharp, 1960). It is also commonly found in adults, usually associated with a hemolytic streptococcus (Lewis, Grey, & Hawlett, 1955) due to a secondary infection from a cold, measles, and other viral infections. Suppurative otitis is also caused by hemophilis influenza (Schuknecht, 1974).

The middle ear space fills with fluid causing the eardrum to bulge outward and thicken if left untreated. The excessive pressure will perforate the eardrum allowing the fluid to discharge into the external auditory canal. Usually the perforation will heal itself with little or no permanent hearing loss. Prior to the discharge of fluid, there is typically a conductive hearing loss. Recovery of hearing is 85–95% when treated early. Treatment consists of antibiotic and decongestant therapy or incision of the eardrum (myringotomy) if needed.

Chronic purulence due to inadequate or ineffective medical therapy can produce fibrotic changes requiring mastoid surgery. In the more extensive cases, sensorineural hearing loss occurs due to ototoxins damaging part of the basal turn of the cochlea (Hulks, 1941; Paparella, Brady, & Hoel, 1970; Paparella, Oda, Hiraide, & Brady, 1972).

b. *Serous otitis media*. This common condition is characterized by the presence of sterile effusion in the middle ear space. The eardrum is retracted and fluid bubbles may be observed. The condition is found in children with hypertrophied adenoid tissue, Eustachian tube obstruction, or allergy. Occasionally it develops when antibiotic therapy has destroyed the bacteria present in a suppurative infection but the fluid remains. Hearing loss rarely exceeds 25 dB with the primary loss in the low frequencies due to the increased stiffness of the ossicular chain. Decongestant therapy is the primary course of treatment. If fluid persists, it may be removed surgically and a ventilation tube placed in the eardrum. The tube allows proper aeration of the middle ear space and hearing loss is reduced (Gunderson & Tonning, 1976; Neel *et al.*, 1977).

Mastoiditis. Mastoiditis is an infection of the mastoid bone air cells. If antibiotic therapy is ineffective, surgery (mastoidectomy) may be indicated. In the early acute stages, the infection may be confined to the mastoid air cells. A simple mastoidectomy may be performed. Since the middle ear is not affected, hearing will remain normal. If the infection becomes chronic and spreads to include the middle ear space, a radical mastoidectomy may be performed. This procedure involves removing the mastoid and the entire ossicular chain. Maximum conductive hearing loss (60 dB) will result. When possible, the surgeon will modify the radical mastoidectomy procedure and save part of the ossicular chain. The extent of the resultant hearing loss will depend upon how well the remaining ossicular structures succeed in providing the function of the ossicular chain.

Cholesteatoma. Cholesteatoma is a squamous epithelial ingrowth that usually enters the middle ear space through a perforation in the upper portion of the eardrum (Schrapnelll's membrane) (Harris, 1961). It may eventually enter the mastoid space. Cholesteatoma causes bone resorption, which can destroy the ossicles, the cochlea, the balance mechanism in the bony labyrinth, and the facial nerve. Surgical removal is often required.

Atresia. Atresia is a congenital malformation of the outer, middle, or inner ear or any combination of these. Gross deformities are not common. The malformations may be vascular, neural, muscular, or osseous. The extent of the hearing loss depends upon the nature and degree of the malformation. Middle ear reconstruction surgery or a hearing aid may be beneficial if the cochlea and auditory nerve are intact (Northern & Downs, 1974).

Otosclerosis. Otosclerosis is a progressive disease of the otic capsule. During the early stages an embryonic form of bone develops that is characteristically spongy and vascular; it usually starts in the oval window, but approximately 20% have a site of origin at the round window. In the advanced stages of the disease, the bone becomes sclerotic and can involve the entire ossicular chain (Wever & Lawrence, 1954). In the advanced stages approximately 25–30% develop a superimposed sensorineural loss (Lempert, 1945). While etiology is still somewhat speculative, the general consensus favors a hereditary etiology (Shambaugh, 1967; Tato *et al.*, 1963).

Otosclerosis can be diagnosed as a clinical entity, a histological entity, or both. The incidence of histological otosclerosis is 5–10% greater than clinical otosclerosis (Lindsay, 1959), since some subjects do not develop a demonstrable hearing loss. The incidence in women is about 6–7 times greater than in men (Davis & Silverman, 1978). Altmann, Glasgold, and MacDuff (1967), however, found the incidence only slightly favored women over men. Women who have the disease may develop an increased loss of hearing with pregnancy. Walsh (1954) estimated 12% of the pregnant women show an increase in hearing loss. Shambaugh (1967) estimated one in four worsen during pregnancy.

Hearing loss due to otosclerosis usually manifests itself after puberty (Davis & Silverman, 1960; Mawson, 1963) but has been seen in children as young as 4 years. The loss is usually bilateral, but is unilateral in 25–35% of the cases (Fowler, 1950). The air conduction thresholds are flat or may show a slightly greater depression in the low frequencies due to the impeded average action of the fixated ossicular chain. During the early stages the hearing loss is only conductive even though the bone thresholds show a symptomatic dip (5 dB at 500 Hz, 10 dB at 1000 Hz, 15 dB at 2000 Hz, 5 dB at 4000 Hz) known as the *Carhart notch* (Carhart, 1952). This bone conduction dip reflects immoblized cochlear fluids causing an inner-ear block. The fixed stapes prevents normal reciprocal movement between the oval and round windows. These depressed bone thresholds return to normal if the stapes is immobilized and the reciprocal movement between the two windows is normal. In the advanced stages, a cochlear impairment is superimposed on the conductive loss (Bellucci, 1958; Carhart, 1960, 1963; Goodhill & Moncur, 1963) and the subject develops a mixed loss. In very advanced cases there may be marked labyrinthine degeneration.

The low-tone loss is due to a mechanical increase in stiffness from the stapedial impedence and a thickening of fibrous tissue at the annular ligament. With increased restriction of stapes mobility, the audiogram configuration is flattened and eventually a high-tone tilt develops due to the increase in mass as the stapes rigidly adheres to the oval window.

Tinnitus is associated with otosclerosis in 80% of the cases in varying degress (Mawson, 1963). It increases in severity during the periods when the disease is progressing rapidly. Mawson (1963) also noted that 25% have vertigo.

Special Surgical Treatments of Otosclerosis

Three surgical procedures have been devised for the treatment of otosclerosis: fenestration, stapes mobilization, and stapedectomy. Stapedectomy is the only procedure currently in use, stapedectomy can also be used in the treatment of the other disorders listed above.

Fenestration. The fenestration, perfected as a one-stage operation by Lempert (1945) involves the construction of a window or fenestra in the ampullated end of the horizontal canal. This fenestra enables the sound to bypass the oval window and travel directly to the perilymph of the semicircular canal. Many complications are associated with this procedure and thresholds usually are not restored to better than 20–25 dB due to the surgical removal of the malleus and incus. The fenestration operation is now obsolete.

Stapes Mobilization. The stapes mobilization, popularized by Rosen (1953), involved jarring loose the stapes from its foci of fixation. Results were often dramatic on the operating table as patients would suddenly report they could hear. The procedure is simple and the hearing results initially good. The major weakness was the tendency for the stapes to refixate so permanent success was only about 20% (McKenzie & Rice, 1962). This operation, too, is now mostly of historical interest (Eimind, 1960; Newby, 1964; Rosen, 1961.)

Stapedectomy. The first stapedectomy procedure used a vein graft to close the oval window after the removal of the stapes. A polyethylene strut connected the vein graft and the incus. Many descriptions and modifications have subsequently been given (Schuknecht, 1962). It is presently the foremost surgical procedure for otosclerosis (Coleman, 1962; McKenzie & Rice, 1962). Schuknecht (1962) estimated that over 90% achieved good hearing restoration when surgery was successful. The air conduction hearing level was restored to within 10 dB of the bone conduction threshold (Newby, 1964).

Over the years, many modifications of stapedectomy have been developed (Schuknecht, 1974). While an air–bone gap of 10 dB or less is achieved after surgery in 80–90% of the cases, less than 70% have this small a gap five years past surgery. Unsuccessful surgery may traumatize a cochlea and result in marked loss of speech discrimination. In some instances, additional injury to the chorda tympani nerve occurs, and equilibrium can be reduced (Harrison, Shambaugh, Derlacki, & Clemis, 1970).

Disorders of the Sensorineural Mechanism

Hereditary Deafness

Generally, hereditary deafness is included in chapters dealing with children who have auditory problems. For this reason, it will be mentioned here only briefly. The major cause of hereditary deafness is genetic inheritance of a dominant, recessive, and/or sex-linked gene. It has been estimated that about half of all children who are deaf have inherited their deafness; actual incidence varies depending upon a number of parameters. Some of these parameters include congenital vs. delayed onset and dominant vs. recessive genotype. Also, there can be an interaction between genotype and environment; some people, for example, have a greater proclivity than others toward developing hearing loss due to environmental noise. While hereditary deafness in young children is usually of severe magnitude involving the sensorineural mechanism, one particular type of hearing impairment does not appear clinically until the teenage years. This condition is called otosclerosis and is a progressive conductive impairment; the rate of progressive hearing loss depends upon genetic determinants and cannot be predicted in advance.

Sudden and Acquired Deafness

Sudden deafness refers to the dramatic onset of a partial or total loss. This can include a disease of the cochlea, the auditory nerve, or the acoustic pathways. Onset may be complete in an hour or in several days. In many instances there is no warning or suspected etiology. The condition may be permanent or temporary, or may fluctuate depending on the etiology. Sudden deafness is commonly associated with cochlear impairments (Hilger & Goltz, 1951; Saltzman & Ersner, 1952; Wilmot, 1959), but can be caused by eighth-nerve tumors (J. Jerger et al., 1958; Van Dishoeck & Bierman, 1957). Acquired deafness refers to partial or total hearing loss that has a longer, more ongoing onset. Hence, the difference between sudden and acquired deafness is the length of the onset period. Some of the etiologic agents common to both sudden and acquired deafness are listed below.

Viral and Bacterial Agents. Viruses and bacteria can directly or indirectly cause hearing loss. Directly, the deafness is due to damage to the nerve ending in the cochlea (Van Dishoech & Bierman, 1957). Cases of unilateral sudden deafness in individuals who were suffering from "head colds" have been reported by Lieberman (1957). Indirectly, the viral infection can cuase otitis media, meningitis, and encephalitis (Ballantyne, 1960; Schuknecht, 1974). The pathologic effect is damage to the stria vascularis, tectorial membrane, scale media, or saccule (Strome, 1977). The organ of Corti and its neural pathways may also be affected (Lindsay, Davey, & Ward, 1960a). Some infections are contracted by the mother and affect the unborn child in utero. Below are some of the more common diseases that can cause deafness:

a. *Rubella* may be contracted by the mother any time during pregnancy and transmitted to the unborn child. Problems are most severe when contracted during the first trimester but the fetus is susceptible throughout the pregnancy (Barr & Lundstrom, 1961; Hemenway, Sanda, & McChesney, 1969). Pathological changes can affect the stria vascularis, organ of Corti, tectorial membrane, and ganglion cells (Lim, 1977). Hearing loss may be accompanied by multiple disabilities including cerebral palsy (Hardy, 1973), cataracts, cardiovascular disorders, and mental retardation (Lim, 1977). Estimates on the incidence of rubella pathology in postrubella children vary from 22% (Lim, 1977) to 35% (Catlin, 1978). Hearing loss is progressive in about 25% of the children studied by Bordley and Alford (1970). Rarely does the disease have any effect on the hearing of adults.

b. *Influenza* rarely causes hearing loss in adults but has in isolated instances. Steward (1961) claimed that toxic exudates inflicted damage on the eighth nerve. Other symptoms would be vertigo, nausea, or nystagmus. Hearing loss when it does occur is usually temporary.

c. *Meningitis* is another cause of deafness. The meningitis *per se* can be caused by an otitis or any other infection that works its way to the meninges. The nature of the hearing loss is independent of the etiology that caused the meningitis. In cases of supporative otitis media or mumps the bactria may travel from the middle ear to the mastoid area and ultimately to the brain through a rupture in the mastoid bone. Many bacteria gain access to the subarachnoid space or the ventricular system through the blood system.

The symptoms of meningitis are well documented in the literature. Incidence is estimated at 18% (Best, 1943) to 10% (Schein & Delk, 1974). Loss of hearing caused by meningitis is sensorineural, often unilateral, and profound (Strome, 1977). There may be inflammation and stiffness of the cochlea, destruction of the eighth nerve, and damage to the stria vascularis, the tectorial membrane, or the hair cells (Lindsay, Procter, & Work, 1960b). Kopetsky (1948) contended that the damage to the organ of Corti was due to the toxic effects of the disease while the damage to the auditory nerve as due to the increased intercranial pressure. Usually the deafness is irreversible. Recruitment is present or absent depending on the retrocochlear involvement. The treatment is supportive antibiotic therapy.

d. *Mumps* has been known to produce deafness since ancient times. Historically, the pathogenesis was ascribed to the labyrinth or the acoustic nerve. The present contention is that the major damage occurs in the cochlea with only minimal damage to the vestibular system. Destruction of the stria vascularis, the organ of Corti, and the tectorial membrane is common and greatest at the base of the cochlea rather than the apex. Lindsay *et al.* (1960a) labeled this condition *endolymphatic labyrinthitis*. Deafness may suddenly appear about a week after the onset of the mumps (Strome, 1977). In addition, there may be tinnitus, vertigo, and vomiting. Deafness is usually unilateral; in fact, mumps is the most common cause of unilateral deafness. Everberg (1960), however, claimed one-fifth of his cases to be bilateral. The audiometric picture is that of profound deafness with perhaps some minimum low-frequency hearing at extreme levels. Vuori, Labikainen, and Peltonen (1963–64) cited some atypical instances when the hearing returned to normal. The overall incidence of hearing loss was given as 0.05/1000 by Everberg (1957) and as 1/372 by Bangs and Bangs (1943), while Hyatt (1962) indicated mumps accounts for 24% of the deaf children in the United States. The incidence for males and females is about equal. Vestibular malfunction is also common (Everberg, 1957, 1960; Lindsay *et al.*, 1960a; Saunders & Lippy, 1959; Steward, 1961). Medical treatment has been unsuccessful; nicotinic acid, vitamin B, and steroids have been ineffectual (Proctor, 1963).

Ototoxic Agents. A broad variety of chemicals and drugs have been implicated as the cause of sensorineural hearing loss (Mawson, 1963; Pararella & Shumrick, 1973). The degree of loss is related to the quantity of the drug ingested; progression of any loss often ceases after the withdrawal of the drug (Stewart, 1961). Factors that influence the ototoxic effect of a drug are the duration over which the drug is administered, other drugs that the patient ingests while receiving the potentially ototoxic drugs, and the ability of the patient's kidneys to remove the drug from the bloodstream (Bergstrom & Thompson, 1976).

a. *Arsenic* and *lead* affect the cochlear and vestibular branches of the eighth nerve. Tinnitus and vertigo are common sequelae (Quick, 1973).

b. *Quinine* and *chloroquine phosphate* are primarily antimalarial drugs is usually reversible (Bergstom & Thompson, 1976). Ingestion of these drugs during pregnancy has led to permanent congenital deafness in the infant (Hart & Nauton, 1964; Schuknecht, 1974).

c. *Salicylates* are the oldest known ototoxic drugs (Bergstom & Thompson, 1976). Ototoxic effects occur when the drug is taken in overdose (Suprapathana, Futrakul, & Campbell, 1979). Tinnitus is usually the first sign of an ototoxic reaction. Hearing loss is usually reversible if the drug intake is reduced or if the drug is withdrawn when tinnitus occurs. Permanant hearing loss, however, has been reported (Jarvis, 1966).

d. *Thalidomide* may cause multiple anomalies of the head and other structures of the unborn infant if ingested by the mother during the second month of pregnancy (Lim, 1977). Anomalies affecting all parts of the ear have been reported.

e. *Streptomycin* shows a propensity for the vestibular branch of the eighth nerve (Davis, Rosenblut, Fernandez, Kimura, & Smith, 1958; Gregg & Lierle, 1961; Ozaki, 1957; Rossi, Corando, DeMicheles, & Mazzio, 1961; Wier, Storey, Curry, & Schloss, 1956), but it also has caused damage to hair cells, the stria vascularis, and the supporting cells of Corti (Davis *et al.*, 1958). Gregg and Lierle (1961) described damage to the ventral cochlear and vestibular nuclei. Symptoms are tinnitus, feelings of fullness in the ear, and loss of hearing (Conway & Birt, 1965; Ozaki, 1957; Varpela, Hietalahti, & Avo, 1969). The hearing loss may be delayed for some time after the drug is administered and the degree of hearing loss is related to the total dosage (Coles, 1959). Treatment involves discontinuing the drug. The incidence of ototoxic reactions is estimated to be about 3.5% (Meurman & Hietalahti, 1960).

f. *Dihydrostreptomycin* can also affect the vestibular mechanism (Shambaugh, Derlacki, Harrison, House, House, Hildyard, Schuknecht, & Shea, 1959), but the major damage is to the cochlea. Harrison (1954) found that approximately 20% who received this drug developed hearing loss. Marcus, Small, and Emanuel (1963) and Varpela *et al.* (1969) found it had detrimental effects on the hearing and vestibular mechanisms of children. The pathogenesis involves the stria vascularis and the hair cells with supporting structures. The cristae within the ampullae of the semicircular canal are affected (Bergstrom & Thompson, 1976). There is marked recruitment (Liden, 1953) and severe irreversible sensorineural pathology. Treatment is abstinence from the drug. Deafness can appear as long as eight months after the drug has been terminated (McGee & Olszewski, 1962).

g. *Kanamycin* is chemically similar to neomycin, but is less toxic to the vestibular system. It causes damage to the hair cells. Patients with poor renal function who cannot easily remove the drug from their blood will be more susceptible to its ototoxic effects. Kanamycin is more ototoxic than dihydrostreptomycin (Abroms, 1977). It is doubtful that neonates incur hearing loss by ingesting kanamycin (Sanders, Eliot, & Cramblett, 1967).

h. *Neomycin* is also a derivative of streptomycin and is toxic to the cochlear hair cells (Kohonen, 1965). Degeneration is progressive for the hair cells and their supporting pillars (Lindsay *et al.*, 1960b). Although hearing loss is usually profound, the vestibular system rarely is involved (Bergstrom & Thompson, 1976). Incidence of ototoxic neomycin reaction is estimated at 8%.

i. *Gentamycin* is a relatively new antibiotic and information is incomplete as to its ototoxic effect. Some researchers estimate ototoxic incidence at 2.3%; however, others contend that gentamycin is more ototoxic than streptomycin. The vestibular system is about twice as sensitive as the auditory system to gentamycin. Severe hearing loss rarely occurs (Bergstrom & Thompson, 1976).

Other drugs that have varying degrees of ototoxicity are viomycin, vanomycin, tobramycin, polymixin B, ethacrynic acid, nitrogen mustard, and chloramphenicol (Bergstrom & Thompson, 1976; Parparella & Shumrick, 1973).

Vascular Disorders. Mattox and Simmons (1977) list hemorrhage, thrombosis, embolism, and vasospasm as the possible types of vascular disorders that affect hearing. Since the first three have long-term effects on the cochlea, destruction of cochlear structures and electrical potentials occurs rapidly and is diffuse. Hearing loss is usually severe with little change of improvement over time. Sheehy (1960) proposed that when hearing loss is reversible, vasospasm was the cause of deafness. There is little evidence, however, to support this contention. Mattox and Simmons (1977) point out that many cases of sudden onset hearing loss can be related to specific factors, and that the diagnosis of vascular disorders is very difficult to make. In their study of 88 patients with undiagnosed sudden onset hearing loss, Mattox and Simmons (1977) found great diversity in the degree of loss and the rate and extent of recovery. Sensitivity at 8000 Hz proved to be an important prognostic indicator for recovery. Recovery was clearly better in ears where threshold at 8000 Hz was better than or equal to the threshold at 4000 Hz, then the degree of recovery was reduced. Histamine, steroid, and vasodilator therapy proved no more effective than no treatment. Mattox and Simmons therefore concluded that much is needed to be learned about sudden onset hearing loss.

Cerebral vascular accidents that cause deafness involve some type of occlusion rather than a hemorrhage. On many occasions the cochlea survives the temporary interruption of its blood supply although little is known about how it does manage to survive (Kimura, Pearlman, & Fernandez, 1959). Complete or partial obstruction of the internal auditory artery usually causes a sudden severe deafness with recruitment, tinnitus, vertigo, and occasionally diplacusis. Recovery can be as sudden as the onset but these people are more likely to have recurrences (Hallberg, 1957).

Acoustic Trauma. The effects of hearing loss from noise exposure is a well-known cause of sudden deafness. An acoustic trauma due to an explosion or some violent pressure release of acoustic energy can inflict sudden auditory damage. Conductive hearing loss will occur if the tympanic membrance is ruptured or the ossicular chain is disrupted. The most likely impairment is a sensorineural hearing loss due to destruction of the hair cells (Bohne, 1976). Acoustic trauma *per se* is a different clinical entity than the gradual hearing loss that evolves after prolonged exposure to intense noise; the latter is called ''noise-induced hearing loss.''

Noise-induced hearing loss. At one time, noise-induced hearing loss was thought to be an artifact of industrial society. Rosen, Bergman, Plester, El-Mofty, and Satti (1962) reported that members of a primitive African tribe who were never exposed to noise showed no traces of hearing loss associated with age. Further investigation revealed, however, that many other aspects of life for these individuals such as diet, social stress, and family structure, were also significantly different and could account for their generally excellent health (Davis & Silverman, 1978).

The severity of a noise-induced hearing loss is related to several variables: (1) the overall intensity level of the noise, (2) the frequency distribution of the energy, (3) the duration of exposure time, and (4) individual susceptibility. If the hearing loss is only a temporary threshold shift then it is called *auditory fatigue*. A permanent threshold shift is the irreversible loss that remains. There is a relationship between temporary threshold shift and permanent threshold shift which is now established. Glorig (1958) indicated that the temporary threshold shift can vary from 0 dB to 35 dB contingent on an individual's susceptibility.

The concept of a *damage-risk criteria* was introduced by Kyter (1950) and subsequently studied by others (Glorig, Ward, & Nixon, 1961; Rosenblith & Stevens, 1953) to determine which combinations of frequency and intensity inflict damage on the auditory mechanism. The detrimental effects of sound were studied in four octave bands, i.e., 300–600, 600–1200, 1200–2400, and 2400–4800 Hz. The threshold shifts at 1000 and 2000 Hz were related to the 1200–2400 band. Most studies concur that any sound beyond 85 dB is a potential damage risk to the auditory mechanism.

Noise-induced hearing losses are sensorineural, with the site in the cochlea. The damage is to the hair cells. The threshold shift starts at 4000 Hz or in some instances 6000 Hz. If the hearing loss increases then the dip widens and deepens. Once the loss spreads to 2000 Hz speech intelligibility becomes involved and the hearing loss interferes with communication. The impairment may continue until most of the high frequencies are seriously impaired or lost beyond 500 Hz. Tinnitus and dysacusis can be marked (Tonndorf, 1976).

The Occupational Health and Safety Administration (OSHA, 1974) developed regulations for industrial hearing conservation programs. Although the regulations are still subject to review, OSHA has determined that exposure to noise in excess of 90 dBA and 8 hours is hazardous to hearing. The permissible exposure period is reduced by half for each 5 dB increment beyond the 90 dB level. This damage-risk criteria should protect between 80% and 90% of the individuals exposed to the noise from noise-related risks to health. When exposure is in excess in the damage-risk criteria, efforts must be made to reduce the level of the noise at its source or to reduce the noise in transmission to the worker. When neither is feasible, hearing protectors may be worn by the exposed employees. Hearing portectors attenuate the noise before it reaches the eardrum and can decrease the noise level as much as 10 to 30 dB. To determine the status of employees' hearing, OSHA requires that all employees undergo a baseline audiogram within the first 90 days of employment, and then have annual hearing retests. Protection management basically involves absence from the noise-producing environment although the protective value of such absence after a certain period of time has been questioned. Most of the hearing loss is permanent although there may be some improvement after the remission of the temporary threshold shift aspect. Noise-induced hearing loss is unique insofar as it is totally untreatable and completely avoidable.

Meniere's Disease. Meniere's disease, also known as endolymphatic hydrops, is characterized by a classic triad of symptoms: (1) acute episodes of vertigo, (2) roaring tinnitus, and (3) unilateral sensorineural hearing loss (Alford, 1972). Attack frequency is highly variable; the attacks can occur years apart. Characteristically, the attacks last from 30 minutes to 3 hours (Pulec, 1976), but longer episodes do occur. The principal histopathological change is a vastly dilated endolymphatic system including the cochlea duct (Hallpike & Cairns, 1938; Hallpike & Wright, 1940). Etiology of Meniere's disease is not known with certainty but has been the object of speculation for many years. Allergy, syphilis, hypoadrenalism, myxedema, and vascular insufficiency are only some of the suggested causes (Pulec, 1976). One current theory is failure of the endolymphatic sac to reabsorb endolymph at a normal rate causing an increase in the volume of fluid (Arenberg, Marovitz, & Shambaugh, 1970; Gussen, 1971).

The audiological pattern of Meniere's disease typically shows a low-frequency sensorineural hearing loss that fluctuates in the early stages of the disease. Loudness recruitment is present and most site of lesion tests indicate a cochlear impairment. There is often extremely poor speech discrimination. The excessive endolymph in the cochlear duct (hence endolymphatic hydrops) increases the stiffness of the basilar membrane and reduces the efficiency of low-tone transmission. Between attacks, the volume of endolymph presumably is reduced and hearing may improve.

Torok (1977) proposed that symptomatic treatment be used to control Meniere's disease. Most forms of therapy seem successful to some extent. Vasodilator therapy was advocated by Sheehy (1960). Histamine and nicotinic acid have also been used (Pulec, 1976). Although several surgical procedures have been tried, the endolymphatic subarachnoid shunt has gained favor (House & Greenfield, 1969). The shunt relieves the pressure in the endolymphatic sac and hearing is stabilized or improved in about half of the patients (Pulec, 1976).

Eighth-Nerve Tumors. Cushing (1917) defined these tumors as inercranial tumors of the eighth nerve. Some versions of this kind of tumor identified by Boyd (1953) are cerebellopentine angle tumor, acoustic neuronoma perineural fibroma, fibroblastoma, Schirannoma, neurinoma, and neurilemmoma. Histologically, almost all eighth-nerve tumors arise from the vestibular branch and from the Schwann sheath cells of the cranial nerve. Rarely does the tumor arise in the cochlear branch of the eight nerve (Nager, 1964). The tumors may first be detected in the third or fourth decade of life (Weaver & Northern, 1976). Growth is slow and the tumor frequently remains a firm, encapsulated mass for many years. Eighth-nerve tumors are a common finding at autopsy. Most occur in only one ear, but Grinker, Bucy, & Sahs (1960) claimed that bilateral neuromas can also be associated with Von Recklinghausen's disease in younger patients.

Schuknecht (1974) found three major ways in which the tumor affects auditory and vestibular functioning: (1) The tumor destroys the fibers of the auditory and vestibular nerves causing a noticeable loss of speech discrimination, with pure tone threshold minimally affected; the foregoing is a classic audiological sign. Vestibular response is also reduced. (2) The tumor destroys the structures of the membranous labyrinth, possibly causing complete loss of the organ of Corti and subsequently total deafness. (3) The tumor results in alterations of the chemical composition of the fluids in the inner ear, resulting in a flat audiogram and loudness recruitment.

Schuknecht (1974) indicated that the sequence of symptoms is relatively consistent. Hearing loss and balance problems are usually the first symptoms to appear. Tinnitus and/or headaches may also be present. In some instances, cerebellar symptoms such as incoordination of the lower limbs and nystagmus may occur. Furthermore, additional cranial nerves may become involved; the trigeminal and facial nerves are usually affected first (Glasscock & Hays, 1973).

A complete battery of auditory tests is necessary to assist in the diagnosis. Typically, the audiogram shows a high-frequency loss; the degree of loss depends on the size of the tumor. Recruitment supposedly is absent (J. Jerger, 1962) in many cases because of retrocochler involvement. The perstimulatory tone decay test reveals excessive auditory adaptation (Carhart, 1957; J. Jerger, Allen, Robertson, & Harford, 1958). Continuous and interrupted Bekesy threshold tracings give Type III or Type IV audiograms (J. Jerger, 1960, 1962). A forward–backward Bekesy trace discrepancy may be present (J. Jerger *et al.*, 1972a). Discrimination loss is disproportionately severe, especially when compared to the air-conduction thresholds. Acoustic reflexes may be absent or may occur at elevated levels, and reflex decay may be present. No one audiometric test is completely reliable but considering the development of retrocochlear and central auditory tests described earlier, the audiologist's contribution to a diagnosis is significant.

In addition to auditory testing, otologic diagnostic procedures should include vestibular testing, X-ray polytomography, and dye contact studies of the internal auditory canal. Many surgical procedures have been developed and early detection has reduced the mortality rate (Glasscock, 1969).

Presbycusis. This term refers to a progressive hearing loss due to the degenerative or aging process. Although Nixon, Glorig, and High (1962), Belal and Stewart (1974), and Belal (1975) found tympanic membrane thickening and ossicular joint calcification caused a slight conductive component, the loss is considered primarily sensorineural. Lowell and Paparella (1977) cautioned that the term presbycusis should not be broadly applied to all individuals who develop sensorineural hearing loss in the later years of life. Genetic and familial hearing losses should be distinguished by personal and family history and are distinct entities from presbycusis.

Four types of presbycusis have been differentiated: (1) Sensory presbycusis is due to an atropic lesion of a few millimeters of the organ of Corti (Johnson & Hawkins, 1972). Hair cells and their supporting structures are involved and a sharp high-frequency hearing loss is present. (2) Neural presbycusis is related to destruction of cochlear neuron fibers with little destruction of hair cells. The result is speech discrimination that is much poorer than would be predicted from the pure tone audiogram. This condition is called *phonemic regression* (Gaeth, 1948). Other central nervous system problems may also be present in these patients. (3) Metabolic presbycusis is a slow progressive process due to atrophy of the stria vascularis. It is characterized by a mild-to-moderate flat hearing loss with good speech discrimination (Schuknecht & Ishii, 1966). (4) Cochlear conductive presbycusis is believed "to be due to a disorder in motion mechanics of the cochlear duct" (Schuknecht, 1974). Hearing loss is progressive in the high-frequency range. Speech discrimination will depend on the slope of the pure tone loss. Glorig and Davis (1961) note that many cases of presbycusis do not manifest loudness recruitment and cite this as evidence of a cochlear conductive loss; in the latter there is an absence of hair cell damage. The audiometer patterns associated with different types of presbycusis are discussed by Gacek and Schuknecht (1969) in detail.

The onset of presbycusis can occur as early as the twenties (Alexander, 1954) and generally is measurable in men at 32 years and in women five years later at 37 years (Corso, 1959). Spoor's (1967) analysis of eight different sources of presbycusis indicates that slight hearing loss is measurable for men age 35 years and women age 40 years. Hearing degenerates at a slow progressive rate. Sataloff and Menduke (1957) indicate that hearing loss stabilizes at age 65 years while Spoor's data suggest increased loss through age 85 years. Men have a slightly greater loss than women (Klotz & Kilbane, 1962). The result of the neural deterioration is to restrict the normal spatial and temporal assimilation of the acoustic signal. Information cannot be transmitted and interpreted rapidly. Presbycusis patients have marked difficulty with speech tests based on accelerated rhythm, interrupted voice, and longer or reversed sentences (Bocca, 1958).

The typical but not exclusive pattern of presbycusis is a gradual sloping sensorineural hearing loss (Bergman, 1967). Speech discrimination may vary from good to poor and may decrease with increasing age (J. Jerger, 1973). Amplification may benefit many individuals but introduce distortion and aggravate the problem in others. Intelligibility is aided if the speaker slows down his rate so the listener can integrate the material. Speech reading may also be of benefit. There is no more of a cure for presbycusis than there is for any other neural deterioration due to aging.

Conclusion

The amount and scope of technical and medical advances in the areas of audiology and otology during the past 15 years have been remarkable. The field of audiology has made significant strides in the development of tests for locating the site of the lesion and improving the efficiency of older tests. Considerable advances have been made in the development and use of central hearing tests. The field of otology has also undergone a radical metamorphosis. Middle ear surgery is close to miraculous and antibiotic drugs have nearly eliminated middle ear infections. There have also been major advances in understanding the anatomy and physiology of the auditory mechanism. From this knowledge evolve the important improvements in diagnostic and therapeutic methodology. The future can only bring more gratifying results.

A number of other improvements have occurred in the last decade. There is universal agreement on audiometer calibration. Terminology is also relatively standardized so that terms are less ambiguous and deceptive. Further understanding is needed for air- and bone-conduction testing at the more intense levels where inadvertent intervention from other sensory receptors clouds the validity (Nober, 1964). Attention was drawn to these "cutile" thresholds in the joint American Speech and Hearing Association–American Academy of Ophthalmology and Otolarygology designation of audiograms. There has been a noteworthy refinement of current tests and development of new ones to isolate the malfunctioning of specific areas of the auditory mechanism. Improved tests for the selection of hearing aids are being developed to ascertain the type of aid, ear to fit, wisdom of binaural fitting, and other associated problems.

Longitudinal research is emerging about the permanence of hearing improvements from new surgical techniques. Medical treatment of inner ear pathologies is currently being explored. Progress in audiology and otology has been so abundant that there has almost been a metamorphosis in some areas each decade. One refinement typically leads to yet another. Virtually all areas warrant further study, so the student has every opportunity to indulge his imagination and creative talents.

References

Abroms, I. Nongenetic hearing loss. In B. F. Jaffe (Ed.), *Hearing loss in children*. Baltimore: University Park Press, 1977.

Alberti, P., & Kristensen, R. The clinical application of impedance audiometry. *Laryngoscope*, 1970, *80*, 735–746.

Alexander, L. W. Diagnostic and etiological considerations of deafness in older persons. *Journal of the American Geriatric Society*, 1954, *2*, 386–395.

Alford, B., Meniere's disease: Criteria for diagnosis and evaluation of therapy for reporting. Report of subcommittee on equilibrium and its measurement. *Transactions of the American Academy of Ophthalmology and Otolaryngology*, 1972, *76*, 1462–1464.

Alford, B., Jerger, J., Coates, A., Peterson, C., & Weber, S. Neurophysiology of facial nerve testing. *Archives of Otolaryngology*, 1973, *97*, 214–219.

Altmann, F., Glasgold, A., & MacDuff, J. P. The incidence of otosclerosis as related to age and sex. *Annals of Otology, Rhinology, & Laryngology*, 1967, *76*, 377–392.

American Speech and Hearing Association. Guideline: For manual puretone threshold audiometry. *Journal of American Speech and Hearing Association*, 1978, *20*, 297–301.

American Standards Association. *Audiometers for general diagnostic purposes*. Z24.5. New York: American Standards Association, 1951.

Anderson, H., & Barr, B. Conductive high-tone hearing loss. *Archives of Otolaryngology*, 1971, *93*, 599–605.

Anderson, H., Barr, B., & Wedenberg, E. Intra-aural reflexes in retrocochlear lesions. In C. Hamberger & J. Wersall (Eds.), *Nobel Symposium: 10 Disorders of the skull base region*. Stockholm, Almqvist and Wiskell, 1969, pp. 49–55.

Anderson, H., Barr, B., & Wedenberg, E. Early diagnosis of VIIIth-nerve tumors by acoustic reflex tests. *Acta Otolaryngologica, Supplement*, 1970, *263*, 232–237. (a)

Anderson, H., Barr, B., & Wedenberg, E. The early detection of acoustic tumors by the stapedius reflex test. In G. Wolstenholme & J. Knight (Eds.), *Sensorineural hearing loss*. Churchill, London, 1970. (b)

Arenberg, I., Marovitz, W., & Shambaugh, G. The role of the endolymphatic sac in the pathogenesis of endolymphatic hydrops in man. *Acta Otolaryngologica*, Supplement, 1970, *275*.

Ballantyne, J. C. *Deafness*. Boston: Little, Brown, 1960.

Bangs, H., & Bangs, J. Involvement of the central nervous system in mumps. *Acta Medical* (Scand.), 1943, *113*, 487.

Barr, B., & Lundstrom, R. Deafness following maternal rubella. *Acta Otolaryngologica*, 1961, *53*, 413–423.

Beedle, R., & Harford, E. An investigation of the relationship between the acoustic reflex growth and loudness growth in normal and pathological ears. *Journal of American Speech and Hearing Association*, 1970, *12*, 435.

Beery, Q. C., Andrus, W. S., Bluestone, C. D., & Cantekin, E. I. Tympanometric pattern classification in relation to middle ear infusions. *Annals of Otology, Rhinology, & Laryngology*, 1975, *84*, 1–9.

Bekesy, G. A new audiometer. *Acta Otolaryngologica*, 1947, *35*, 411–422.

Belal, A. Presbycusis: Physiological or pathological. *Journal of Laryngology*, 1975, *89*, 1011–1025.

Belal, A., & Stewart, T. Pathological changes in the middle ear joints. *Annals of Otology, Rhinology, & Laryngology*, 1974, *83*, 159.

Bellucci, R. A guide for stapes surgery based on a new surgical classification of otosclerosis. *Laryngoscope*, 1958, *68*, 741–759.

Bergman, M. Effects of aging on hearing. *Maico Audiological Library Series*, 1967, 2, 6.

Bergstrom, L., & Thompson, P. Ototoxicity. In J. Northern, (Ed.), *Hearing Disorders*, (Vol. 12) Boston: Little, Brown, 1976. pp. 136–152.

Berlin, C. I., & McNeill, M. R. Dichotic listening. In N. J. Lass (Ed.), *Contemporary Issues in Experimental Phonetics*. New York: Academic Press, 1976.

Best, H. Deafness and the deaf in the United States. New York: Macmillan, 1943.

Bocca, E. Binaural hearing: Another approach. *Laryngoscope*, 1955, 65, 1164–1171.

Bocca, E. Clinical aspects of cortical deafness. *Laryngoscope*, 1958, *68*, 301–309.

Bocca, E., & Calearo, C. Central hearing process. In J. Jerger (Ed.), *Modern developments in audiology*. New York, Academic Press, 1963.

Bocca, E., Calearo, C., & Cassinari, B. A new method for testing hearing in temporal lobe tumors. *Acta Otolaryngologica*, 1954, *44*, 219–221.

Bohne, B. A. Mechanisms of noise damage in the inner ear. In D. Henderson, R. P. Hamernik, D. S. Dosanjh, & J. H. Mills (Eds.), *Effects of noise on hearing*, New York: Raven Press, 1976.

Bordley, J. E., & Alford, B. R. The pathology of rubella deafness. *International Audiology*, 1970, *9*, 58–67.

Boyd, W. *A textbook of pathology: An introduction to medicine* (6th Ed.). Philadelphia: Lea and Febiger, 1953.

Brooks, D. Secretive otitis media in school children. *Journal of International Audiology*, 1970, *9*, 141.

Brooks, D. N. The role of the acoustic impedance bridge in impedance screening. *Scandinavian Audiology*, 1974, *3*, 99–104.

Brooks, D. N. Middle ear impedance measurements in screening. *Audiology*, 1977, *16*, 288–293.

Butler, R. A., Diamond, I. T., & Neff, W. D. Role of auditory cortex in discrimination of changes in frequency. *Journal of Neurophysiology*, 1957, *20*, 108–120.

Calearo, C. *Proceedings of the 5th Congressional International Society of Audiology*, Bonn, Germany, 1960.

Carhart, R. Clinical application of bone conduction audiometry. *Archives of Otolaryngology*, 1950, *51*, 798–808.

Carhart, R. Bone conduction advances following fenestration surgery. *Translations of the American Academy of Ophtholamology and Otolaryngology*, 1952, *56*, 621–629.

Carhart, R. Clinical determination of abnormal auditory adaptation. *Archives of Otolaryngology*, 1957, *65*, 32–39.

Carhart, R. Audiometry in diagnosis. *Laryngoscope*, 1958, *68*, 253–277.

Carhart, R. Assessment of sensorineural response in otosclerosis. *AMA Archives of Otolaryngology*, 1960, *71*, 141–149.

Carhart, R. Atypical audiometric configuration associated with otosclerosis. *Annals of Otology, Rhinology, & Laryngology*, 1963, *71*, 744–758.

Carhart, R., & Jerger, J. F. Preferred method for clinical determinations of pure-tone thresholds. *Journal of Speech & Hearing Disorders*, 1959, *24*, 330–345.

Carhart, R., & Porter, L. S. Audiometric configuration and prediction of threshold for spondees. *Journal of Speech & Hearing Disorders*, 1971, *14*, 486–495.

Catlin, F. Etiology and pathology of hearing loss in children. In F. N. Martin (Ed.), *Pediatric audiology*, Englewood Cliffs, N.J.: Prentice-Hall, 1978.

Coleman, B. H. Stapedectomy: Experiences with the fat graft and steel pin operation. *Journal of Larngology, Otology, and Rhinology*, 1962, *76*, 100–113.

Coles, R. R. A case of vestibular failure after streptomycin. *Journal of Laryngology*, 1959, *73*, 555–559.

Conway, N., & Birt, B. D. Fetal ear damage due to streptomycin. *British Medical Journal*, 1965, *2*, 260–263.

Cooper, J. C., Jr., Gates, G. A., Owen, J. H., & Dickison, H. D. An abbreviation impedance bridge technique for school screening. *Journal of Speech & Hearing Disorders*, 1975, *40*, 260–269.

Corso, J. F. Age and sex differences in pure-tone thresholds. *Journal of Acoustical Society America*, 1959, *31*, 498–507.

Cushing, H. W. *Tumors of the Nervous Acousticus*. Philadelphia: Saunders, 1917.

Davis, H. Slow cortical response evoked by acoustic stimuli. *Acta Otolaryngolica*, 1965, *59*, 179–85.

Davis, H. Slow electrical responses of the human cortex. *Proceedings of the American Philosophical Society*, 1968, *112*, 150–156.

Davis, H., & Goodman, A. C. Subtractive hearing loss, loudness recruitment, and decruitment. *Annals of Otology, Rhinology, & Laryngology*, 1966, *75*, 87–94.

Davis, H., & Silverman, S. R. *Hearing and deafness*. New York: Holt, Rinehart and Winston, 1960.

Davis, H., & Silverman, S. R. *Hearing and deafness*. (4th Ed.). New York: Holt, Rinehart and Winston. 1978.

Davis, H., Rosenblut, B., Fernandez, C., Kimura, R., & Smith, C. A. Modification of cochlear potentials produced by extensive venous obstruction. *Laryngoscope*, 1958, *68*, 596–627.

Dirks, D., & Malmquist, C. Comparison of frontal and mastoid bone conduction thresholds in various conduction lesions. *Journal of Speech & Hearing Disorders*, 1969, *12*, 725–746.

Djupesland, G. Use of impedance indicator in diagnosis of middle ear pathology. *Journal of International Audiology*, 1969, *8*, 570–578.

Eagles, E. L., Wishik, S. M., & Doerfler, L. G. Hearing sensitivity and ear distress in children: A prospective study. *Laryngoscope*, Supplement, 1967, 1–274.

Egan, J. P. Articulation testing methods. *Laryngoscope*, 1948, *58*, 955–991.

Egan, J. P., & Hake, H. W. On the masking pattern of a simple auditory stimulus. *Journal Acoustical Society of America*, 1950, *22*, 622–630.

Eimind, K. The reaction of the cochlea to stapes mobilization. *Acta Otolaryngologica*, 1960, *52*, 453–460.

Elbrond, O. Defects of the auditory ossicles in ears with intact tympanic membrane. *Acta Otolaryngologica*, Supplement, 1970, 264.

Everberg, G. Deafness following mumps. *Acta Otolaryngologica*, 1957, *48*, 397–403.

Everberg, G. Unilateral total deafness in children: Clinical problems with a special view to vestibular function. *Acta Otolaryngologica*, 1960, *52*, 253–269.

Fletcher, H. Auditory patterns. *Revised Modified Physics*, 1940, *12*, 47–65.

Fletcher, H., & Steinberg, J. C. Articulation testing methods. *Bell System Technical Journal*, 1929, *8*, 806–854.

Fowler, E. P. A method for the early detection of otosclerosis. *Acta Otolaryngologica*, 1936, *24*, 731–741.

Fowler, E. P. Sudden deafness. *Annals of Otology, Rhinology, & Laryngology*, 1950, *59*, 980–987.

Fowler, E. P. Loudness recruitment: Definition and clarification. *Archives of Otolaryngology*, 1963, *78*, 748–753.

Fowler, E. P. Some attributes of "loudness recruitment" and "loudness decruitment". *Annals of Otology, Rhinology, & Laryngology*, 1965, 2, 500–506.

Gacek, R. R., & Schuknecht, H. F. Pathology of presbycusis. *International Audiology*, 1969, *8*, 199–209.

Gaeth, J. Study of phonemic regression in relation to hearing loss. Unpublished master's thesis, Northwestern University, Chicago, 1948.

Geisler, C. D., Frishkopf, L. S., & Rosenblith, W. A. Extracranial responses to acoustic clicks in man. *Science*, 1958, *128*, 1210–1211.

Glasscock, M., III. Middle fossa approach to the temporal bone: An otologic frontier. *Archives of Otolaryngology*, 1969, *90*, 15–27.

Glasscock, M., & Hays, J. Pitfalls in the diagnosis of acoustic and other cerebellopontine angle tumors. *Laryngoscope*, 1973, *83*, 1038.

Glorig, A., Jr. *Noise and your ear*. New York: Grune and Stratton, 1958.

Glorig, A., & Davis, H. Age, noise, and hearing loss. *Annals of Otology, Rhinology, & Laryngology*, 1961, *70*, 556–571.

Glorig, A., Jr., Ward, W., & Nixon, J. Damage risk criteria and noise-induced hearing loss *AMA Archives of Otolaryngology*, 1961, *74*, 413–423.

Goetzinger, C. P., Proud, G. O., & Embrez, J. E. Masking and bone conduction. *Acta Otolaryngologica*, 1962, *54*, 287–291.

Goldstein, D. P., Hayes, C. S., & Peterson, J. L. A comparison of bone conduction thresholds by conventional and Rainville methods. *Journal of Speech & Hearing Research*, 1962, *5*, 244–255.

Goldstein, R., & Rodman, L. B. Early components of the averaged evoked responses to rapidly repeated auditory stimuli. *Journal of Speech & Hearing Research*, 1967, *10*, 697–705.

Goldstein, R., Goodman, A., & King, R. B. Hearing and speech in infantile hemiplegia before and after left hemispherectomy. *Neurology*, 1956, *6*, 869–875.

Goodhill, V., & Moncur, J. The low frequency air-bone gap. *Laryngoscope*, 1963, *73*, 850–867.

Green, D. The modified tone decay test (MTDT) as a screening procedure for eighth nerve lesions. *Journal of Speech & Hearing Disorders*, 1963, *28*, 31–36.

Greenwood, D. D. Auditory masking and the critical band. *Journal of Acoustical Society America*, 1961, *33*, 484–502. (a)

Greenwood, D. D. Critical band width and the frequency coordinates of the basilar membrane. *Journal of Acoustical Society America*, 1961, *33*, 1344–1356. (b)

Gregg, J. D., & Lierle, D. M. Ototoxicity of certain antibiotic drugs. *General Practice*, 1961, *23*, 94–100.

Greisen, O., & Rasmussen, P. Stapedius muscle reflexes and otoneurological examinations in brainstem tumors. *Acta Otolaryngologica*, 1970, *70*, 366–370.

Grinker, R. R., Bucy, P. C., & Sahs, A. L. *Neurology* (5th Ed.). Springfield, Ill.: Charles C. Thomas, 1960.

Gundersen, T., & Tonning, F. M. Ventilating tubes in the middle ear: Long-term observations. *Archives of Otolaryngology*, 1976, *102*, 198–199.

Gussen, R. Meniere's disease: New temporal bone findings in two cases. *Laryngoscope*, 1971, *81*, 1695–1707.

Hallberg, O. E. Sudden deafness and its relation to atherosclerosis. *Journal of American Medical Association*, 1957, *165*, 1649–1652.

Hallpike, C. S., & Cairns, H. Observations on the pathology of Meniere's syndrome. *Journal of Laryngological Otolaryngology*, 1938, *53*, 625–655.

Hallpike, C. S., & Wright, A. J. On the histological changes in the temporal bones of a case of Meniere's disease. *Journal of Laryngological Otolaryngology*, 1940, *55*, 59–65.

Hardy, J. B. Fetal consequences of maternal viral infection in pregnancy. *Archives of Otolaryngology*, 1973, *98*, 218–227.

Harper, A. Acoustic impedance as an aid to diagnosis in otology. *Journal of Laryngological Otolaryngology*, 1961, *75*, 614–620.

Harris, I. Tympanosclerosis—A revised clinicopathologic entity. *Laryngoscope*, 1961, *71*, 1488–1533.

Harrison, H. H. Ototoxicity of dihydrostreptomycin. *Quarterly Bulletin Northwest University Medical School*, 1954, *28*, 271–273.

Harrison, W., Shambaugh, G., Jr., Derlacki, E., & Clemis, J. The perilymph fistula problem. *Laryngoscope*, 1970, *80*, 1000.

Hart, C., & Naunton, R. The ototoxicity of chloroquine phosphate. *Archives of Otolaryngology*, 1964, *80*, 407.

Heilman, K., Hamner, L., & Wilder, B. An audiometric defect in temporal lobe dysfunction. *Neurology*, 1973, *23*, 384–386.

Hemenway, W. G., Sando, I., & McChesney, D. Temporal bone pathology following maternal rubella. *Archives of Otolaryngology*, 1969, *193*, 287–300.

High, W. G., Fairbanks, G., & Glorig, A. Scale for self-assessment of hearing handicap. *Journal of Speech & Hearing Disorders*, 1964, *29*, 215–230.

Hilger, J. A., & Goltz, N. F. Some aspects of inner ear therapy. *Laryngoscope*, 1951, *61*, 695–717.

Hirsh, J. J., Davis, H., Silverman, S. R., Reynolds, E. G., Eldbert, E., & Benson, R. W. Development of materials for speech audiometry. *Journal of Speech & Hearing Disorders*, 1952, *17*, 321–337.

Hood, J. D. Auditory fatigue and adaptation in the differential diagnosis of end organ disease. *Annals of Otology, Rhinology, & Laryngology*, 1955, *64*, 507–518.

Hood, J. D. The principles and practice of bone conduction audiometry: A review of the present position. *Laryngoscope*, 1960, *70*, 1211–1228.

House, H., & Greenfield, E. Five-year study of wire loop-absorbable gelatin sponge technique. *Archives of Otolaryngology*, 1969, *89*, 420–421.

Hughson, W., & Thompson, E. A. Correlation of hearing acuity for speech with discrete frequency audiograms. *Archives of Otolaryngology*, 1942, *36*, 526–540.

Hughson, W., & Westlake, H. D. Manual for program outline for rehabilitation of aural casualties both military and civilian. *Transactions of the American Academy of Ophthalmology and Otolaryngology*, Supplement, 1944, *48*, 1–15.

Hulks, J. Bone conduction changes in acute otitis media. *Archives of Otolaryngology*, 1941, *33*, 333–350.

Hyatt, H. W., Sr. Complications of mumps. *General Practice*, 1962, *25*, 124–126.

Jackson, C., & Jackson, C. *Diseases of the nose, throat and ear*. Philadelphia: W. B. Saunders, 1959.

Jarvis. J. F. A case of unilateral permanent deafness following acetylsalicylic acid. *Journal of Laryngology, Otolaryngology, Rhinolaryngology*, 1966, *80*, 318.

Jerger, J. F. Bekesy audiometry in analysis of auditory disorders. *Journal of Speech & Hearing Research*, 1960, *3*, 275–287.

Jerger, J. F. Hearing tests in otologic diagnosis. *Journal of the American Speech and Hearing Association*, 1962, *4*, 139–145.

Jerger, J. Auditory test, for disorders of the central auditory mechanism. In W. Fields, & B. Alford (Eds.), *Neurological aspects of auditory and vestibular disorders*. Springfield, Ill.: Charles C Thomas, 1964, 77–93.

Jerger, J. Clinical experience with impedance audiometry. *Archives of Otolaryngology*, 1970, *92*, 311–324.

Jerger, J. Diagnostic audiometry. In J. Jerger (Ed.), *Modern development in audiology* (Vol. 2) New York: Academic Press, 1973, pp. 75–115.

Jerger, J. Impedance terminology. *Archives of Otolaryngology*, 1975, *101*, 589–590.

Jerger, J., & Harford, E. The alternate and simultaneous balancing of pure tones. *Journal of Speech & Hearing Disorders*, 1960, *3*, 15–30.

Jerger, J. F., & Herer, G. Unexpected dividends in Bekesy audiometry. *Journal of Speech & Hearing Disorders*, 1961, *26*, 390–391.

Jerger, J., & Jerger, S. Critical evaluation of SAL audiometry. *Journal of Speech & Hearing Research*, 1965, *8*, 103–125.

Jerger, J., & Jerger, S. Diagnostic significance of PB word functions. *Archives of Otolaryngology*, 1971, *93*, 573–580.

Jerger, J., & Jerger, S. Audiological comparisons of cochlear and eighth nerve disorders. *Annals of Otology, Rhinology, & Laryngology*, 1974, *83*, 275–285. (a)

Jerger, J., & Jerger, S. Diagnostic value of Bekesy comfortable loudness tracings. *Archives of Otolaryngology*, 1974, *99*, 351–360. (b)

Jerger, J., & Jerger, S. Clinical validity of central auditory tests. *Scandinavian Audiology*, 1975, *4*, 147–163. (a)

Jerger, J., & Jerger, S. A simplified tone decay test. *Archives of Otolaryngology*, 1975, *102*, 403–407. (b)

Jerger, J., & Tillman, J. A new method for the clinical determination of sensorineural acuity level (SAL). *Archives of Otolaryngology*, 1969, *71*, 948–953.

Jerger, J. F., Allen, G., Robertson, D., & Harford, E. Hearing loss of sudden onset. *Archives of Otolaryngology*, 1958, *73*, 350–357.

Jerger, J. F., Sheed, J., & Harford, E. On the detection of extremely small changes in sound intensity. *Archives of Otolaryngology*, 1959, *69*, 200–211.

Jerger, J., Speaks, C., & Trammell, J. L. A new approach to speech audiometry. *Journal of Speech & Hearing Disorders*, 1968, *33*, 318–328.

Jerger, J., Jerger, S., & Mauldin, L. The forward–backward discrepancy in Bekesy audiometry. *Archives of Otolaryngology*, 1972, *72*, 400–406. (a)

Jerger, J., Jerger, S., & Mauldin, L. Studies in impedance audiometry. I. Normal and sensorineural ears. *Archives of Otolaryngology*, 1972, *96*, 513–523. (b)

Jerger, J., Anthony, L., Jerger, S., & Mauldin, L. Studies in impedance audiometry. III. Middle ear disorders. *Archives of Otolaryngology*, 1974, *99*, 165–172. (a)

Jerger, J., Burney, P., Mauldin, L., & Crump, B. Predicting hearing loss from the acoustic reflex. *Journal of Speech & Hearing Disorders*, 1974, *39*, 11–22. (b)

Jerger, S., & Jerger, J. Extra- and intra-axial brain stem auditory disorder. *Audiology*, 1975, *14*, 93–117.

Jerger, S., & Jerger, J. Diagnostic value of crossed vs. uncrossed acoustic reflexes: Eighth nerve and brain stem disorders. *Archives of Otolaryngology*, 1977, *103*, 445–453.

Jerger, S., Neely, J. G., & Jerger, J. Recovery of crossed acoustic reflexes in brain stem auditory disorder. *Archives of Otolaryngology*, 1975, *101*, 329–332.

Jewitt, D., Romano, M., & Williston, J. Human auditory evoked potentials: Possible brain stem components detected on the scalp. *Science*, 1970, *167*, 1517–1518.

Johnson, E. W. Auditory test results in 110 surgical confirmed retrocochlear lesions. *Journal of Speech & Hearing Disorders*, 1965, *30*, 307–317.

Johnson, E. W. Confirmed retrocochlear lesions: Auditory test results in 163 patients. *Archives of Otolaryngology*, 1966, *84*, 247–254.

Johnson, E. W. Auditory findings in 200 cases of acoustic neuromas. *Archives of Otolaryngology*, 1968, *88*, 50–55.

Johnson, L., & Hawkins, J., Jr. Sensory and neural degeneration with aging, as seen in micro-disections of the human inner ear. *Annals of Otology, Rhinology, & Laryngology*, 1972, *81*, 179–193.

Juhn, S. K., Huft, J. S., & Paparella, M. M. Lactate dehydrogenase activity and isoenzyme patterns in middle ear effusions. *Annals of Otology, Rhinology, & Laryngology*, 1973, *82*, 192—195.

Karja, J., & Palva, A. Reverse frequency—sweep Bekesy audiometry. *Acta Otolaryngologica*, Supplement, 1970, *263*, 225–228.

Katz, J. The use of staggered spondaic words for assessing the integrity of the central auditory nervous system. *Journal of Auditory Research*, 1962, *2*, 327–337.

Katz, J. The SSW test: An interior report. *Journal of Speech & Hearing Disorders*, 1968, *33*, 132–146.

Katz, J. Audiologic diagnosis: Cochlea to cortex. *Menorah Medical Journal*, 1970, *1*, 25–38.

Keith, R. W. Impedance audiometry with neonates. *Archives of Otolaryngology*, 1973, *97*, 465–467.

Keith, R. W. Middle ear function in neonates. *Archives of Otolaryngology*. 1975, *101*, 376–379.

Kimura, R., Pearlman, H. B., & Fernandez, C. Experiments on temporary obstruction of the internal auditory artery. *Laryngoscope*, 1959, *69*, 591–613.

Klotz, R. E., & Kilbane, M. Hearing in an aging population, preliminary report. *New England Journal of Medicine*, 1962, *266*, 277–280.

Kobrak, H. G., Lindsay, J. R., & Pearlman, H. B. Experimental observations on the question of auditory fatigue. *Laryngoscope*, 1941, *51*, 798–810.

Kohonen, A. Effect of some ototoxic drugs upon the pattern and innervation of cochlear sensory cells in the guinea pig. *Acta Otolaryngologica*, Supplement, 1965, *208*, 1.

Kopetsky, S. J. *Deafness, tinnitus, and vertigo*. New York: Thomas Nelson and Sons, 1948.

Kryter, K. The effects of noise on man. *Journal of Speech & Hearing Disorders*, 1950, *1*, 36–37.

Kryter, K., & Ades, H. Studies on the function of the higher acoustic nerves centers in the cat. *American Journal of Psychology*, 1943, *56*, 501–536.

Kupperman, G. L., & Mendel, M. I. Threshold of the early components of the averaged electroencephalic response determined with tone pips and clicks during drug-induced sleep. *Audiology*, 1974, *13*, 379–390.

Lempert, J. Lemport Fenestra Novalis with Mobile Stapple. *Archives of Otolaryngology*, 1945, *41*, 1–41.

Lewis, R. S., Grey, J. D., & Hawlett, A. B. Acute otitis media. *British Medical Journal*, 1955, *66*, 142–146.

Liden, G. Loss of hearing following treatment with dihydrostreptomycin or streptomycin. *Acta Otolaryngologica*, 1953, *43*, 551–572.

Liden, G., Peterson, J., & Bjorkman, G. Tympanometry. *Acta Otolaryngologica*, Supplement, 1970, *263*, 218–224. (a)

Liden, G., Peterson, J., & Bjorkman, G. Tympanometry. *Archives of Otolaryngology*, 1970, 92, 248–257. (b)

Liden, G., Peterson, J. L., & Harford, E. R. Simultaneous recording of changes in relative impedance and air pressure during acoustic and nonacoustic elicitation of the middle ear reflexes. *Acta Otolaryngologica*, Supplement, 1970, *263*, 208–217. (c)

Lieberman, A. Unilateral deafness. *Laryngoscope*, 1957, *67*, 1237–1265.

Lightfoot, C. The M-1 test of bone conduction hearing. *Laryngoscope*, 1960, *70*, 1552–1559.

Lilly, D. A comparison of acoustic impedance data obtained with Madsen and Zwislocki instruments. *American Speech and Hearing Association*. 1970, *12*, 441.

Lim, D. Histology of the developing inner ear: Normal anatomy and developmental abnormalities. In B. F. Jaffe (Ed.), *Hearing loss in children*. Baltimore: University Park Press, 1977.

Lindsay, J. R., Davey, P. R., & Ward, P. H. Inner ear pathology in deafness due to mumps. *Annals of Otolaryngology*, 1960, *69*, 918–935. (a)

Lindsay, J. R., Procter, L. R., & Work, W. P. Histopathologic inner ear changes in deafness due to neomycin. *Laryngoscope*, 1960, *70*, 383–392. (b)

Lowell, S. E., & Paparella, M. M. Presbycusis: What is it? *Laryngoscope*, 1977, *87*, 1710–1717.

Lynn, G. E., & Gilroy, J. Audiological abnormalities in patients with temporal lobe tumors. *Journal of Neurological Science*, 1972, *12*, 167–184.

Lynn, G. E., & Gilroy, J. Effects of brain lesions on the perception of monotic and dichotic speech stimuli. *Proceedings of a Symposium on Central Auditory Processing Disorders*. M. D. Sullivan (Ed.). Omaha: University of Nebraska Medical Center, 1975.

Lynn, G. E., & Pirkey, W. P. Measurement of the sensory-neural acuity level (SAL) in conductive hearing-loss cases. *Journal of Auditory Research*, 1962, *2*, 323–326.

Marcus, R. E., Small, H., & Emanuel, B. Ototoxic medication in premature children. *Archives of Otolaryngology*, 1963, *77*, 198–204.

Martin, F. N., & Forbis, N. K. The present status of audiometric practice: A follow-up study. *Journal American Speech and Hearing Association*. 1978, *20*, 531–541.

Mattox, D. E., & Simmons, R. B. Natural history of sudden sensorineural hearing loss. *Annals of Otology, Rhinology, & Laryngology*, 1977, *86*, 463–481.

Matzker, J. Two new methods for the assessment of central auditory functioning in cases of brain disease. *Annals of Otology, Rhinology, & Laryngology*, 1959, *68*, 1185–1197.

Mawson, S. R. *Diseases of the ear, Part IV*. Baltimore: Williams and Wilkins, 1963.

McGee, T. M., & Olszewski, J. Streptomycin sulfate and dihydrostreptomycin toxicity: Behavioral and histopathological studies. *Archives of Otolaryngology*, 1962, *75*, 295–311.

McKenzie, W., & Rice, J. C. Schuknecht's operation for middle ear deafness. *Lancet*, 1962, *1*, 943–944.

Metz, O. The acoustic impedance measured on normal and pathological ears. *Acta Otolaryngologica*, Supplement, 1946, *63*, 1–131.

Metz, O. Influence of the patulous eustachian tube on the acoustic impediance of the ear. *Archives of Otolaryngology*, Supplement, 1953, *109*, 105–112.

Meurman, O. M., & Hietalahti, J. Modern treatment of pulmonary tuberculosis and its effect on hearing. *Acta Otolaryngologica*, Supplement, 1960, *158*, 351–355.

Meyer, D. R., & Woolsey, C. N. Effects of localized cortical damage destruction upon auditory discrimination conditioning in the cats. *Journal of Neurophysiology*, 1952, 15, 149–162.

Michael, L. A. The SAL test in the prediction of strapedectomy results. *Laryngoscope*, 1963, *73*, 1370–1376.

Morales-Garcia, C., & Poole, J. P. Masked speech audiometry in central deafness. *Acta Otolaryngologica*, 1972, *74*, 307–316.

Nager, G. Association of bilateral VIIIth nerve tumors with meningiomas in von Recklinghausen's disease. *Laryngoscope*, 1964, *74*, 1220–1261.

Neel, H. B., Keating, L. W., & McDonald, T. J. Ventilation in secretory otitis media. Effects on middle ear volume and eustachias tube function. *Archives of Otolaryngology*, 1977, *103*, 228–231.

Newby, H. *Audiology*. New York: Appleton-Century-Crofts, 1964.

Niemeyer, W., & Sesterhenn, G. Calculating the hearing threshold from the stapedius reflex threshold for different sound stimuli. *Audiology*, 1974, *13*, 421–427.

Nixon, J. C., Glorig, A., & High, W. S. Changes in air and bone conduction thresholds. *Journal of Laryngology*, 1962, *76*, 288–298.

Nober, E. H. GSR magnitudes for different intensities of shock, conditioned tone and extinction tone. *Journal of Speech & Hearing Research*, 1958, *1*, 316–324.

Nober, E. H. Pseudo-auditory bone conduction thresholds. *Journal of Speech & Hearing Disorders*, 1964, *29*, 469–476.

Nober, E. H. Vibrotactile sensitivity of deaf children. *Laryngoscope*, 1967, *87*, 2128–2146.

Nober, E. H. Cutile air and bone conduction thresholds of the deaf. *Exceptional Children*, 1970, 571–579.

Noffsinger, D., Kurdziel, S., & Applebaum, E. L. Value of special auditory tests in the latero-medical inferior pontine syndrome. *Annals of Otology, Rhinology, & Laryngology*, 1975, *84*, 384–390.

Northern, J. L., & Downs, M. P. *Hearing in children*. Baltimore: Williams and Wilkins, 1974.

Occupational Health and Safety Administration, *Federal Register*, October 24, 1974, *39*, 37773–37778.

Olsen, W., & Noffsinger, D. Masking level differences for cochlear and brain stem lesions. *Annals of Otology, Rhinology, & Laryngology*, 1976, *85*, 820–825.

Olsen, W., Noffsinger, D., & Kurdziel, S. Speech discrimination in quiet and in white noise by patients with peripheral and central lesions. *Acta Otolaryngologica*, 1975, *80*, 375–382.

Olsen, W., Noffsinger, O., & Carhart, R. Masking level differences encountered in clinical populations. *Audiology*, 1976, *15*, 287–301.

Ozaki, T. Prevention of adverse effects of streptomycin on the ear. *Archives of Otolaryngology*, 1957, *66*, 673–678.

Palva, T., & Palva, A. Masking in audiometry. *Acta Otolaryngologica*, 1962, *76*, 593–595.

Palva, T., Karja, J., Palva, A. Foward vs. reversed Bekesy tracings. *Archives of Otolaryngology*, 1970, *91*, 449–452.

Paparella, M. M., & Shumrick, D. A. *Otolaryngology*, Philadelphia: W. B. Saunders, 1973.

Paparella, M., Brady, D., & Hoel, R. Sensorineural hearing loss in chronic otitis media and mastoiditis. *Transactions of the American Academy of Ophthalmology and Otolaryngology*, 1970, *74*, 108–115.

Paparella, M., Oda, M., Hiraide, F., & Brady, D. Pathology of sensorineural hearing loss in otitis media. *Annals of Otology, Rhinology, & Laryngology*, 1972, *81*, 632–647.

Paradise, J. L., Smith, C. G., & Bluestone, C. D. Tympanometric detection of middle ear effusion in infants and young children. *Pediatrics*, 1976, *58*, 198–210.

Payne, E. E., & Paparella, M. M. In J. L. Northern (Ed.), *Hearing Disorders*. Boston: Little, Brown, 1976.

Proctor, D. F. *The nose, paranasal sinuses, and ears in childhood*. Springfield, Ill.: Charles C Thomas, 1963.

Pulec, J. L. Meniere's disease. In J. L. Northern (Ed.), *Hearing Disorders*. Boston: Little, Brown, 1976.

Quick, C. A. Chemical and drug effects on inner ear. In M. M. Paparella, & D. A. Shumrick (Eds.), *Otolaryngology II:* Ear. Philadelphia: W. B. Saunders, 1973, pp. 391–406.

Rainville, M. J. New methods of masking for the determination of bone conduction curves. *Translation Beltone Institute Hearing Research*, 1959, *11*, 1–10.

Rapin, I., Schimmel, H., Tourk, L. M., Krasnegor, N. A., & Pollak, C. Evoked responses to clicks and tones of varying intensity in waking adults. *Electroencephalography and Clinical Neurophysiology*, 1966, *21*, 335–344.

Renvall, V., Liden, G., Jungert, S., & Nilsson, E. Impedance audiometry as a screening method in school children. *Scandinavian Audiology*, 1973, *2*, 133–137.

Robertson, E., Peterson, J., & Lamb, L. Relative impedance measurements in young children. *Archives of Otolaryngology*, 1968, *88*, 162–168.

Rosen, S. Mobilization of the stapes to restore hearing in otosclerosis. *New York Journal of Medicine*, 1953, *22*, 2650–2653.

Rosen, S. Assessment of the techniques of stapes surgery. *Journal of American Medical Association*, 1961, *178*, 1144–1146.

Rosen, S., Bergman, M., Plester, D., El-Mofty, A., & Satti, M. H. Presbycusis study of a relatively noise-free population in the Sudan. *Annals of Otology, Rhinology, & Laryngology*, 1962, *71*, 727–743.

Rosenblith, W., & Stevens, K. Handbook of acoustic noise control: Noise and man. *Wright Patterson Air Development Center of Technology Report*, 1953, *2*, 52–2041.

Rossi, G., Corando, D. E., DeMicheles, G., & Mazzio, R. P. Deafness and vestibular impairment caused by strepto- and dihydrostreptomycin. A clinical study of 209 cases. *Acta Medical (Scand.)*, 1961, *169*, 169–180.

Ruhm, H., Walker, E., & Flanigin, H. Acoustically-evoked potentials in man: Mediation of early components. *Laryngoscope*, 1967, *77*, 806–822.

Saltzman, M., & Ersner, M. The ear stroke. *Eye, Nose, and Throat Monthly*, 1952, *31*, 372–374.

Sanchez-Longo, L. P., Forster, F. M., & Auth, T. L. A clinical test for sound localization and its applications. *Neurology*, 1957, *7*, 655–663.

Sanders, D. Y., Eliot, D. S., & Cramblett, H. G. Retrospective study for possible kanamycin ototoxicity among neonatal infants. *New England Journal of Medicine*, 1967, *271*, 949–951.

Sanders, J. W., Josey, A. F., & Glasscock, M. E. Audiologic evaluation in cochlear and eight nerve disorders. *Archives of Otolaryngology*, 1974, *100*, 283–289.

Sataloff, J., & Menduke, H. Presbycusis. *Transactions of the American Academy of Opthalmology and Otolaryngology*, 1957, *61*, 141–146.

Saunders, W. H., & Lippy, W. H. Sudden deafness and Bell's palsy: A common cause. *Annals of Otology, Rhinology, & Laryngology*, 1959, *68*, 830–837.

Schein, J. D., & Delk, M. T. *The deaf population of the United States*. Silver Springs, Md.: National Association of the Deaf, 1974.

Schuknecht, H. F. Presbycusis. *Laryngoscope*, 1955, *65*, 402–419.

Schuknecht, H. Perceptive hearing loss. *Laryngoscope*, 1958, *68*, 429–439.

Schuknecht, H. F. (Ed.). *Otosclerosis*. Henry Ford Hospital, International Symposium. Boston: Little, Brown, 1962.

Schuknecht, H. *Pathology of the ear*. Cambridge, Mass.: Harvard University Press, 1974.

Schuknecht, H., & Ishii, T. Hearing loss caused by atrophy of the stria vascularis. *Japanese Journal of Otology*, 196, *69*, 1825.

Selters, W. A., & Brackmann, D. E. Acoustic tumor detection with brain stem electric response audiometry. *Archives of Otolaryngology*, 1977, *103*, 181–187.

Shambaugh, G. E., Jr. *Surgery of the ear* (2nd Ed.). Philadelphia: W. B. Saunders, 1967.

Shambaugh, G. E., Derlacki, E. L., Harrison, W. H., House, H., House, W., Hildyard, V., Schuknecht, H., & Shea, J. J. Dihydrostreptomycin deafness. *Journal of American Medical Association*, 1959, *170*, 1657–1660.

Sharp, H. S. The management of acute otitis media. *British Journal of Clinical Practice*, 1960, *14*, 279–282.

Sheehy, J. Vasodilator therapy in sensorineural hearing loss. *Laryngoscope*, 1960, *70*, 885–913.

Shimizu, H. Evoked response in VIIIth nerve lesions. *Laryngoscope*, 1968, *78*, 2140–2152.

Shurin, P. A., Pelton, S. I., & Finkelstein, J. Tympanometry in the diagnosis of middle-ear effusion. *New England Journal of Medicine*, 1977, *296*, 412–417.

Smith, B. B., & Pesnick, D. M. An auditory test for assessing brain stem integrity: Preliminary report. *Laryngoscope*, 1972, *82*, 414—424.

Sparks, R., Goodglass, H., & Nickel, B. Ipsilateral vs. contralateral extinction in dichotic listening resulting from hemipheric lesions. *Cortex*, 1970, *6*, 249–260.

Speaks, C. Dichotic listening: A clinical or research tool? In N. Sullivan (Ed.), *Central auditory processing disorders*. Omaha: University of Nebraska Press, 1975.

Spoor, A. Presbycusis values in relations to noise induced hearing loss. *International Audiology*, 1967, *6*, 47–57.

Stewart, J. P. (Ed.) *Logan Turner's diseases of the nose, throat and ear*. Bristol: John Wright and Sons, 1961.

Strome, M. Sudden and fluctuating hearing losses. In B. F. Jaffe (Ed.), *Hearing loss in children*. Baltimore: University Park Press, 1977.

Studebaker, G. Clinical masking and air-bone conducted stimuli. *Journal of Speech & Hearing Disorders*, 1964, *29*, 231–235.

Surapathana, L., Futrakul, P., & Campbell, R. Salicylism revisited: Unusual problems in diagnosis and management. *Clinical Pediatrics*, 1979, *9*, 658–661.

Tato, J. M., DeLozzio, C. B., & Valencia, J. I. Chromosomal study in otosclerosis. *Acta Oto–Laryngologica*, 1963, *56*, 265–270.

Terkildsen, K. Clinical application of impedance measurements with a fixed frequency technique. *International Audiology*, 1964, *3*, 147–155.

Thomas, K. Objective determination of the middle ear pressure. *Acta Otolaryngologica*, Supplement, 1961, *158*, 212–216.

Tillman, T. Clinical applicability of the SAL test. *Archives of Otolaryngology,* 1963, *78,* 20–32.

Tillman, T. W. The assessment of sensorineural acuity. In B. Graham (Ed.), *Sensorineural hearing processes and disorders,* Boston: Little, Brown, 1967.

Tillman, T. W. Special hearing tests in otoneurologic diagnosis. *Archives of Otolaryngology,* 1969, *89,* 25–30.

Tillman, T. W., & Gretis, E. S. Masking of tones by bone-conducted noise in normal and hearing-impaired listeners. *Abstracts of American Speech and Hearing Association Convention Program,* 1969, *69.*

Tonndorf, J. Bone conduction; studies in experimental animals. *Acta Otolaryngologica,* Supplement, 1966, *213,* 1–132.

Tonndorf, J. Relationship between the transmission characteristics of the conductive system and noise-induced hearing loss. In D. Henderson, R. P. Hamernik, D. S. Dosanjh, & J. H. Mills (Eds.), *Effects of noise on hearing.* New York: Raven Press, 1976.

Torok, N. Old and new in Meniere's disease. *Laryngoscope,* 1977, *87,* 1870–1877.

Van Dishoeck, H. A., & Bierman, T. Sudden perceptive deafness and viral infections. *Annals of Otology,* 1957, *66,* 963–980.

Van Wagoner, R. S., & Goodwine, S. Clinical impressions of acoustic reflex measures in an adult population. *Archives of Otolaryngology,* 1977, *103,* 582–584.

Varpela, E., Hietalahti, J., & Aro, M. J. T. Streptomycin and dyhydrostreptomycin medication during pregnancy and their effect on the child's inner ear. *Scandinavian Journal of Respiratory Diseases,* 1969, *50,* 101–109.

Vuori, M., Labikainen, E. A., & Peltonen, T. In J. R. Lindsay (Ed.), *Yearbook of the ear, nose, and throat.* Chicago: Yearbook Medical Publishers, 1963–1964.

Walsh, T. E. The effect of pregnancy on the deafness due to otosclerosis. *Archives of Otolaryngology,* 1954, *154,* 1407–1409.

Walton, W. K. *A tympanometry—ASHA model for identification of the hearing impaired child.* Windsor, Conn.: Capital Region Education Council, May 1975.

Ward, W. D. Auditory fatigue and masking. In J. Jerger (Ed.), *Modern developments in audiology.* New York: Academic Press, 1963.

Weaver, M., & Northern, J. L. The acoustic nerve tumor. In J. L. Northern (Ed.), *Hearing disorders,* Boston: Little, Brown, 1976.

Webster, A. Acoustical impedance and the theory of horns and of the phonograph. *Proceedings of the National Academy of Science,* 1919, *5,* 275–282.

Wever, E. G., & Lawrence, M. *Physiological acoustics.* Princeton: Princeton University Press, 1954.

Wier, J. A., Storey, P. B., Curry, F. J., & Schloss, J. M. Ototoxicity from intermittent streptomycin therapy of pulmonary tuberculosis: A study of 105 patients treated 8–10 months. 1956, *30,* 628–632.

Wilber, L., Goodhill, V., & Hogue, A. Comparative acoustic impedance measurements. *Journal of American Speech and Hearing Association,* 1970, *12,* 435.

Wilmot, T. Sudden perceptive deafness in young people. *Journal of Laryngology,* 1959, *73,* 466–469.

Zwislocki, J. Some measurements of the impedance at the eardrum. *Journal of Acoustical Society of America,* 1957, *29,* 349–356. (a)

Zwislocki, J. Some impedance measurements on normal and pathological ears. *Journal of Acoustical Society of America,* 1957, *29,* 1312–1317. (b)

4

Annette R. Zaner and Janet M. Purn

AUDITORY PROBLEMS IN CHILDREN*

Introduction

Compared with other handicapping conditions, the incidence of handicapping hearing impairment in young children is relatively low. Nevertheless, children with auditory deficits are prominent among those for whom "education" in early infancy is deemed essential. Indeed, early identification and intervention strategies for hearing-impaired youngsters are thought to be so critical for adequate development that proposals for innovative approaches with this population were among the first to be encouraged by grants from the United States Department of Health, Education, and Welfare. In fact, services for infants with hearing impairments were considered such a high priority that the United States granting agencies suggested that no minimum age be specified for including hearing-impaired children in intervention programs. On recommendation from professional consultants, it was strongly urged that "age of identification," rather than chronological age, be considered as the appropriate date for enrollment.

In the early 1950s, language-training programs for very young hearing-impaired children were established through the cooperative efforts of health and education agencies. Such programs were often housed in hospital outpatient facilities and were staffed by professional teams that included audiologists,

*In describing auditory problems it is becoming increasingly difficult to separate peripheral from central impairments. Therefore, although this chapter is included in the section of the book devoted to peripheral disturbances of communication, it includes reference to both peripheral and central problems.

Annette R. Zaner and Janet M. Purn • The Mount Carmel Guild, 17 Mulberry Street, Newark, New Jersey 07102.

speech-language pathologists, otolaryngologists, psychologists, and teachers of the hearing impaired. It was not at all unusual to find 18-month-old hearing-impaired babies routinely included on those rosters. Following the rubella epidemic in the early 1960s, there was even greater general awareness of the need for early intervention, and in the mid and late 1970s, with greater availability of more sophisticated screening and testing techniques and with improved dissemination of public information, the enrollment age decreased even further. Consequently, in 1981 it is quite usual to find 6- to 12-month-old hearing-impaired children, and their parents, regularly attending "school."

It would appear, then, that hearing impairment is viewed as potentially creating the kind of problems that make early intervention most essential. It follows, therefore, that "normal" hearing in children must be among the more critical factors in normal development.

The ability of the human organism to respond normally to a variety of auditory stimuli is dependent upon the adequate functioning of the separate aspects of the auditory system and upon the intact interrelatedness among those parts. Individuals with normal hearing, as measured by a clinical audiometer, are able to hear a variety of pure tones across a wide frequency spectrum, at an intensity level that represents average threshold for a broad segment of our population. Beyond the ability to respond to auditory sensations at average thresholds, however, normal hearing, for the purposes of this chapter, will also include the ability to perceive and process auditory verbal stimuli necessary for the acquisition and competent development of language. An auditory "problem" will include any of those conditions subsequent to a decrement in auditory sensitivity and/or the inability to process auditory stimuli received by the peripheral mechanism. Auditory problems can cause delayed or disordered language acquisition and impaired cognitive development with consequent limitations in social, emotional, and/or vocational growth. Auditory problems may also have a negative effect upon educational or academic achievements. In short, a child with an auditory problem can suffer a handicap far in excess of the primary disorder.

Although we are concerned with auditory problems in children of all ages, the emphasis here will be on those auditory problems affecting children through the age of eight, by which time linguistic development should have reached a sophisticated level. Since it is a bias of this chapter to view auditory and linguistic behaviors as interrelated, emphasis will be placed on descriptions and discussions of auditory problems that occur in children prior to the development of sophisticated language. Furthermore, assessment of auditory function will be described utilizing developmental strategies. That is, the goals of assessment change as development proceeds. During infancy, for example, assessment is undertaken for the purposes of determining "hearing" versus "severe hearing impairment"; later, thresholds of sensitivity are obtained, bilaterally; and during the school years, assessment may also pursue a differential diagnosis between peripheral and central disorders. Both theory and technology have advanced so that assessment of auditory function of considerably higher order than sensitivity levels is now possible. Not only can we now examine central auditory processing behaviors, but researchers are currently providing data that will enable us better to understand the relevance and complexity of what some of those tests are measuring.

This chapter includes a discussion of research related to various requisites for human communication, a description of auditory problems as they relate to type and degree of hearing loss, intervention and assessment techniques and procedures, and a delineation of medical and educational classification systems for hearing impairment. Finally, there is included a presentation of dilemmas that currently concern professionals working with children with auditory problems.

Auditory Requisites for Human Communication

Knowledge of the normal development of audition as a requisite for the development of verbal communication is preliminary to an understanding of the problems that hearing-impaired children face. Although there is documented evidence of the interest in infants' abilities to hear, dating as early as the 1800s (Taylor & Mencher, 1972), it is only within the past 20 years that researchers have begun to report on systematic investigations of the developmental aspects of audition. Although we are yet far from answering all questions about the neural bases for language acquisition, it is now certain that the ability to hear and to process what we hear is critical to such development.

Anatomical/Neurological Considerations

In order for audition to be accomplished, we are assuming the presence of an intact peripheral mechanism. This includes the availability of an external meatus that ends at the tympanic membrane; intact, appropriately articulated ossicles, joining the oval window of the cochlea; a patent Eustachian tube; and a cochlea capable of transducing auditory stimuli into neural impulses that ascend to the auditory cortex.

The human auditory pathways are of particular interest. They ascend both directly and indirectly to cortical levels. The cochlea contains about 15,500 hair cells, which provide input to over 30,000 nerve cells in the spiral ganglion. The cochlear nucleus serves as the first major station to the cortex. Pathways from the cochlear nucleus ascend in both crossed and uncrossed tracts. The majority of the fibers cross to the opposite superior olivary complex and then to the nuclei of the lateral lemniscus. The pathway then ascends to the inferior colliculus, where cell density increases to about one hundred times that seen at the cochlear nucleus. From the inferior colliculus, the pathway ascends to the medial geniculate body, which projects directly to the primary auditory cortices in the temporal lobes.

The auditory cortex is surrounded by auditory association areas, which connect with other sensory association areas involving somatesthesia and vision, and converge in the parietotemporal region. It has been known for some time that this anatomical region plays a primary role in mediating auditory verbal stimuli (Penfield & Roberts, 1959; Wernicke, 1874). Since 1968 there have been reports of asymmetry in the size of the planitemporali of human cortical hemispheres. In the overwhelming majority of subjects this region in the left hemisphere has been reported as significantly larger than the homologus area in the right hemisphere (Geshchwind & Levitsky, 1968). This asymmetry is not seen in infrahuman species.

Since much of our information about the human auditory system was derived from infrahuman species, and because communicative behavior varies widely between species having apparently identical auditory systems (Eisenberg, 1976), the uniquely human attribute of verbal communication is thought to be related to the manner of and capability for central processing of auditory stimuli. The size asymmetry between the right and left temporal lobes of humans supports this hypothesis.

Research suggests, further, that in all species the mechanisms that are present in the neonatal organism are probably those that have proven their adaptability and value to survival (Eisenberg, 1976; Morse, 1972). That is, the stage of mechanism development present at birth is dependent upon the requirements of survival in a particular environment. The ability of human infants to discriminate second and third formant transitions (so critical to speech perception) is therefore here viewed as testimony to the primacy of audition in the development of verbal communication.

Infant Speech Perception

Investigations of the auditory abilities of human infants have documented the presence of a remarkable degree of competence in these very young subjects for dealing with speech or speechlike stimuli (Eimas, Siqueland, Jusczyk, & Vigorit, 1971). Other studies reported a measurable response preference for stimuli that are composed of speech or speechlike sounds (Eisenberg, 1976). Eimas *et al.* found that infants of one month and older are not only responsive to speech sounds, but are able to discriminate between sounds that differ along the voicing continuum. They concluded, as a result of these findings that, "the means by which categorical perception of speech . . . is accomplished may well be part of the biological make-up of the organism" (p. 306).

Morse (1972), in a carefully controlled study of 40- to 54-day-old infants, investigated the ability of that population to discriminate between synthetic speech and nonspeech stimuli. His findings revealed that these infants were able to "discriminate the acoustic cues for place of articulation and intonation" (p. 477). After comparing the infants' responses to speech and nonspeech stimuli, Morse suggested that infants respond in a "linguistically relevant" manner to acoustic cues for place of articulation, in the presence of speech stimuli only.

Eilers and Minifie (1975) investigated the ability of very young infants to discriminate *fricative* sounds. The results of this study were of particular interest because previous studies had included only *stops*. Information on fricative discrimination therefore served to extend knowledge of infants' perceptual skills and it also provided data that could be compared with earlier studies. The infants studied by Eilers and Minifie responded to /s/:/v/ and /s/:/ʃ/ contrasts, but not to /s/:/z/ contrasts. Analysis of the acoustic features that interacted to produce these discriminations led Eilers and Minifie to conclude that "discrimination of consonants can be explained as resulting from infants' sensitivity to acoustic properties of formant transitions" (p. 167).

Miller and Morse (1976), utilizing a cardiac-orienting paradigm, with appropriate controls, demonstrated that three- and four-month-old infants were able to discriminate between phonemic categories involving place of articulation. Attempts to elicit such discriminations within phonemic categories having the same place of articulation were not successful. Miller and Morse compared their findings with those of Morse and Snowdon (1975) who performed similar experiments with Rhesus monkeys. The animals also demonstrated an ability to discriminate between categories. They were further able, however, to make within category discriminations, an ability the human infants did not demonstrate. These findings were thought to be suggestive of the existence of a feature detection mechanism, characteristic of mammalian or primate species. Human infants apparently do not operate in such a manner. Findings in the infant speech perception research suggest rather that phonemic differences having linguistic significance are those that are critical for infant discrimination ability.

Eilers, Wilson, and Moore (1977) utilized a visually reinforced identification of speech discrimination technique to document developmental changes in infants' speech discrimination ability. They noted that infants require exposure to and listening experience with some sounds to show evidence of discrimination.

All of the studies of speech perception reported above indicate that the capability of humans to discriminate the linguistically relevant acoustic cues in speech stimuli is present in infancy. The suggestion is that "some aspects of processing in a speech mode are either a genetically endowed capacity in infants, or they are learned within the first few weeks of life" (Morse, 1977, p. 166).

Sensory Deprivation

Our increasing understanding of the important role of audition, particularly infant audition, has raised interesting questions about the effects of early sensory deprivation upon neural structures. Although such information has long been available for visual deprivation (Brattgard, 1952; Riesen, 1960), data is only now emerging on auditory sensory deprivation. In 1977, Webster and Webster published a report of their findings, indicating changes in the globular cells of the cochlear nucleus and their ascending projections in mice, which resulted from sensory deprivation. This data raises significant and provocative questions about both the peripheral vs. central controversy and about the recently reported high correlations between a clinical history of repeated serous otitis media and subsequent emergence of learning disabilities (Battin, 1978; Lewis, 1976; Northern & Downs, 1978). If further research in the sensory deficit area continues to yield evidence consistent with that reported by the Websters, we must consider the possibility that youngsters whose hearing sensitivity is impaired, even intermittently, may well exhibit potentially irreversible brain stem auditory pathway changes. The significance and ramifications of such a possibility cannot now be fully evaluated. We remain very primitive in our knowledge of how auditory processing is accomplished. The functions of various central nervous system structures are only now beginning to be studied. Nevertheless, the possibility hinted at by the Websters, that there may be a direct relationship between the presence of fewer globular cells and the inability to process auditory information optimally, poses a most intriguing question.

Auditory Problems

Auditory problems in children are those that are caused by auditory sensory deficits and/or auditory association impairments. These causative factors will be explored in this section. Another kind of difficulty that will be described here is thought to be caused by a psychological disorder that is expressed in auditory dysfunction. Following is a description of auditory problems found in children, with special attention to the interrelationship of all the causative factors (see Table I).

1. Sensory Deficits

Traditionally, auditory sensory deficit was described solely in terms of the *nature* and *degree* of the hearing loss. It is now felt that such a limited description is inadequate for determination of resultant auditory problems; the *age* at which the hearing loss occurred must also be considered as a major factor in predicting the effects of the sensory deficit.

Support for consideration of the *age-of-onset* variable in viewing auditory problems can be found in a number of studies dealing with language acquisition. The theory of critical periods, for instance, states that there are certain intervals during which an organism is programmed to utilize most effectively particular types of sensory input, for the purpose of developing specific functions. Lenneberg (1967) hypothesized the notion of critical periods for language acquisition. In reporting on the facility for development of language in hearing-impaired, postmeningitic children, Lenneberg stated that the children whose hearing is lost after the onset of speech, "can be trained much more easily in all language arts . . . [whereas] children deafened before completion of the second year" (p. 155) "show no advantage in language acquisition and development over congenitally hearing-impaired youngsters." Even though this comparison by Lenneberg would seem to negate the importance of auditory linguistic experience during the first two years, his observation called attention to the concept of time constraints on language development. Current thought on the critical period for language acquisition is that it is considerably expanded from Lenneberg's estimation. Fry (1977) suggests that "there is a natural period in the life of the individual for the brain development that language acquisition represents and that is from birth to about four or five years of age" (p. 299).

Another convincing argument against relying solely upon the nature and degree of hearing loss as determinants of auditory problems is that these two variables alone cannot furnish adequate information, should the sensory deficit be expressed in an irregular audiometric configuration. Degree of hearing loss is determined by calculating the pure tone average for frequencies of 500 Hz, 1000 Hz, and 2000 Hz. In the presence of an irregular audiometric configuration, however, the mid-frequency pure tone average can be misleading. For instance, when using the averaging technique to determine the degree of loss, children who

Table I. Relationship of Auditory Problems to Type and Degree of Hearing Loss

Auditory problem	Diagnosis	Site of lesion	Age of onset
I. Sensory Deficits			
A. Severe-to-Profound Cannot hear without amplification; unable to rely primarily on audition for acquisition/development of verbal language; educational progress frequently impeded; segregated school is often an option.	Sensorineural hearing loss	Cochlea to cochlear nucleus	Prenatal to prelingual[a]
B. Moderate May appear to hear "selectively"; speech and language acquisition/development usually delayed and/or impaired; will need amplification; educational progress impeded.	1. Sensorineural hearing loss 2. Conductive hearing loss 3. Mixed hearing loss	Cochlea to cochlear nucleus Outer ear to oval window Outer ear to cochlear nucleus	Prenatal through childhood
C. Mild-to-Moderate May appear "inattentive"; speech and language acquisition/development may be delayed and/or impaired; hearing may fluctuate; misdiagnosis may be a factor; educational progress impeded.	1. Sensorineural hearing loss 2. Conductive hearing loss 3. Mixed hearing loss	Cochlea to cochlear nucleus Outer ear to oval window Outer ear to cochlear nucleus	Prenatal through childhood

D. Mild Hearing loss of slight degree; may fluctuate; speech and language acquisition/development may be subtly affected.	1. Sensorineural hearing loss 2. Conductive hearing loss	Cochlea to cochlear nucleus Outer ear to oval window	Prenatal through childhood Postnatal through childhood
II. Association Impairments Impaired ability to process auditory information, often in the presence of normal sensitivity.	Central auditory processing disorder	Cochlear nucleus through auditory cortex	Prenatal through childhood
III. Psychological Disorders Hearing loss is affected, with no organic auditory pathology.	Nonorganic auditory disorder	None	School age

[a] Prelingual refers to that period in a child's development prior to the emergence of expressive language, at approximately 18 months of age.

exhibit irregular audiograms (Fig. 1), or audiograms that are characterized by significantly poorer thresholds for high frequencies (Fig. 2), may fall into the normal limits for hearing sensitivity and none the less be handicapped because of the auditory deficit. In such cases, even speech reception threshold, which usually correlates well with pure tone average loss, may be consistent with normal limits. Where there are such variations of configuration and severity, therefore, it is imperative that we consider results of tests of speech discrimination administered at conversational levels (about 40 dB HL) in conjunction with other audiometric findings (McCandless, 1976). When irregular audiograms are found, such speech discrimination scores are thought essential predictors of auditory functioning.

Additional factors which can contribute to the impact of the sensory deficit include the child's cognitive capacity and physical and emotional status, the age at which identification was accomplished, the etiology of the hearing loss, and so forth. Although all of these factors have independent and interactive implications, the interaction between *degree* of loss and *age* of onset is thought to be most predictive of concomitant language delay (McConnell, 1973).

A. Severe-to-Profound Deficits. Children who exhibit severe-to-profound prelingual auditory sensory deficits are able to hear only the most intense sounds. In many cases, these youngsters perceive only the very low frequency, or vibratory components of complex auditory stimuli such as speech or music. Audition is so

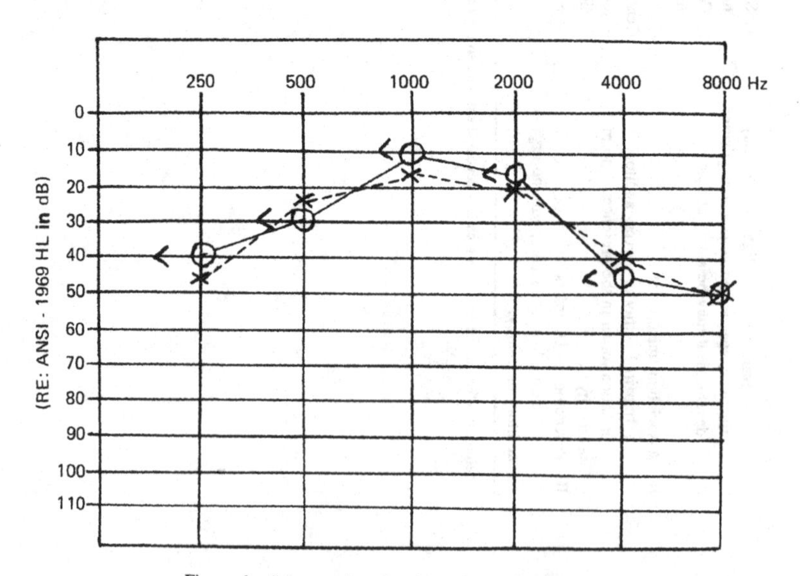

Figure 1. An example of an irregular audiogram.

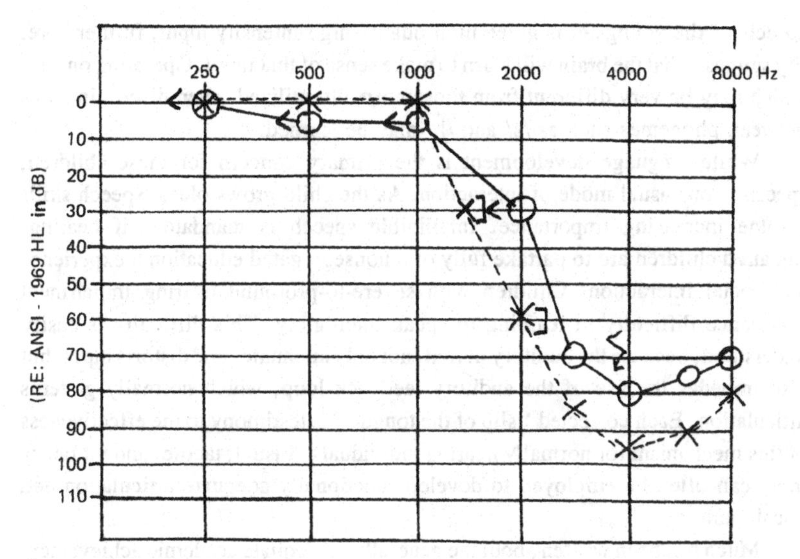

Figure 2. An example of an audiogram showing significantly poorer thresholds for high frequencies.

seriously impaired that such hearing losses usually preclude the utilization of the auditory channel as the prime input for language acquisition.

Visual input supplements and, in some cases, supplants audition, for language acquisition. Because vision is a directed sense, however, it is not entirely adequate for this task. It is, in fact, most difficult for children to receive and process new visual information in any manner comparable to that in which auditory information is handled. That is, a linguistically rich environment is almost always available to a child with normal hearing, whereas special attention is necessary for visual language (i.e., lipreading) input. In addition, visual language input is often ambiguous. Individual speaker differences and the rapidity of speech movements contribute to the inferiority of vision as a source of linguistic input (Rodel, 1978). According to Jeffers and Barley (1971), "Language is an auditory phenomenon, one designed for hearing, not viewing" (p. 16).

Despite these limitations, however, many children with severe-to-profound hearing impairments do acquire competent verbal language systems. Fry (1977) points out that by employing criteria, or cues, which differ from those usually processed, many individuals are able to make distinctions between stimuli, which should be "impossible" because of the limitations imposed by sensory deficits. For instance, by using alternative auditory information (i.e., vowel transitions rather than consonant determinants) children do apparently learn to discriminate speech. If the youngster is given high-quality high-intensity input, furthermore, Fry suggests that the brain will learn to make sense of that input. Operating on cues which may be very different from those normally utilized, even discriminations between phonemes such as /s/ and /ʃ/ may be learned.

While language development is the primary concern for these children, speech is our usual mode of interaction. As the child grows older, speech skills assume increasing importance. Intelligible speech is mandatory if hearing-impaired children are to partake fully of a nonsegregated educational experience and social interaction. Children with severe-to-profound hearing impairment experience difficulty in learning to speak intelligibly. This difficulty is easily understood, because the auditory deficit, not only attenuates and distorts input, but also impedes the use of the auditory feedback loop, which normally governs articulation. Each corrected "slip of the tongue" is testimony to the effectiveness of this mechanism for normally hearing individuals. Visual, tactile, and vibratory input can often be employed to develop functionally adequate articulation and vocal quality.

Much has been written about the generally inadequate academic achievement of children with severe hearing impairments. Because many academic skills are language determined, the academic difficulties encountered by these children are most probably a reflection of the very severe language delay that their hearing impairment causes. There has been much concern about the "thought processes" of children with severe-to-profound hearing impairment. Bangs (1968) cites the following anonymous contribution from a relatively unsophisticated, but remarkably perceptive individual.

> As a totally deaf person, the more I think of it the more ludicrous all this discussion on "how the deaf think" appears to me. Surely your writers are not suggesting (or are they) that God manufactured a special "brain box" for the deaf alone? The way I see it the deaf appear to think differently only because of the lack of education as compared to a hearing person in the same social and age group. (p. 149)

Hearing aids are mandatory for severely hearing-impaired youngsters, and many of them are able to achieve some functional ability to discriminate speech when they are afforded early and intensive speech and language stimulation and amplification (Northern & Downs, 1978).

Severe-to-profound auditory sensory deficits are most often sensorineural in nature and are caused by lesions in the cochlea, although the cochlear nucleus is sometimes implicated. Sensorineural losses are usually characterized by differential sensitivity loss of high-frequency sounds and by speech discrimination impairment.

Common etiologies include pre- or postnatal, bacterial, and viral infections; genetic or familial factors; syndromes; and ototoxicity. Many of the common etiologic factors are known to produce central nervous system damage or other related deficits that may have serious implications for language acquisition. Therefore, etiology is always a significant factor in determining the total effect of severe-to-profound auditory sensory deficit.

B. Moderate Deficits. Children who exhibit moderate auditory sensory deficits are able to hear speech at loud conversational intensity levels. Even though moderately hearing-impaired children characteristically have difficulty understanding and discriminating even high-intensity speech, in most cases the auditory channel remains the primary avenue for language acquisition. Therefore, children with moderate hearing impairments derive great benefit from amplification, and are usually more appropriately placed in nonsegregated educational settings that offer ancillary support services.

Prelingual onset of hearing loss and possibly delayed identification are perhaps more critical for children with moderate impairments than those with severe-to-profound deficits. Such children will often be missed in neonatal screenings (Mencher, 1974); they will show evidence of language development, albeit delayed and/or deviant; and they appear (to most people concerned with their welfare) to hear. Only when speech and language delay becomes very obvious does parental concern become intensified, finally motivating audiologic investigation.

Children with moderate hearing impairments often demonstrate defective articulation. Vocal quality is usually adequate, and speech is generally intelligible. These children are very amenable to speech and language stimulation and, when provided with adequate and appropriate amplification, usually learn to speak with good intelligibility.

Academic problems are often the major source of difficulty for children with moderate hearing impairments. Because language acquisition is usually delayed but not obviously aberrant, the youngsters' problems may escape the teacher's attention. As a result, the kind of classroom modification that is essential if such a youngster is to prosper is often not effected. Furthermore, necessary supplemental or ancillary services may not be made available.

All of our enthusiasm for amplification noted above is often misplaced. The appalling disrepair of hearing aids worn by school-aged children has been amply documented (Gaeth & Lounsberry, 1966; Schell, 1976). In many instances, if a child is noted to be wearing a hearing aid, the functionality of that instrument is assumed. In fact, few teachers of mainstreamed children routinely check hearing aids. This problem is most crucial during the very early academic years, when the youngster is unable either to maintain the aid independently, or to express dissatisfaction with an aid that is nonfunctional.

An additional concern, but one that has received only limited attention in educational settings, is the effect of room acoustics upon the hearing-impaired child's ability to process amplified speech. If one were to ask a hearing-impaired adult to describe the most problematic situations for amplification, the adult would invariably mention the large room with hard walls and floors, in which a large group is trying to hear a single speaker in a noisy background. Is there a more accurate description of the average classroom? It is imperative that educators become aware of the very real contribution that acoustic environment makes in determining the success or failure of the hearing-impaired child. Probably because of both the low incidence of hearing loss and the high cost of appropriate modifications in classroom acoustics, such awareness may be slow in coming. However, without these modifications, the educational and vocational fate of many children is seriously jeopardized (Olsen, 1977; Ross, 1972).

Moderate auditory sensory deficits may be the result of sensorineural, conductive, or mixed pathology. As described earlier, sensorineural losses are caused by lesions in the cochlea or cochlear nucleus. Conductive losses result in a decrease in sound intensity without a decrease in discrimination ability. They are caused by lesions in the external or middle ear. A mixed loss occurs when there are both conductive and sensorineural components.

Etiologies include those mentioned earlier for sensorineural impairments; middle ear pathologies such as serous otitis media, ossicular disarticulation, and congenital malformation of ossicles or external auditory meatus; and conditions that predispose children to middle ear pathology, such as cleft palate or other craniofacial disorders.

C. Mild-to-Moderate Deficits. Children who sustain mild-to-moderate auditory sensory deficits are able to hear speech of conversational intensity levels. However, they often cannot discriminate speech presented at such levels without utilizing visual cues. Unfortunately, without amplification, speech at conversational levels does not provide these children with the high intensity, high quality, and high frequency of auditory input that is necessary for optimal language acquisition and development.

With appropriate amplification and speech and language stimulation, children with mild-to-moderate hearing losses acquire language through auditory channels. They are subject to late identification because the effects of mild-to-moderate hearing loss are often subtle and may not be noticed during infancy and early childhood. Although speech and language skills appear only mildly delayed, comprehensive evaluation often reveals more significantly delayed linguistic behavior. This language delay will often become more pronounced as the youngster progresses academically, and as the demands on linguistic and cognitive functions become more stringent. Children with mild-to-moderate hearing impairment are also subject to all of the amplification and acoustic environment problems noted earlier.

Etiologic and diagnostic factors include all of those mentioned in the previous categories.

D. Mild Deficits. Children with mild hearing impairments are often the last to be identified because they exhibit the most subtle problems. These children hear and often understand speech of conversational intensity levels, but without amplification this "understanding" requires considerable effort, and often results in fatigue and irritiability. In the past, when amplification systems were relatively unsophisticated and when our knowledge of language acquisition was less precise, such youngsters usually did not receive amplification. "Preferential seating" was thought to be adequate compensation for the sensory deficit. We are now of the firm opinion that most children with mild losses need amplification, professional attention to speech and language stimulation, and ancillary academic services.

Etiologic and diagnostic factors include all of those mentioned earlier. It is imperative that we mention again the mild, often intermittent, conductive hearing loss associated with recurrent serous otitis media. Many investigators are now suggesting that these mild problems result in long-term language and learning disorders (Lewis, 1976; Northern & Downs, 1978).

II. Association Impairments

Children with auditory association problems are those who have impaired ability to *process* auditory information, often in the presence of normal auditory sensitivity. These children may have delayed or deviant language development, an inability to make auditory figure–ground discriminations, difficulty maintaining auditory attention, hypersensitivity to loud sounds, or difficulty "listening" (Willeford, 1977).

Central auditory processing disorders are currently receiving much attention, particularly as possible components of learning disability. Although the etiology of auditory processing disorders in children is often obscure, they are assumed to be caused by dysfunction somewhere along the auditory pathways from the cochlear nucleus through the auditory cortex.

Most original research in this area was performed on adult patients (Berlin, Lowe-Bell, Janetta, & Kline, 1972; Bocca, Callearo, Cassinari, & Miglinvacca,

1955; Lynn & Gilroy, 1972). Willeford (1977), however, has worked extensively with school-aged children, and reports a significant incidence of central auditory processing problems among those with learning disabilities. Habilitative or rehabilitative measures are being developed that focus on increasing the redundancy of speech through frequent repetition and restatement, reducing competing auditory and visual stimuli, and vigilant monitoring of attention by requiring frequent feedback from the child (Dempsey, 1978; Willeford, 1977).

III. Psychological Disorders

Nonorganic hearing loss is a *psychological* rather than an *auditory* problem. Such a psychological problem exists when voluntary thresholds indicate reduced auditory sensitivity and when this reduction does not reflect a true sensory deficit. Such a loss may be consciously affected by the child (malingering) or, less commonly, may be the result of an unconscious mechanism.

Often associated with academic pressures, nonorganic hearing losses can be relatively minor, transient experiences that are resolved during audiologic evaluation. More rarely, however, they can be symptoms of more serious psychological disorders. Through audiologic evaluation, the diagnostician is almost always able to ascertain the functional nature of the problem.

Common features include a loss identified during school screening, but previously unsuspected by the parents, significant disparity between responses to speech and pure tone stimuli, and unusual variability in pure tone thresholds (Topp, 1977). Counseling for both the child and the parents is essential immediately following audiologic assessment and appropriate referrals should be made when indicated.

Assessment

Assessment of the auditory mechanism in children is one of the most challenging tasks facing the audiologist. Each youngster brings a unique combination of social, physical, emotional, and cognitive capabilities to the test situation. The audiologists' role must therefore include the identification of all factors that contribute to the youngster's overall status, and the determination of the most appropriate clinical tools for the assessment procedure.

The past twenty years have been a period of tremendous growth in pediatric audiology. We now have available a wide variety of behavioral and objective measures of hearing sensitivity, which are used in identification and assessment programs. The following section discusses identification or screening programs, and current trends in the assessment of auditory function in children. Assessment techniques will be discussed in operant terms, and have been separated into behavioral and objective measures. Emphasis will be placed upon identification and assessment procedures as determined by developmental status.

Identification (Screening)

Technological advances that brought precision to audiologic measures, increased understanding of language acquisition and its prerequisites, and social commitment to the early education of handicapped youngsters have combined to argue for the earliest possible identification of hearing-impaired children. Identification programs are designed to separate rapidly and reliably from an asymptomatic population those who demonstrate some disorder which requires intervention from those who do not (Northern & Downs, 1978). Public health officials have listed the following criteria, as those which determine the appropriateness of a disorder for mass screening:

1) Occurrence frequent enough or consequence serious enough to warrant mass screening . . .
2) Amenability to treatment or prevention which will forestall or change the expected outcome . . .
3) Availability of facilities for diagnosis and treatment . . .
4) Cost of screening tool reasonably commensurate with benefit to public . . .
5) A screening tool that validly differentiates disease from non-disease . . .
6) Acceptance by the public. (North, 1976)

Current screening programs include two aspects: The identification of hearing loss and the identification of middle ear dysfunction.

Hearing screening programs have been in existence since the "fading numbers" test of the 1940s (Goldstein, 1933; McFarlan, 1927) and are one of the few mass screening programs to meet the tests of time and effectiveness (Klein, 1977).

Neonatal Screening

Possibly in response to the rubella epidemic, infant hearing screening programs proliferated in the 1960s. These programs, typified by that reported by Downs and Sterritt (1967), were staffed by trained volunteers who utilized a high-frequency, high-intensity stimulus and observed the behavioral responses of infants. Critical evaluation of these screening programs revealed problems. In 1970, Ling, Ling, and Doehring cited the "inherent complexities" of infant screening and reported that the observer's judgments of response presence were influenced by knowledge of stimulus presentation. Goldstein and Tait (1971) reported that commonly employed neonatal screening programs did not meet their objectives and were problematic in cases of delayed onset of hearing loss or progressive hearing loss. They suggested that fewer children, who are selected via a high-risk register, be evaluated more intensively. At about the same time, the National Joint Committee on Newborn Screening issued a statement urging intensive research into "stimuli, response patterns, environmental factors, status at the time of testing and behavior of observers" (Joint Committee on Infant Hearing Screening, 1971, p. 79). In 1973, the Joint Committee recommended the use of a high-risk register containing the following:

A. History of hereditary childhood hearing impairment.
B. Rubella or other nonbacterial intrauterine fetal infection (e.g., cytomegalovirus infection, Herpes infection).
C. Defects of ear, nose or throat. Malformed, low-set or absent pinnae; cleft lip or palate (including submucous cleft); any residual abnormality of the otorhinolaryngeal system.
D. Birthweight less than 1500 g.
E. Any free or indirect serum bilirubin concentration which is judged to be potentially toxic. (Joint Committee on Infant Hearing Screening, 1974)

Infants who demonstrated any of the risk factors were to be referred for in-depth audiologic evaluation, and were to be followed longitudinally to identify delayed onset, or progressive hearing loss. In 1976, Feinmesser and Tell reported a longitudinal study of neonatal screening on 17,731 infants. Evaluating their program, which used an extended high-risk register, they concluded that use of a restricted register similar to the one recommended by the Joint Committee was both economically feasible and efficient in indentifying severely hearing-impaired infants.

In 1974, members of the Joint Committee (Mencher, 1976) approved the experimental use of behavioral screening techniques, with established criteria, in addition to the high-risk register. Behavioral screening measures currently under investigation include the "arousal" test and the "Crib-O-Gram." The "arousal" test, proposed by Mencher (1974) presents a 90–100 dB SPL narrow band (3000 Hz) or white noise stimulus to a sleeping infant to elicit a generalized body response. The "Crib-O-Gram" is an automated, computerized screening device that uses a sensor under the infant's mattress to detect responses to cycled stimuli (3000 Hz narrow band noise) over a 24-hour period (Simmons, 1974).

At the present time, three states, New Jersey, California, and Massachusetts have laws requiring neonatal screening. The high-risk register is used in California and Massachusetts and will be implemented in New Jersey.

Toddler Screening

Hearing-screening procedures for toddlers (aged 12–30 months) remain most problematic. Northern and Downs (1978) suggest that well-baby centers, pediatricians, and audiologists utilize orienting responses to noise makers, as was originally described by Ewing and Ewing (1944). Although such procedures are certainly of some value in the audiology facility where stimuli can be adequately controlled, their use in other settings is questionable. Environmental conditions, the availability of a trained examiner, and the difficulty in presenting controlled and consistent stimuli in such situations provide formidable obstacles to valid screening. Possibly middle ear function screening, which will be discussed, in combination with acoustic reflex measures will be more effective with this population. Further research in these areas is needed.

Preschool Screening

Preschool-aged children are usually quite amenable to behavioral audiometry techniques, and therefore regular audiometric screening procedures can be used. It is recommended that 3- to 5-year-old children receive an individual ear screening utilizing pure tones presented at 20 dB HL re: ANSI, 1969, at 500, 1000, 2000, 3000, 4000 Hz (New Jersey Speech and Hearing Association, 1978). Children who fail to respond at any frequency are referred to evaluation following a rescreening within the same session, if possible. Preschool-aged children should be screened annually. It is appropriate to note that the Verbal Auditory Screening for Children test, which was designed for preschool children (Griffing, Simonton, & Hedgecock, 1967) has proven ineffective in identifying children with mild or high-frequency hearing losses (Mencher & McCulloch, 1970) and is not recommended for use.

School-Aged Screening

Elementary school children are screened most efficiently and effectively using the individual pure tone sweep check. These children are usually able to respond verbally, or via hand raising, although play techniques may be substituted if appropriate. Screening parameters and referral criteria are as reported for preschool-aged children. School-aged children should receive annual screening in grades K–3, and semiannual screening thereafter. Any child who is exhibiting academic difficulties or for whom there is suspicion of hearing loss should be included in the annual screening, as should those at high risk for hearing loss.

Middle Ear Function

The advent of middle ear function measures generated much interest in their possible use as tools for mass screening. In 1975, Lescouflair published a thought-provoking critique of hearing screening programs, emphasizing that the goals of the programs were unattainable due to equipment, environment, and examiner limitations. He specifically observed that pure tone audiometric screenings do not always detect children who exhibit otologic pathology warranting referral. Pure tone screenings performed in noisy environments are particularly vulnerable to false negatives in the case of conductive impairment, which is usually expressed in a low-frequency hearing loss. Impedance/admittance measures provide direct evaluation of middle ear function and can be valuable in such situations. Many investigations have shown high correlations between impedance and otoscopy. However, there have also been reports of significant overreferrals resulting from screenings that employ only one tympanogram (Lewis, Dugdale, Canty, & J. Jerger, 1975). These overreferrals may be related to the natural course of serous otitis media, the most common cause of childhood conductive hearing loss. Serous otitis media often fluctuates and is in many cases subject to spontaneous remission (Brooks, 1976). The time interval between the impedance screening and the otoscopic examination may in part explain the overreferrals. It is also known that although the flat, or type B tympanogram is highly correlated with the presence of effusion in the middle ear, other tympanograms are highly variable in their ability to predict the presence or absence of middle ear effusion (Orchik, Dunn, & McNutt, 1978). An open-ended screening technique, utilizing serial typanograms was proposed in 1975 by Lewis *et al.* The open-ended screening concept has yet to be evaluated on large numbers of children of varying ages and socioeconomic status. However, it has good potential for reducing overreferrals caused by transient middle ear pathology. Serial tympanograms can also serve to document the natural course of the disease process.

In 1976, a large group of professionals from medicine, audiology, and related disciplines met to discuss impedance as a mass screening tool. The report of the task force formed during the International Symposium on Impedance Screening for Children (1978) did not recommend the implementation of mass impedance screening programs at the present time. The task force cited current limitations in our knowledge of the epidemiology and natural course of serous otitis media, a lack of sufficient validating studies on the effectivness of pharmaceutical therapy regimens, and incomplete data on the reliability and validity of impedance tests with varying populations and different types of equipment. They strongly urged that intensive research be directed toward these areas. Programs already in existence have the potential to contribute invaluable data, and it was recommended that the following procedures be implemented in all ongoing screenings for preschool- and school-aged children:

1. A combination of tympanometry and acoustic reflex measurement should be used.
2. A 105 dB HL signal should be used to elicit an acoustic reflex in the contralateral mode, and a 105 dB SPL signal in the ipsilateral mode, or both.
3. Tones between 1000 Hz and 2000 Hz should be used as the eliciting stimulus. Whether broad-band noise or pure tone is preferable as an elicitor for this purpose remains to be established. Whichever stimulus is used must be specified and described.
4. Acoustic reflex measurements can be obtained either with the ear canal air pressure that results in minimum acoustic impedance, or with the ear canal air pressure equal to ambient pressure. The condition should be specified.
5. A 220 Hz probe tone should be used for tympanometry; however probe tones up to 3000 Hz are acceptable.
6. For typanometry an air pressure range of 400 to + 100 mm (H_2O) is preferable. A range of 300 to +100 mm is acceptable.
7. Failure on the initial screen is denoted by either an absent acoustic reflex or an abnormal tympanogram. An abnormal tympanogram is defined as one that either:
   a) is "flat" or "rounded," i.e., without a definite peak, or
   b) has a "peak" at, or more negative than, −200 mm (H_2O).

   It is noted that flat or rounded tympanograms are more highly correlated with middle ear effusions than are tympanograms with peaks at negative-pressure readings.
8. Any child failing the initial screening should be retested in 4 to 8 weeks. Any child who has an acoustic reflex *and* a "normal" tympanogram on the initial screening passes and is "cleared."

The following schema is recommended for various screening findings:

Classification	*Initial Screen*	*Retest*	*Disposition*
1	Passes AR and tymp.	Not indicated	Cleared
2	Fails AR and/or tymp.	Fails AR and/or tymp.	Referred
3	Fails AR and/or tymp.	Fails AR *or* tymp.	Referred
4	Fails AR and/or tymp.	Passes both	At Risk

Classifications 1–3 should constitute the majority of children in a given population. Referral, when indicated, should be made to an appropriate medical facility in the community. Classification 4 constitutes a group of children that requires special monitoring by the agency responsible for the screening program. These "at risk" children should be tested periodically to determine the possible need for future referral. Whenever possible children who initially have an acoustic reflex and pass tympanometry should be tested periodically since long term studies are needed in order to further refine pass/fail criteria and management.

Certain populations have significantly higher incidences of middle ear pathology, with correspondingly more serious consequences. In recognition of these problems, impedance screening is warranted and recommended for the following groups:

Known sensorineural hearing loss

Developmentally delayed

Cleft Palate or other cranio-facial anomaly

Down's Syndrome Children (International Symposium on Impedance Screening for Children, 1978, pp. 570–573)

Identification programs are again at the crossroads. It is evident that more sophisticated measures will soon be available to larger numbers of children than ever before. If our research experience with infant hearing screening is carried over to impedance screening, we can expect the accumulation of new data on many populations to provide the impetus for more reliable and valid test procedures and protocols. There is much yet to be done. However, it is clear that we are entering a new era, in which objective measures will receive greater attention, innovative screening programs for hearing and middle ear problems will be implemented, and some of the effects of hearing loss will be minimized.

Assessment (Evaluation)

The assessment of hearing in young children, once thought to require a "grandmotherly" appearance and affect on the part of the audiologist, has shifted considerably from the days when it was thought to be possible to obtain thresholds only on children older than five years (Ewing & Ewing, 1944). We now routinely evalute infants and very young or handicapped children, and fully expect to obtain accurate and valid information both about hearing sensitivity and middle ear function.

Audiologic assessment should be undertaken when a child fails a routine hearing screening, or whenever there is any suspicion that hearing sensitivity is less than normal. The purpose of a hearing evaluation is to document the existence, nature, degree, and likely effects of a hearing problem. The audiologist looks for the presence of an auditory sensory deficit, a central auditory processing disorder, or a nonorganic hearing problem. The evaluative procedures involve the use of behavioral measures in which the auditory stimulus elicits some overt reponse or a change in the youngster's behavior, and object measures which utilize electrophysiologic monitoring to record nonovert responses to auditory stimuli. When evaluating central auditory function, tests are employed that measure the child's ability to process auditory information. All assessment procedures should be dependent upon the child's developmental status and maturational level.

Auditory Sensory Deficits

Behavioral Measures. Behavioral techniques may be divided into those that are *reflexive* and those that are *conditioned*. Reflexive techniques require that the observer note under which conditions specific reflexes were elicited. Conditioned techniques are those in which a behavior elicited in response to an auditory stimulus is differentially reinforced. When assessing the hearing sensitivity of the very young infant (0–4 months), the audiologist relies heavily upon reflexive responses to intense signals. The reflexes elicited include the Moro, or startle, the auropalpebral, and arousal (Hodgson, 1978). The data obtained must be carefully evaluated, taking into consideration the possible existence of recruitment. These reflexive measures provide, at best, a qualitative description of hearing sensitivity.

Two conditioning strategies, one dealing with visual reinforcement of orienting responses and the other dealing with tangible reinforcement of motor responses, are relatively recent developments in hearing assessment of young children. While more traditional behavioral techniques are indeed entirely adequate for evaluating the hearing of many children, these conditioning strategies have proven to be most effective for assessment with those youngsters who are not amenable to conventional testing procedures because they are chronologically, developmentally, and/or in some other way, inaccessible.

Conditioned Orienting Reflex (COR) audiometry (Suzuki & Ogiba, 1961) has become a major tool of the clinical audiologist. This technique was modified and broadened to Visual Reinforcement Audiometry by Liden and Kankkonen (1969). Visual Reinforcement Audiometry utilizes a visual stimulus (lights, slides, animated puppets) to reinforce responses to auditory stimuli. The technique was originally described as effective for children between 12 and 30 months of age. Subsequent studies demonstrated that Visual Reinforcement Audiometry increased the number of responses to sound in children as young as five months (Moore, Wilson, & Thompson, 1977) and documented the technique's effectiveness for obtaining thresholds in Down's Syndrome children (Greenberg, Wilson, Moore, & Thompson, 1978). The later study showed a systematic relationship between consistency of response and mental age. A cutoff equivalent of 10 months on the Bailey Scales of Infant Development was found to be most predictive of success in obtaining threshold. Reliable thresholds for noise stimuli were obtained on 81% of the Down's Syndrome subjects in one session.

Dix and Hallpike (1947) introduced the Peep Show technique, which was later modified and refined (Guilford & Haug, 1952). Effective for youngsters aged three to six years (Hardy, 1965), the Peep Show also uses a visual reinforcer for a motoric response to auditory stimuli. In the United States, the Peep Show has been replaced by a variety of different play audiometry techniques.

Tangible Reinforcement Operant Conditioning Audiometry (TROCA) (Lloyd, Spradlin, & Reid, 1968) offers a tangible reinforcer for a motor response, such as pressing a button, to auditory stimuli. The technique is considerably strengthened when the child chooses his own reinforcer (candy, cereal, beads, tokens, etc.) from an array of possibilities. Originally proposed for the assessment of mentally retarded youngsters, Tangible Reinforcement Operant Conditioning Audiometry has also been used extensively with normal children.

Conditioned play audiometry is probably the most widely utilized behavioral hearing assessment technique for children older than 24 months. Play audiometry makes some play activity (dropping blocks, stringing beads, building towers, or whatever the audiologist's or child's ingenuity can create) contingent upon the presentation of an auditory signal (Barr, 1954, 1955). Because this technique attaches some "significance" to abstract pure tone stimuli, children will often continue the "game" long enough for the completion of air and bone thresholds.

All of the conditioned behavioral procedures require interaction between the child and the examiner. Many children will enter the test situation unafraid and eager to participate. Others, however, will be timid, frightened, or even terrified by the strange rooms and new faces. If play audiometry is to be successful, the audiologist must be able to deal with all eventualities. Only a strong understanding of the principles of child development will allow uniform success.

Zaner (1974) points out that three developmental principles appear most relevant to the auditory test situation: (1) separation—maternal separation and its concomitant childhood anxieties, (2), psychological differentiation, and (3) the emergence of basic trust. To test the hearing of youngsters who do not "cooperate" it is necessary to separate them from whatever is preventing their response. Often ingenuity will allow the clinician to guide the child from whatever stimulus prevents entry into the situation. Separation does not necessarily in this case imply physical distance, but the child's ability to behave autonomously, even if seated upon his or her mother's lap.

The differentiation concept dictates that organisms have been influenced by all earlier experiences and that earlier experiences have potentially greater effects on later development. When there is an early sensory deficit, the later appearance of an expected system may be delayed, distorted, or not forthcoming. Sensory deprivation can affect the differentiation process, causing uneven performance. For example, the verbal, apparently happy child may walk into the room and proceed to have hysterics if you ask mother to sit "over there."

The establishment of basic trust has been referred to as as the infant's first developmental task (Erikson, 1963) and is determined by the kind of environment in which the infant grows. Children who have acquired basic trust during infancy are less likely to experience severe separation anxiety later on when seen for audiologic assessment. Children who, because of illness or other reason, are separated from their mothers during these critical development periods, experience intense anxiety and do not establish basic trust as anticipated (Zaner, 1974).

Familiarity with developmental principles and their possible behavioral manifestations is clearly prerequisite to successful play audiometry.

Objective Measures. Objective measures of hearing sensitivity consist of electrophysiologic monitoring of responses to auditory stimuli. It should be noted that whereas the monitoring process is objective, the review of the results is subject to interpretation. Although different responses are being investigated, only impedance/admittance measures have attained widespread clinical acceptance. Impedance/admittance measures now routinely provide information about middle ear and Eustachian tube function (Bluestone, 1974), intactness of the tympanic membrane, the patency of ventilation tubes (Northern & Downs, 1978), and via acoustic reflex measures, hearing sensitivity (J. Jerger, Burney, Mauldin, & Crump, 1974; Niemeyer & Sesterhenn, 1974).

Feldman (1978) divides the impedance/admittance battery into three segments: tympanometry, acoustic reflex measures, and Eustachian tube function tests. Tympanometry measures the sound transmission characteristics of the tympanic membrane and middle ear ossicles as a function of air pressure variations in the external canal. Tympanograms have been classified (J. Jerger, 1970) and described according to pressure, amplitude, and shape (Feldman, 1975). The accuracy of tympanometry for the detection of middle ear pathology has been studied intensively. Haughton (1977) reported 80% agreement between tympanometry and findings at myringotomy in an ear, nose, and throat (ENT) clinic population. Cooper, Gates, Owen, and Dickson (1975) reported 84% agreement between impedance measures and otoscopy. Audiometry detected only 24% of conductive losses. Bess, Schwartz, and Redfield (1976) also found impedance superior to audiometry in identifying middle ear pathology in cleft palate children. Paradise, Smith, and Bluestone (1976) concluded that the use of tympanometry "should result in improved detection of middle ear effusion and other middle ear anomalies" (p. 201). Orchik *et al.* (1978) reitereated J. Jerger's (1970) statement that tympanograms alone are not good predictors of effusion, and stated that prediction of middle ear effusion based upon single tympanograms is difficult at best. The use of serial tympanograms in conjunction with reflex measures was suggested to improve predictive abilities.

Acoustic reflexes are discernible as a time-linked impedance change in response to auditory stimuli. The reflex can be elicited either ipsilaterally or contralaterally and occurs in response to high-intensity stimuli. The acoustic reflex test battery has been utilized in the detection of middle ear pathology, cochlear pathology, retrocochlear pathology, central pathology, and to infer the existence of nonorganic hearing loss. The most promising application of acoustic reflex measures for children lies in the prediction of hearing sensitivity from acoustic reflexes to noise and pure tones (Jerger *et al.*, 1974; Niemeyer & Sesterhenn, 1974). The procedure is based upon the increased size of critical bands for loudness summation (Schwartz & Sanders, 1976) and the common differential high-frequency loss of sensitivity in sensorineural losses. Keith (1977) evaluated the accuracy of predictions based upon acoustic reflex thresholds to pure tone and broad band noise. He found that the Niemeyer and Sesterhenn formula was accurate within 15 dB for 82% of his subjects. The Jerger weighted formula yielded results similar to those found using the Niemeyer and Sesterhenn formula. This procedure is felt to be more reliable for children than for adults because of smaller variations in ear canal volume (Jerger *et al.*, 1974). Although the technique is not yet widely accepted, it shows tremendous promise for children and may be further refined with continued investigation.

Tympanometry has also found application as a test of Eustachian tube function. The Eustachian tube controls middle ear aeration and determines middle ear pressure. Tympanometry can therefore provide valuable information about the adequacy of tubal function (Bluestone, 1974). Seidemann and Givens (1977) proposed a tympanometric procedure to determine Eustachian tube patency in an otologically normal population. Their protocol called for evaluation of change not only in middle ear pressure, but also in middle ear function. Various other techniques (inflation/deflation, Toynbee, and Valsalva) have been proposed for evaluation of Eustachian tube function. However, no one technique has gained universal acceptance (Feldman, 1978).

Impedance/admittance measures have had a profound effect on the practice of audiology. Once limited to inferring the existence of middle ear dysfunction from pure tone audiometry, audiologists are now able to measure and describe middle ear function directly. Impedance/admittance systems routinely identify middle ear dysfunction in very young and severely hearing-impaired children, a task that heretofore required a physician skilled in pneumatic otoscopy.

The availability of impedance/admittance stimulated research into the effects of mild, but repeated episodes of otitis media. Virgil Howie (1975), a pediatrician, noted significant differences in verbal intelligence quotients between a group he defined as "otitis prone" (children who suffer more than six bouts of otitis media within first two years of life) and a control group. As early as 1962, Eisen reported probable adverse effects of early auditory sensory deprivation. Holm and Kunze (1969) compared children who had histories of chronic otitis media with those who did not and found significant delays in language skills, but no difference in visual or motor skills. In 1973, Kaplan, Fleshman, Bender, Baum, and Clark reported that the onset of otitis media during the years critical to language development and the number of episodes the youngster undergoes play an important role in delaying verbal development. In a carefully controlled study Lewis (1976) reported similar findings in an Australian population. Downs (1977) noted that such children are candidates for the title, "Irreversible Auditory Learning Disaster." Battin (1978) reported her observations of such children and concluded that amplification should be considered for children exhibiting losses of 15 dB HL or greater.

Other electrophysiologic measures are currently in research stages. Measures that have been or are being investigated for pediatric application include: respiration audiometry, heart rate audiometry, electrocochleography, and electric response audiometry.

Respiration measures are based upon Howell's (1928) observation that sensory stimuli change the rate or amplitude of respiration. Respiratory changes following auditory stimulation were recorded on a pneumograph, evaluated, and thresholds determined. Although Bradford (1975) reported success using an accelerometer for both human and animal subjects, more recent evidence suggests that respiration measures are not clinically feasible. Hayes and J. Jerger (1978) reported that a high response identification rate could only be achieved at the expense of a high false negative rate. This technique was therefore judged inadequate for clinical hearing measurement.

Heart rate measures never attained acceptance by large numbers of clinicians. Most of the research in heart rate changes following sensory stimulation was in fact performed by psychologists (Eisenberg, 1975, 1976). Heart rate is most efficiently measured via the cardiotachometer, which measures the time interval between heart beats to determine heart rate. Eisenberg indicates that the presence of a response is documented when: (1) the distribution of heart rate during time sample following stimulus differs significantly from distribution of rate for time sample immediately prior to stimulus onset; and (2) the differences between pre- and poststimulus rates differ significantly in kind or degree from those that characterize contiguous nonstimulus samples. Clinical studies utilizing heart rate change following auditory stimuli in children include: Bartoshuk (1962a, b, 1964), who found cardiac acceleration to auditory signals in neonates; Keen (1965), who also found acceleration following auditory stimulation in infants; Schulman and Wade (1970), who evaluated average responses to 34 dB SPL noise bands in high-risk infants; and Schulman (1970), who found heart rate effective for the determination of auditory response levels for children between 3 and 13 years of age. Eisenberg (1976) suggests that heart rate research not be limited to determining auditory sensitivity, but that the technique be applied to the larger questions of determining developmental auditory function.

Electric response audiometry refers to the use of evoked electrical potentials for identification of auditory responses. Auditory evoked potentials are described by: site of origin, response wave form, response latency, and amplitude (Skinner & Glattke, 1977). The auditory evoked potentials currently being investigated include: electrocochleography, early/brainstem, middle, late, and very late responses.

Electrocochleography measures electric energy originating within the cochlea or auditory nerve (Glattke, 1978). The potentials that are seen include: the action potential of the auditory nerve, the summating potential and the cochlear microphonic (Davis, 1976a,b). The action potential has been shown to be the best indicator of peripheral auditory function (Glattke, 1978). Application of electrocochleography has been limited in the United States by concern about the effects of electrode placement. The most effective electrode placements are closest to the generators of the response and require minor surgery for placement on the promontory. Clinical techniques are therefore a compromise between ideal recording sites and the amount of risk to the patient. The three placements now in use are: transtympanic, intrameatal, and extrameatal, or outside of the canal. Transtympanic placements appear to record the largest potentials, they require perforation of the tympanic membrane, and in children they require the use of general anesthesia with all of its associated risks. Intrameatal and surface techniques yield small potentials and have been suggested for use only as a screening tool (Glattke, 1978). Interpretation of electrocochleographic findings is based upon: response threshold, response wave form, growth function of response amplitude, and rate of change of response latency (Aran, Portmann, Portmann, & Pelerin, 1972).

In summary, the action potential (AP) response is an excellent indicator of cochlear and auditory nerve integrity. The quantification and interpretation of responses remains an inexact science. However, it appears that analysis of cochlear potential (CP) and summating potential (SP) components may soon allow the separation of sensory from neural hearing losses. Electrocochleography has excellent potential for clinical use. Current limitations include: the invasive nature of the procedure and the need for further study of the validity of the technique when low-frequency tone pips are used.

Brain stem, or early evoked potentials are also receiving intensive attention. They are particularly applicable to children because they utilize surface electrodes to record responses at the vertex. The early response can be observed as a series of seven peaks with latencies between 4–8 msec. The origins of various peaks have been placed anatomically via comparisons with animals. Following this theoretical construct, the first peak is assumed to arise from the cochlear nerve, the second peak from the cochlear nuclei of the brain stem, and the fifth peak from the associated inferior colliculus (Goldenberg & Derbyshire, 1975). Davis (1976a) summarized the clinical attributes of brain stem evoked potentials: the wave form is consistent and easily recordable, latency is optimal, the response is not confused with the cochlear microphonic, nor is it masked by muscle reflexes. Brain stem evoked potentials are not affected by sedation, which is in fact recommended for young children to reduce myogenic artifacts.

Clinical studies of brain stem auditory evoked potentials have demonstrated that the brain stem responses are reliably recorded from premature infants, and are not subject to habituation. The threshold for auditory sensitivity as measured via brain stem evoked potentials is considerably lower than that obtained using heart rate. Schulman-Galambos and Galambos (1975) also reported shifts in latency, correlated with age, which they relate to maturational changes in the auditory system. Mokotoff, Schulman-Galambos, and Galambos (1977) evaluated 81 infants and children whose responses to more traditional audiologic evaluative techniques were questionable. In 23% of these cases, the original audiologic impression was not confirmed. Hume and Cant (1977) reported on a series of 88 children under the age of three years who were evaluated via brain stem evoked potentials. They obtained 78 reliable decisions, 73% of which were later verified by other test procedures. The brain stem auditory evoked potential has been shown to have limited variability and demonstrated reliability (Hecox & Galambos, 1974). It remains one of the most promising objective measures of hearing for children (J. Jerger & Hayes, 1976) and adults.

The middle and late responses (latencies between 10–50 msec and 50–500 msec, respectively) are known as encephalic responses (McCandless, 1977). Middle responses are thought to originate in the upper brain stem, or lower cortical areas, while the late responses reflect cortical activity. Both middle and late responses can give frequency specific information about hearing sensitivity. There remains some controversy regarding the origin of the middle responses. Several investigators (Cody, Jacobson, Walker, & Bickford, 1964; Kiang, Crist, French, & Edwards, 1963; Mast, 1965) reported evidence supporting a myogenic origin, while others (Goldstein, 1965; Ruhm, Walker, & Flanigan, 1967) support a neuroelectric origin. Electroencephalographic measures have been extensively studied and reported upon in the literature. These tests reflect very complex nervous system changes, which are influenced by a myriad of variables. Research continues to define the variables and measure the contributions of various components to the measurement procedure. Response criteria must be stringently stated, as intraobserver agreement has been reported as only fair (Rose, Keating, Hedgecock, Schreur, & Muller, 1971). The effects of subject state and/or sedation are problematic in late responses. Although many patients yield electroencephalic responses that are in good agreement with behavioral thresholds, these measures are often not effective with the young, hyperactive, or emtionally disturbed child, for whom other measures also fail.

In summary, only impedance/admittance measures have received widespread clinical application. Other electrophysiologic measures, although promising for the future, are presently in research stages. Both brain stem measures and electrocochleography appear to have the best potential, and may in the future provide answers for those children who are not accessible through the more traditional behavioral methods.

Auditory Association Impairments

Contemporary diagnostic audiology practice is no longer limited to measuring and describing peripheral auditory sensitivity. Techniques that permit direct assessment of central auditory processing capabilities are now readily available. It must be cautioned, however, that most of the tests currently employed in assessing *central* function have been developed on and are largely appropriate for adult subjects. Nevertheless, a variety of tests are being used in clinical settings for evaluating children.

Almost all of the research that generated the current central auditory test battery was performed on adults who sustained surgically or radiologically confirmed brain lesions (Lynn & Gilroy, 1972). Early findings with temporal lobe patients, who evidenced virtually no hearing loss or speech discrimination problems when they were tested using standard procedures, were found to have significantly reduced discrimination abilities when a distorted auditory signal was presented to the ear contralateral to the lesion (Berlin, *et al.*, 1972). These results were consistent with previous findings which demonstrated that, before the effects of the temporal lesions on auditory function became evident, it was necessary to distort the test signal significantly (Bocca, Callearo, & Cassinari, 1954). Studies such as these point to awareness that, as a result of the complexity and redundancy of function of the central auditory mechanism, significant signal distortion is necessary before the abilities of the system are taxed. Central auditory processing tests are designed to ferret out the subtle manifestations of such disorders, therefore, by introducing either distortion of or competition to a primary auditory signal. Following is a description of the different tests available. They can be categorized generally as tests utilizing distorted signals or signals with competing messages.

Distortion of the speech signal can be accomplished in a number of ways. Bocca and Callearo (1963) found that low pass filtering of speech, using an 800 Hz cutoff caused a depression in discrimination scores in the ear contralateral to the lesion and effectively differentiated the temporal lobe lesion group from the controls. J. Jerger (1960) reported similar findings when evaluating patients with lesions in the temporal lobes and in the brain stem. More recently, Lynn and Gilroy (1978) found that "temporal lobe lesions almost always cause a contralateral ear deficit with monaural low pass filter distorted tests" (p. 213).

Time-altered speech is another method of distortion that has generated interest as a potential clinical procedure for the identification of central auditory processing disorders. Callearo and Lazzaroni (1957) were among the earliest investigators of time-compressed speech. Methodological problems caused difficulty in replication of their early work. Technological advances that later allowed control of frequency during time compression of speech, however, sparked renewed interest in this technique. Beasley, Maki, and Orchik (1976) provided normative data for young adults and children aged 4–8 years. They used both the Phonetically Balanced Kindergarten (PBK) word lists and the Word Intelligibility by Picture Identification (WIPI) test as stimuli. The WIPI was significantly easier and yielded higher scores for all subjects. Other investigators have focused upon the use of time-compressed speech as a correlate of aging (Konkle, Beasley, & Bess, 1977), and for the detection of cortical lesions (Kurdziel, Noffsinger, & Olsen, 1976). Time-compressed speech has great potential for implementation as a clinical tool. Current research is designed to increase the normative data pool, and to investigate alternative stimuli, such as sentences, which will tax the pattern detection ability of the brain.

At present, Katz's (1962, 1968) Staggered Spondaic Word Test (SSW) and the battery of tests proposed by Willeford (1976) have generated most of the clinical and research interest in central auditory processing problems in children. The SSW is a dichotic task that utilizes time-staggered spondiac words in both competing and noncompeting conditions. Katz (1977) reports that the test has been administered to more than 10,000 patients who sustained a variety of lesions, as well as to a large group of normals. The SSW, therefore, is one of the very few measures for which adequate normative data exists. At present, the norms available suggest that the test can be used reliably with patients between 11–60 years of age. Children younger than 11 exhibit some variability in scores and usually show an error pattern demonstrating a lateral effect favoring the right ear (Brunt, 1978). Some caution must be exercised when interpreting Staggered Spondaic Word Test scores with younger children.

The battery of tests proposed by Willeford includes competing sentences, filtered speech (monaural), binaural fusion, and rapidly alternating speech tasks. The competing sentence and filtered speech tests are sensitive to cortical dysfunction, while binaural fusion and rapidly alternating speech measure the integrity of brain stem function. Norms are available for children aged five and greater. Willeford cautions, however, that norms should be established within each clinical facility, in order to control for socioeconomic and other variables, which should not be generalized from the published norms generated in Colorado Springs. White (1977), utilizing the Willeford battery, reported normative data for a group of children aged 5 to 10 years. White found several variances from the published norms, particularly on the binaural fusion task, for which test scores on her normal population showed a wide range and large standard deviation. These findings support Willeford's suggestion of the need for individual clinic norms.

The interpretation of central auditory processing test results remains challenging. Dempsey (1977) poses several interesting theories about central auditory processing tests, and the basic perceptual tasks that they may represent. Filtered speech tests frequently yield poor scores binaurally for children who demonstrate learning problems. Dempsey notes that such children are often "verbally inept" and may simply be unable to deal with this difficult task. The binaural fusion task, which upon initial examination appears to measure brain stem integrity, may in fact reflect a cortical inability to deal with filtered symbolic material. Dempsey also notes that syntactic variations in response to rapidly alternating speech tasks, usually considered acceptable, may deserve further investigation. These errors may in fact reflect an inability to process rapidly presented syntactic material, usually considered a cortical function. Rees (1973) raised similar questions about what central auditory processing tasks are measuring. She noted that all of these tasks are in fact language based, and therefore should be considered language rather than auditory processing tasks. Dempsey and Rees pose serious questions, which remain unanswered. Speaks (1978) states that the results of dichotic listening tasks, when administered to adults, must be viewed with considerable reservation and applied cautiously. For children, whose developing and maturing neurological systems complicate the picture tremendously, the ultimate in caution is warranted.

Psychological Disorder

Nonorganic hearing loss occurs when there is any conscious or unconscious attempt to exaggerate hearing thresholds. Children are usually quite unsophisticated in such attempts and the alert clinician usually experiences little difficulty in identifying nonorganic loss. The evaluation of a child suspected to demonstrate a nonorganic loss begins, as do all evaluations, with the case history and observation of the child's behavior. In many cases, the audiologist will find that the "hearing loss'' was first identified during a school screening, and that the child's parents, teachers, and physician had seen no evidence of a hearing, or middle ear problem prior to the screening. The child will usually come quickly when called from the waiting area, and will often respond appropriately to questions and commands when presented outside the test area.

Upon entering the formal test situation, however, the child may ask for repetitions, or request that the examiner speak louder. If the existence of a nonorganic loss is suspected, the audiologist should alter the usual test procedure and begin with impedance/admittance tests, and proceed immediately to speech audiometry. Because this procedure presents two tests that are most likely unfamiliar to the child, it often reduces the possibility of the nonorganic loss appearing in subsequent pure tone audiometry.

If the nonorganic loss is not suspected during the early stages of the evaluation, a significant discrepancy between thresholds for pure tone and speech should alert the examiner to its possible existence. Inordinately high speech discrimination scores, obtained when stimuli are presented at very low sensation levels, will serve both to confirm the examiner's impression of nonorganic loss and to document the child's ability to hear and understand low-intensity speech. Following reinstruction, the pure tone audiogram should be repeated, using an ascending technique and very small intensity increments. If the loss is not resolved during this procedure, alternative signals such as calibrated warble tones or narrow-band noise may be substituted for pure tones.

Special tests for nonorganic loss are not usually required for children, both because the youngsters are usually quite easily evaluated using the techniques mentioned earlier, and because children are usually unable to maintain such a loss consistently or for any significant period of time. However, when required, special tests such as the Stenger (Taylor, 1949), Bekesy Ascending–Descending Gap Evaluation (Hood, Campbell, & Hutton, 1964), and tests that utilize stapedial reflex thresholds for pure tones and noise should be employed.

The results of evaluation of children who demonstrate nonorganic hearing loss should be discussed in detail with the child's parents and with the referring agent. When appropriate, referrals for professional counseling should be made.

Classification Systems

Children with hearing impairments are classified within two basic systems: *medical* and *educational*. Medical classification usually includes information consisting only of decibel threshold level, diagnosis, and site of lesion. Educational classification, which varies somewhat from state to state, utilizes operational definitions of hearing impairment, and contains information relative to the child's auditory function.

Medical Classification

A. Degree of Loss

The following scale of hearing impairment is after Goodman (1965).

Within Normal Limits: −10 to 26 dB[1]
Mild Hearing Loss: 27 to 40 dB
Moderate Hearing Loss: 41 to 55 dB
Moderately Severe Hearing Loss: 71 to 90 dB
Profound Hearing Loss: greater than 90 dB

B. Diagnosis/Site of Lesion

The following diagnostic categories are derived from audiologic data and reflect site of lesion.

1. *Conductive Hearing Loss:* A conductive hearing loss occurs when there is a disruption of sound transmission through the outer and/or middle ear.

2. *Sensorineural Hearing Loss:* A sensorineural hearing loss occurs when there is a disruption of sound that has been converted to electrical impulses, from the cochlea through the eighth nerve pathways.

3. *Mixed Hearing Loss:* A mixed hearing loss occurs when there is interference with the transmission of sound and, subsequently, electrical impulses, caused by disruption within both conductive and sensorineural mechanisms.

Conductive, sensorineural, and mixed hearing loss can result from disease, injury, or malformation within the various auditory structures. Thus, using a medical classification system, hearing loss is described by combining the degree of loss and the diagnosis (i.e.: severe, mixed hearing loss; profound, sensorineural hearing loss).

The medical classification system was designed for the purposes of identifying potentially remediable problems. Therefore, the emphasis is appropriately placed on the degree and nature of the loss, and on the diagnostic and etiologic factors. The limitations of such classification are important to mention. Because the

[1]All dB levels refer to pure tone average thresholds, re: ANSI, 1969. It should be noted that although Goodman's scale refers to I.S.O., 1964, the scale still pertains.

scaling system is derived from an averaging process, it fails to account for hearing losses that may be irregular or confined to either end of the frequency spectrum. In such cases, the "average" dB loss can be misleading. Even though speech audiometry is routinely included in the diagnostic battery, speech discrimination scores are not considered in the classification system. In other words, the auditory *deficit* is emphasized, and the organism's *residual* capacity is not considered.

Educational Classification

Classification systems for educational purposes have been designed to help describe how much effect the auditory sensory deficit has had on language acquisition and development, since linguistic competence is prerequisite to educational progress. Theoretically, then, there is a direct relationship between the auditory sensory deficit and the child's potential educational success.

Educational classification includes the following categories (McConnell, 1973).

> 1. The *deaf*—those whose hearing loss is so severe at birth and in the prelingual period . . . that it precludes the normal, spontaneous development of spoken language.
> 2. The *partially hearing*—those whose hearing loss in the prelingual period or later is not of sufficient severity to preclude the development of some spoken language, and those who have normal hearing in the prelingual period but acquire hearing loss later. (The term "partially hearing" which is used extensively in the United Kingdom, and the term "hard of hearing," which is more frequently used in the United States, are interchangeable.) (p. 352)

The major criticism of this kind of classification system is that it is not operational enough. That is, it does not yet take into account individual differences and abilities that can contribute greatly to the educational success or failure of a particular child.

Recent legislation (P.L.94-142), together with greater understanding and awareness on the part of special educators, has served to promote a trend toward improved classification categories (Rules & Regulations pursuant to N.J. Chap. 28, New Jersey Administrative Code, 1978). Because in-depth hearing evaluations are now mandatory for all children with auditory problems, and because audiologists, by virtue of training and experience, are the only professionals equipped to perform these evaluations and interpret their results, they are in a unique position to further this trend. The task is to combine the medical and educational classification systems in such a way that the abilities and needs of individual children with auditory problems are considered. For instance, the child with a profound, bilateral, prelingual sensorineural hearing loss may function educationally and socially (as a result of early identification, appropriate amplification, and excellent stimulation) as a *partially hearing* rather than a *deaf* youngster.

The current interest in classifying children who exhibit central auditory processing difficulties is evidence of audiological input into the development of such new categories. Since the organic bases of these disorders are as yet undetermined, central auditory processing problems are best described by the functional limitations they impose.

Dilemmas and Perplexities

Programs for the identification and management of children with auditory problems have improved considerably in recent years. Nevertheless, with added knowledge about hearing and the effects of hearing loss, with greater sophistication about instrumentation for testing and for amplification purposes, and with improved means for dissemination of this information, there are many questions with which those professionals dealing with these children must grapple. Following is a brief discussion about some of those dilemmas and perplexities.

Amplification

In most discussions about auditory problems in children, the concept of the criticality of early amplification is universally stated. There is certainly no controversy about the age at which a child should be fitted with hearing aids; the earlier the better. Controversy does exist, however, on the question of how to go about providing the amplification. At one extreme, it is suggested that young children require trial periods with amplification before firm decisions can be made about the recommendation of specific hearing aids (Madell, 1976). At the other extreme, it is recommended that the *loaner* concept of ongoing hearing aid evaluations for young children be abandoned. Instead, it is suggested that audiometric data be utilized in order to determine requisite electroacoustic properties for the immediate provision of amplification (Zaner, 1976).

Perhaps more appropriate than either of these are the "speech spectrum" methods orginally described by Gengel, Pascoe, and Shore (1971) and later modified by Erber (1973). By calculating the differential between aided threshold levels and the computed average level of noise bands in the speech spectrum, these procedures are used to estimate the amount and sensation level of the speech spectrum that is probably audible to a child, with amplification. The objective of these procedures is to provide amplification throughout the speech spectrum at optimal sensation levels that have been precisely determined.

Schwartz and Larson (1977) compared the two speech spectrum methods with the traditional approach, involving measurement of aided versus unaided thresholds. They found that results with the traditional approach were often misleading. In their sample, children with severe-to-profound losses often did not receive spectral information at sensation levels sufficient for processing. With the speech spectrum approach, however, they were able to provide more adequate evaluations of amplification for speech reception purposes.

The use of speech spectrum methods requires some modification of standard audiometric instrumentation in order to be considered clinically feasible. The advantages of the use of the procedures, however, would seem to justify the investment in time and equipment. Perhaps the answer to the dilemma about amplification in young children lies in the use of speech spectrum methods for hearing aid evaluations.

Central Processing

One of the more perplexing issues confronting educators has been the difficulty in classifying children who evidence learning problems that appear to be related to specific inabilities to process language. Cognizant of the relationship between language and hearing, educators and psychologists have routinely referred such children for audiological evaluation. In the past, it was only possible, however, for the audiologist to test the peripheral mechanism and determine whether or not hearing *sensitivity* was adequate. Tests for auditory *association* were not available. More recently, the audiologist has at his or her disposal a wide variety of tests of central auditory function.

Because the results of these tests seem geared toward answering many heretofore unanswerable questions about central auditory processing, educators have started to request that they be administered to large numbers of learning-disabled children. Audiology clinics have rallied to this demand, and tests of central auditory processing are now widely used.

The use of these tests, however, has not necessarily diminished the perplexity. That is, there remain many unanswered questions about using these test materials to diagnose the learning problems under question. These unresolved questions argue for a conservative approach in the administration of central auditory processing tests and a cautious interpretation of findings.

Diagnosis

Emerging from a medical model, the audiological categorization of auditory problems orginally included information restricted to the nature and degree of the sensory deficit. Especially with children, such strict adherence to a description of hearing "loss" helped create dilemmas when such information was reported to other than medical sources. Schools, for instance, are more interested in descriptions of residual *abilities*, in children with auditory problems, than in technical reports of their *deficits*. Schools also require information concerning the specific educational needs of their hearing-impaired students. They want to know, that is, what are the recommendations for day-to-day classroom involvement in order for a hearing-impaired youngster to achieve optimally.

Motivated by these questions, and faced with a greater number of referrals for audiological evaluations as distinct from medical examinations, audiologists now seem to be writing reports of their evaluations that include functional descriptions of the child's auditory behavior, along with the medical-type diagnosis. It is anticipated that as the writing of individual education plans becomes a more universal activity, the reporting of audiological information solely by nature and degree of loss will become less prevalent.

Early Identification

The key to effective intervention for children with auditory problems lies in early identification and management. Because there is not yet an objective, effective, widely available test to determine hearing impairment in very young or difficult-to-test children, early identification has not necessarily become the rule. Therefore, despite the fact that early identification has been universally acclaimed as a critical factor in dealing with the habilitation or rehabilitation of children with hearing impairments, too often such youngsters are not identified early enough.

The reason that tests for early identification are not readily administered is probably based on a number of interrelated factors. First, there is still professional controversy about the administration of neonatal hearing screenings. The controversy involves questions about which instrumentation to use, who should do the testing, which responses to accept, and whether or not any of the screening procedures currently in use are valid. Although recommendations have been made that might lessen the controversial issues, there appears to be apathy on the part of the various concerned professionals about initiating new screening programs and maintaining programs already in existence.

Second, young children who may exhibit transient or recurrent middle ear pathology are often not referred for audiological studies. Without such a routine universal referral system, identification of a hearing impairment in those children is often neglected.

Third, there is not as yet effective legislation that would mandate either the development of appropriate procedures or the routine utilization of currently accepted tests and measurements. Adequately funded legislation can be seen as an appropriate impetus toward the establishment of an effective early identification system.

Lastly, early identification of hearing impairments in children is in fact an expensive undertaking. The reluctance to invest in an identification program that is not necessarily cost effective is probably related to low-incidence factors in hearing impairment. That is, is it appropriate to spend a great deal of money in order to identify the anticipated one out of every thousand severely hearing-impaired child? Or, is it reasonable to expect that money be invested in identifying the predicted five or six out of every thousand children with hearing impairments that are less than severe? The cost is surely high, but until such time as we can agree that the investment is appropriate, we will continue to be faced with an irresolvable dilemma over the gap between our knowledge about the importance of early identification and actual practice.

References

Aran, J. M., Portmann, M., Portmann, C., & Pelerin, J. Electrocochleogramme chez l'adulte et chez l'enfant. *Audiology,* 1972, *11,* 77–89.

Bangs, T. *Language and learning disorders in the pre-academic child, with curriculum guide.* New York: Appleton-Century-Crofts, 1968.

Barr, B. Pure tone audiometry for pre-school children, a preliminary report. *Acta Otolaryngologica,* (Stockholm) Suppl. *110,* 1954.

Barr, B. Pure tone audiometry for pre-school children. *Acta Otolaryngologica,* Suppl. *121,* 1955.

Bartoshuk, A. K. Response decrement with repeated elicitation of human neonatal cardiac acceleration to sound. *Journal of Comparative Physiological Psychology,* 1962, *55,* 9–13. (a)

Bartoshuk, A. K. Human neonatal cardiac acceleration to sound: habituation and dishabituation. *Perceptual Motor Skills,* 1962, *15,* 15–27. (b)

Bartoshuk, A. K. Human neonatal cardiac responses to sound: a power function. *Psychonomic Science,* 1964, *1,* 151–152.

Battin, R. Developmental effects of early middle ear pathology. *Texas Journal of Audiology and Speech,* 1978, *3,* 11–12.

Beasley, D., Maki, J., & Orchik, D. Children's perception of time—compressed speech on two measures of speech discrimination. *Journal of Speech and Hearing Disorders,* 1976, *41,* 216–225.

Berlin, C. I., & Lowe, S. S. Temporal and dichotic factors in central auditory testing. In J. Katz (Ed.), *Handbook of clinical audiology* (1st Ed.). Baltimore: Williams and Wilkins, 1972.

Berlin, C. I., Chase, R. A., Dill, A., & Hagepanos, T. Auditory findings in temporal lobectomized patients *Journal of the American Speech and Hearing Association,* 1965, *7,* 386. (Abstract)

Berlin, C. I., Lowe-Bell, S. S., Janetta, P. S., & Kline, D. G. Central auditory deficits after temporal lobectomy. *Archives of Otolaryngology,* 1972, *96,* 4–10.

Bess, F., Schwartz, D., & Redfield, N. Audiometric, impedance and otoscopic findings in children with cleft palates. *Archives of Otolaryngology,* 1976, *102,* 465–469.

Bluestone, C. Eustachian tube function. In J. Jerger (Ed.), *Impedance audiometry*. New York: American Electromedics, 1974.

Bocca, E., & Callearo, C. Central hearing processes. In J. Jerger (Ed.), *Modern developments in audiology*. New York: Academic Press, 1963.

Bocca, E., Callearo, C., & Cassinari, V. A new method for testing hearing in temporal lobe tumours. *Acta Otolaryngologica*, 1954, *44*, 219–221.

Bocca, E., Callearo, C., Cassinari, V., & Miglinvacca, F. Testing cortical hearing in temporal lobe tumors. *Acta Otolaryngologica*, 1955, *45*, 289–304.

Bradford, L. *Psychological measures of the audio vestibular system*. New York: Academic Press, 1975.

Brattgard, S. O. The importance of adequate stimulation for the chemical composition of retinal ganglion cells during early post natal development. *Acta Radiologica*, Suppl. *96*, 1952, 1–80.

Brooks, D. School screening for middle ear effusion. *Annals of Otology, Rhinology, & Laryngology*, 1976, *85*, (Suppl. 25, Number 2, Part 2), 223–228.

Brunt, M. The staggered spondaic word test. In J. Katz (Ed.), *Handbook of clinical audiology* (2nd ed.). Baltimore: Williams and Wilkins, 1978.

Callearo, C., & Lazzaroni, A. Speech intelligibility in relation to the speed of the message. *Laryngoscope*, 1957, *67*, 410–419.

Cody, D. T. R., Jacobson, J. L., Walker, J. C., & Bickford, R. G. Averaged myogenic and cortical potentials to sound in man. *Annals of Otology*, 1964, *73*, 763–777.

Cooper, J. C., Gates, G., Owen, J., & Dickson, H. An abbreviated impedance bridge technique for school screening. *Journal of Speech and Hearing Disorders*, 1975, *40*, 260–269.

Davis, H. Brainstem and other responses in electric response audiometry. *Annals of Otology, Rhinology, and Laryngology*, 1976, *85*, 3–13. (a)

Davis, H. Principles of electric response audiometry. *Annals of Otology, Rhinology, and Laryngology*, 1976, *85*, (Suppl. 28, Part 3). (b)

Dempsey, C. Some thoughts concerning alternate explanations of central auditory test results. In R. Keith (Ed.), *Central auditory dysfunction*. New York: Grune and Stratton, 1977.

Dempsey, C. Materials distributed at a colloquium in Newark, N.J., May 23, 1978.

Dix, M. R., & Hallpike, C. S. The peep show. *British Medical Journal*, 1947, *2*, 719–723.

Downs, M. The expanding imperatives of early identification. In F. Bess (Ed.), *Childhood deafness: Causation, assessment and management*. New York: Grune and Stratton, 1977.

Downs, M. & Sterritt, G. A guide to newborn and infant hearing screening programs. *Archives of Otolaryngology*, 1967, *85*, 37–44.

Eilers, R., & Minifie, F. Fricative discrimination in early infancy. *Journal of Speech and Hearing Research*, 1975, *18*, 158–167.

Eilers, R., Wilson, W., & Moore, J. Developmental changes in speech discrimination in infants. *Journal of Speech and Hearing Research*, 1977, *20*, 766–780.

Eimas, P. D., Siqueland, E. R., Jusczyk, P., & Vigorit, J. Speech perception in infants. *Science*, 1971, 303–306.

Eisen, N. H. Some effects of early sensory deprivation on later behavior: the quondom hard of hearing child. *Journal of Abnormal Psychology*, 1962, *65*, 338–342.

Eisenberg, R. Cardiotachometry. In L. Bradford (Ed.), *Physiological measures of the audio vestibular system*. New York: Academic Press, 1975.

Eisenberg, R. B. *Auditory competence in early life*. Baltimore: University Park Press, 1976.

Erber, N. Body baffle and real ear effects in the selection of hearing aids for deaf children. *Journal of Speech and Hearing Disorders*, 1973, *38*, 224–231.

Erikson, E. *Childhood and society* (2nd ed.). New York: Norton, 1963.

Ewing, I. R., & Ewing, A. W. G. The ascertainment of deafness in infancy and early childhood. *Journal of Laryngology and Otology*, 1944, *59*, 309–338.

Feinmesser, M., & Tell, L. Neonatal screening for detection of deafness. *Archives of Otolaryngology,* 1976, *5,* 297–299.

Feldman, A. Acoustic impedance–admittance measurements. In L. J. Bradford (Ed.), *Physiological measures in the audio vestibular system.* New York: Academic Press, 1975.

Feldman, A. Acoustic impedance–admittance. In J. Katz (Ed.), *Handbook of clinical audiology* (2nd ed.). Baltimore: Williams and Wilkins, 1978.

Fry, D. Language development and the deaf child: a psycholinguistic approach. In F. Bess (Ed.), *Childhood deafness: Causation, assessment and management.* New York: Grune and Stratton, 1977.

Gaeth, J. H., & Lounsberry, E. Hearing aids and children in elementary schools. *Journal of Speech and Hearing Disorders,* 1966, *31,* 283–289.

Gengel, R., Pascoe, D., & Shore, I. A frequency response procedure for evaluating and selecting hearing aids for severely hearing impaired children. *Journal of Speech and Hearing Disorders,* 1971, *36,* 341–353.

Geschwind, N., & Levitsky, W. Human brain: left right asymmetries in temporal speech region. *Science,* 1968, *161,* 186–187.

Glattke, T. Electrocochleography. In J. Katz (Ed.), *Handbook of clinical audiology* (2nd ed.). Baltimore: Williams and Wilkins, 1978.

Goldenberg, R., & Derbyshire, A. Averaged evoked potentials in cats with lesions of auditory pathways. *Journal of Speech and Hearing Research,* 1975, *18,* 420–429.

Goldstein, M. A. *Problems of the deaf.* St. Louis: The Laryngoscope Press, 1933.

Goldstein, R. Early components of the AER. *Acta Otolaryngologica* (Stockholm), 1965, Suppl. *206,* 127–128.

Goldstein, R., & Tait, C. Critique of neonatal hearing evaluation. *Journal of Speech and Hearing Disorders,* 1971, *36,* 1–18.

Goodman, A. Reference zero levels for pure tone audiometer. *Journal of the American Speech and Hearing Association,* 1965, *7,* 262–263.

Greenberg, D., Wilson, W., Moore, J., & Thompson, G. Visual reinforcement audiometry (VRA) with young Down's Syndrome children. *Journal of Speech and Hearing Disorders,* 1978, *43,* 448–458.

Griffing, T. S., Simonton, K. M., & Hedgecock, L. D. Verbal auditory screening for preschool children. *Transactions of the American Academy of Ophthalmology and Otolaryngology,* 1967, *71,* 105–111.

Guilford, F., & Haug, C. Diagnosis of deafness in the very young child. *Archives of Otolaryngology,* 1952, *55,* 101–106.

Hardy, J. B. Studies of hearing in neonates. In The young deaf child, identification and management. *Acta Otolaryngologica* (Stockholm), 1965, Suppl. 206, 34–36.

Haughton, P. Validity of tympanometry for middle ear effusions. *Archives of Otolaryngology,* 1977, *103,* 505–513.

Hayes, D., & Jerger, J. Response detection in respiration audiometry. *Archives of Otolaryngology,* 1978, *104,* 183–185.

Hecox, K., & Galambos, R. Brain stem auditory evoked responses in human infants and adults. *Archives of Otolaryngology,* 1974, *99,* 30–33.

Hodgson, W. R. Testing infants and young children. In J. Katz (Ed.), *Handbook of clinical audiology* (2nd ed.). Baltimore: Williams and Wilkins, 1978.

Holm, V. A., & Kunze, L. H. Effect of chronic otitis media on language and speech development. *Pediatrics,* 1969, *43,* 833–839.

Hood, W. H., Campbell, R. A., & Hutton, C. L. An evaluation of the Bekesy ascending descending gap. *Journal of Speech and Hearing Research,* 1964, *7,* 123–132.

Howell, W. H. *A textbook of physiology* (10th ed., revised). Philadelphia: Saunders, 1928.

Howie, V. The "otitis prone" condition. *American Journal of Disabilities in Children*, 1975, *129*, 676–678.

Hume, A., & Cant, B. Diagnosis of hearing loss in infancy by electric response audiometry. *Archives of Otolaryngology*, 1977, *103*, 416–418.

International symposium on impedance screening for children. *Pediatrics*, 1978, *62*, 570–573.

Jeffers, J., & Barley, M. *Speechreading*. Springfield, Ill.: Charles C Thomas, 1971.

Jerger, J. Observation in auditory lesions in the central auditory pathways. *Archives of Otolaryngology*, 1960, *71*, 797–806.

Jerger, J. Clinical experience with impedance audiometry. *Archives of Otolaryngology*, 1970, *92*, 311–324.

Jerger, J., & Hayes, D. The cross check principle in pediatric audiology. *Archives of Otolaryngology*, 1976, *102*, 614–620.

Jerger, J., Burney, P., Mauldin, L., & Crump, B. Predicting hearing loss from the acoustic reflex. *Journal of Speech and Hearing Disorders*, 1974, *39*, 11–22.

Joint Committee on Infant Hearing Screening: Statement (Sept. 16, 1970). *Journal of the American Speech and Hearing Association*, 1971, 13, 79.

Joint Committee on Infant Hearing Screening: Supplementary statement (Summer, 1973). *Journal of the American Speech and Hearing Association*, 1974, 16, 160.

Kaplan, G., Fleshman, K., Bender T., Baum, C., & Clark, P. Long term effects of otitis media—a ten year cohort study of Alaskan Eskimo children. *Pediatrics*, 1973, *52*, 577–585.

Katz, J. The use of staggered spondaic words for assessing the integrity of the central auditory nervous system. *Journal of Auditory Research*, 1962, *2*, 327–337.

Katz, J., The SSW test: an interim report. *Journal of Speech and Hearing Disorders*, 1968, *33*, 132–146.

Katz, J. The staggered spondaic word test. In R. Keith (Ed.), *Central auditory dysfunction*. New York: Grune and Stratton, 1977.

Keen, R. E., Chase, H., & Graham, F. K. Twenty four hour retention by neonates of a habituated heart rate response. *Psychonomic Science*, 1965, *2*, 265–266.

Keith, R. An evaluation of predicting hearing loss from the acoustic reflex. *Archives of Otolaryngology*, 1977, *103*, 419–424.

Kiang, N., Crist, A. H., French, M. A., & Edwards, A. B. Postauricular electric response to acoustic stimuli in humans. *Quarterly Progress Report* (M.I.T.), 1963, *2*, 218–225.

Klein, J. *Epidemiology and natural history of middle ear disease*. Paper presented at the international symposium on impedance screening for children. Vanderbilt University School of Medicine, Nashville, Tennessee, June 20–22, 1977.

Konkle, D., Beasley, D., & Bess, F. Intelligibility of time-altered speech in relation to chronological aging. *Journal of Speech and Hearing Research*, 1977, *20*, 108–115.

Kurdziel, S. A., Noffsinger, D., & Olsen, W. Performance by cortical lesion patients on 40 and 60% time-compressed material. *Journal of the American Audiology Society*, 1976, *2*, 3–7.

Lenneberg, E. H. *Biological foundations of language*. New York: John Wiley & Sons, 1967.

Lescouflair, G. Critical view on audiometric screening in school. *Archives of Otolaryngology*, 1975, *101*, 469–473.

Lewis, N. Otitis media and linguistic incompetence. *Archives of Otolaryngology*, 1976, *102*, 387–390.

Lewis, N., Dugdale, A., Canty, A., & Jerger, J. Open-ended tympanometric screening: a new concept. *Archives of Otolaryngology*, 1975, *101*, 722–725.

Liden, G., & Kankkonen, A. Visual reinforcement audiometry. *Acta Otolaryngologica* (Stockholm), 1969, *67*, 281–292.

Ling, D., Ling, A., & Doehring, D. Stimulus, response and observer variables in the audiotory screening of newborn infants. *Journal of Speech and Hearing Research*, 1970, *13*, 9–18.

Lloyd, L. L., Spradlin, J. E., & Reid, M. J. An operant audiometric procedure for difficult to test patients. *Journal of Speech and Hearing Disorders*, 1968, *33*, 236–245.

Lynn, G. E., & Gilroy, J. Neuro-audiological abnormalities in patients with temporal lobe tumors. *Journal of Neurological Science*, 1972, *17*, 167–184.

Lynn, G., & Gilroy, J. Effects of brain lesions on the perception of monotic and dichotic speech stimuli in central auditory processing disorders. In M. Sullivan (Ed.), *Proceedings of a conference held at the University of Nebraska Medical Center*, May 23–24, 1974, (3rd printing). Omaha: University of Nebraska, 1978.

Madell, J. Hearing aid evaluation procedures with children. In M. Rubin (Ed.), *Hearing aids: current developments and concepts*. Baltimore: University Park Press, 1976.

Mast, T. Muscular versus cerebral sources for the short latency human evoked responses to clicks. *Journal of Applied Physiology*, 1965, *20*, 725–730.

McCandless, G. Special considerations in evaluating children and the aging for hearing aids. In M. Rubin (Ed.), *Hearing aids: current developments and concepts*. Baltimore: University Park Press, 1976.

McCandless, G. Electrophysiologic measurement in the assessment of the young child. In F. Bess (Ed.), *Childhood deafness: causation, assessment and management*. New York: Grune and Stratton, 1977.

McConnell, F. Children with hearing disabilities. In L. Dunn (Ed.), *Exceptional children in the schools* (2nd ed.). New York: Holt, Rinehart and Winston, 1973.

McFarlan, D. The voice test of hearing. *Archives of Otolaryngology*, 1927, *5*, 1–29.

Mencher, G. Infant hearing screening: the state of the art. *Maico Audiological Series*, 1974, *12*, #7.

Mencher, G. T. *Early identification of hearing loss, proceedings of the Nova Scotia conference*. Basil, Switzerland: S. Karger, 1976.

Mencher, G. T., & McCulloch, B. F. Auditory screening of kindergarten children, using the VASC. *Journal of Speech and Hearing Disorders*, 1970, *35*, 241–247.

Miller, C., & Morse, P. The "heart" of categorical speech discrimination in young infants. *Journal of Speech and Hearing Research*, 1976, *19*, 578–589.

Mokotoff, B., Schulman-Galambos, C., & Galambos, R. Brain stem auditory evoked responses in children. *Archives of Otolaryngology*, 1977, *103*, 38–43.

Moore, J., Wilson, W., & Thompson, G. Visual reinforcement of head-turn responses in infants under 12 months of age. *Journal of Speech and Hearing Disorders*, 1977, *42*, 328–334.

Morse, P. The discrimination of speech and non speech stimuli in early infancy. *Journal of Experimental Child Psychology*, 1972, *14*, 477–490.

Morse, P. Infant speech perception. In D. Sanders (Ed.), *Auditory perception of speech*, Englewood Cliffs, N.J.: Prentice-Hall, 1977.

Morse, P. A., & Snowdon, C. T. An investigation of categorical speech discrimination by Rhesus monkeys. *Perception & Psychophysics*, 1975, *17*, 9–16.

New Jersey Administrative Code. Title 6, Education. N.J.A.C. 6:28-1.1 et seq. adopted August 2, 1978.

New Jersey Speech and Hearing Association, *Guidelines for hearing screening*, 1978.

Niemeyer, W., & Sesterhenn, G. Calculating the hearing threshold from the stapedial reflex for different sound stimuli. *Audiology*, 1974, *13*, 421–427.

North, F. A. Chapter 4. In W. Frankenburg & B. W. Camp (Eds.), *Pediatric screening tests*. Springfield, Ill.: Charles C Thomas, 1976.

Northern, J. L., & Downs, M. P. *Hearing in children* (2nd ed.). Baltimore: Williams and Wilkins, 1978.

Olsen, W. Acoustics and amplification in classrooms for the hearing impaired. In F. Bess (Ed.), *Childhood deafness: Causation, assessment and management*. New York: Grune and Stratton, 1977.

Orchik, D., Dunn, J., & McNutt, L. Tympanometry as a predictor of middle ear effusion. *Archives of Otolaryngology*, 1978, *104*, 4–6.

Paradise, J., Smith, C. L., & Bluestone, C. Tympanometric detection of middle ear effusion in infants and young children. *Pediatrics*, 1976, *58*, 198–210.

Penfield, W., & Roberts, L. *Speech and brain mechanisms*. Princeton: University Press, 1959.

Rees, N. Auditory processing factors in language disorders: the view from Procrustes' bed. *Journal of Speech and Hearing Disorders*, 1973, *38*, 304–315.

Riesen, A. H. Brain and behavior: session 1. Symposium, 1959, #4: Effects of stimulus deprivation on the development and atrophy of the visual sensory system. *American Journal of Orthopsychiatry*, 1960, *30*, 23–36.

Rodel, M. Visual and auditory training for children. In J. Katz (Ed.), *Handbook of clinical audiology* (2nd ed.). Baltimore: Williams and Wilkins, 1978.

Rose, D., Keating, L., Hedgecock, L., Schreur, S., & Muller, K. Aspects of acoustically evoked responses. *Archives of Otolaryngology*, 1971, *94*, 347–350.

Ross, M. Classroom acoustics and speech intelligibility. In J. Katz (Ed.), *Handbook of clinical audiology* (1st ed.). Baltimore: Williams and Wilkins, 1972.

Ruhm, H., Walker, E., & Flanigan, H. Acoustically-evoked potentials in man: mediation of early components. *Laryngoscope*, 1967, *77*, 806–822.

Schell, Y. Electroacoustic evaluation of hearing aids worn by public school children. *Audiology and Hearing Education*, 1976, *2*(6), 7–15.

Schulman, C. A. Heart rate response habituation in high risk premature infants. *Psychobiology*, 1970, *6*, 690–694.

Schulman, C., & Wade, G. The use of heart rate in the audiological evaluation of non verbal children: II Clinical trials on an infant population. *Neuropaediatrie*, 1970, *2*, 187–196.

Schulman-Galambos, C., & Galambos, R. Brain stem auditory evoked responses in premature infants. *Journal of Speech and Hearing Research*, 1975, *18*, 456–465.

Schwartz, D., & Larson V. A comparison of three hearing aid evaluation procedures for young children. *Archives of Otolaryngology*, 1977, *103*, 401–406.

Schwartz, D. M., & Sanders, J. Critical bandwidth and sensitivity prediction in the acoustic stapedial reflex. *Journal of Speech and Hearing Disorders*, 1976, *41*, 244–255.

Seidemann, M., & Givens, G. Tympanometric assessment of Eustachian tube potency in children. *Journal of Speech and Hearing Disorders*, 1977, *42*, 487–498.

Simmons, F. B., & Russ, F. Automated newborn screening: the crib-o-gram, *Archives of Otolaryngology*, 1974, *100*, 1–7.

Skinner, P. Electroencephalic response audiometry. In J. Katz (Ed.), *Handbook of clinical audiology* (2nd ed.). Baltimore: Williams and Wilkins, 1978.

Skinner, P., & Glattke, T. J. Electrophysiologic response audiometry: state of the art. *Journal of Speech and Hearing Disorders*, 1977, *42*, 170–198.

Speaks, C. Dichotic listening: a clinical or research tool in central auditory processing disorders. In M. Sullivan (Ed.), *Proceedings of a conference held at the University of Nebraska Medical Center*, May 23–24, 1974 (3rd printing). Omaha: University of Nebraska, 1978.

Suzuki, T., & Ogiba, Y. Conditioned orientation audiometry. *Archives of Otolaryngology*, 1961, *74*, 192–198.

Taylor, G. J. An experimental study of tests for the detection of auditory malingering. *Journal of Speech and Hearing Disorders*, 1949, *14*, 119–130.

Taylor, D., & Mencher, G. Neonate response: the effect of infant state and auditory stimuli. *Archives of Otolaryngology*, 1972, *95*, 120–124.

Topp, S. Abnormal central auditory processing. In B. Jaffe (Ed.), *Hearing loss in children*. Baltimore: University Park Press, 1977.

Webster, D. B., & Webster, M. Neonatal sound deprivation affects brain stem auditory nuclei. *Archives of Otolaryngology*, 1977, *103*, 392–396.

Willeford, J. Central auditory function in children with learning disabilities. *Audiology and Hearing Education,* 1976, *2*(2), 12–20.

Willeford, J. Assessing central auditory behavior in children: a test battery approach. In R. Keith (Ed.), *Central auditory dysfunction.* New York: Grune and Stratton, 1977.

Wernicke, C. *Der aphasische Symptomenkomplex. Eine psychologische Studie auf anatomische Basis.* Breslau: Cohn & Wright, 1874.

White, E. Children's performance on the SSW and Willeford battery: interim clinical data. In R. Keith (Ed.), *Central auditory dysfunction.* New York: Grune and Stratton, 1977.

Zaner, A. Differential diagnosis of hearing impairment in children: developmental approaches to clinical assessment. *Journal of Communication Disorders,* 1974, *7,* 17–30.

Zaner, A. Hearing aids for infants and children: noninstrumental selection criteria. In M. Rubin (Ed.), *Hearing aids: Current developments and concepts.* Baltimore: University Park Press, 1976.

PART II

CENTRAL DISTURBANCES

5

Christy L. Ludlow

RECOVERY AND REHABILITATION OF ADULT APHASIC PATIENTS

RELEVANT RESEARCH ADVANCES

Introduction and Overview

If we compare the amount of knowledge we have acquired in aphasia with the amount of knowledge we still need to acquire, the latter must be regarded as the more sizable portion. In contrast with some of the other disorders speech–language pathologists deal with, at least the cause of this communication disorder is known. However, although the cause of aphasia can be assumed to be brain damage affecting the hemisphere dominant for language (usually the left), there is a great deal more we don't know. We have as yet to understand the answers to the following questions:

- What brain mechanisms are responsible for the different symptoms seen in aphasic patients?
- What is the neurological basis for language recovery when it does take place?
- Are aphasic symptoms due to the language remaining in the left hemisphere or due to the limited ability of the right hemisphere to perform?
- What is the most effective method of language therapy?
- What are the factors that can predict an individual patient's prognosis for language recovery?

Christy L. Ludlow • Communicative Disorders Program, National Institute of Neurological and Communicative Disorders and Stroke, Bethesda, Maryland 20205.

As this chapter will demonstrate, the answers to each of these questions are not yet available. However, there has been a dramatic increase in interest in aphasia in the last decade, and some new knowledge is now available. This chapter will review some of the knowledge that has been gained to answer the following questions.

- What are the main syndromes of aphasia and the symptoms of language pathology associated with each?
- How can we measure degree of language impairment in aphasia?
- How can we determine a patient's aphasic syndrome?
- When does the greatest amount of recovery take place?
- What is the recovery process for different aphasia syndromes?
- What is the most realistic goal of language therapy: to try to teach an aphasic patient what language he or she has lost, or to improve the patient's ability to use what language remains?
- Are some treatment methods more effective than others for certain aphasic patients?

Syndromes of Aphasia

It has often been remarked that there are as many classification systems for delineating the different syndromes of aphasia as there are major writers in the field. This was particularly true between 1960 and 1970 when classification systems were described by Luria (1964); Wepman and Jones (1964); Schuell, Jenkins, and Jimenez-Pabon (1964); and Jakobson (1961). In addition, there was renewed interest in earlier descriptions of syndromes (Geschwind, 1972; Head, 1926; Weisenberg & McBride, 1935). The proliferation of classification systems not only hampered communication between aphasiologists but also was frequently the topic of lively debates at scientific and professional meetings. In more recent years, the dust seems to have settled and there is some agreement concerning at least five different types of aphasia: Broca's aphasia, Wernicke's aphasia, anomic aphasia, conduction aphasia, and transcortical sensory aphasia. This agreement has come about largely due to the contributions of Dr. Norman Geschwind, Dr. Frank Benson, and Dr. Harold Goodglass, and their colleagues at the Boston Veterans Administration Hospital who provided detailed descriptions of the symptoms associated with each of the syndromes. In 1972, Geschwind's article in *Scientific American* summarized the postulated neurological bases for many of these syndromes, and Goodglass and Kaplan's publication in the same year, *The Assessment of Aphasia and Related Disorders* provided detailed descriptions as well as test profiles of each of these five syndromes. The following syndrome descriptions are based on Goodglass and Kaplan (1972, pp. 54–73) and include references where possible to the names used for the same syndromes by other writers.

Broca's Aphasia. The most noted symptom of this syndrome is the nonfluent telegraphic speech produced by these patients. In contrast with the fluent production of verbal stereotypies (see Alajouanine, 1956), all propositional speech is slow, labored, awkward in articulation with very simple sentence structure and contains a limited vocabulary. The following is an example of telegraphic speech. "Me . . . my wife . . . went . . . school, no, speech, speech, speech therapy. Oh, I don't know, I went . . . and work, work." Relative to their expressive disorder, these patients exhibit good auditory comprehension of single words although they have marked difficulties decoding complex syntactic structures. Thus, their comprehension disorder is similar to their expressive problem although sometimes less marked (Caramazza & Zurif, 1976). Their reading and writing abilities are similar to their oral–aural disorders and they often have spelling errors. Their repetition of single words and phrases is better than their spontaneous speech as is their writing of single words or phrases to dictation. They often recognize their errors and will try to correct them. Generally, they are highly motivated in therapy although they may become appropriately depressed. They will work long hours in therapy and use what language they have remaining to them quite effectively to meet their daily needs (Holland, 1978a).

During the early stages of recovery, these patients may only produce verbal stereotypies. Propositional speech may emerge slowly beginning as single words with a slow labored rate of production and frequent sound distortions. Once propositional speech begins to predominate the prognosis is improved (Alajouanine, 1956).

In this syndrome, the lesion usually involves the third frontal convolution of the left hemisphere, and although the lesion may extend beyond this region the anterior Broca's area is usually involved.

Broca's aphasia has also been termed anterior aphasia, nonfluent aphasia, verbal aphasia (Head, 1926), efferent motor aphasia (Luria, 1964), Group III (Schuell *et al.*, 1964), expressive aphasia (Weisenberg & McBride, 1935), and syntactic aphasia (Wepman & Jones, 1964).

Wernicke's Aphasia. The patients present fluent speech usually devoid of specific meaning, with fewer nouns than verbs, and many demonstrative and nondefined pronouns, e.g., "I like to do those things all the time with him, just like that." These patients have both literal paraphasias involving sound substitutions such as "crepe raketer" for "tape recorder" and verbal paraphasias involving semantic substitutions such as ''chair'' for ''table.'' Speech articulation is effortless and these patients often talk excessively (Kerschensteiner, Poeck, and Brunner, 1972), but continuously have word-finding difficulties. Repetition of others is errored with many paraphasic errors and frequent insertions.

These patients' difficulties in auditory comprehension are more severe than their expressive disorders. They are usually unaware of their errors during the early postonset period, making treatment difficult until they become more aware of their difficulties and begin to inhibit their output. These patients do not begin to attempt to self-correct until after they have received language therapy. Writing to dictation is severely impaired and reading comprehension poor. Responses during testing are often quick, but errored and inconsistent.

The range in degree of severity is great with more severe forms having almost no auditory comprehension and only jargon as speech output. In the milder cases, speech is fluent and intelligible but devoid of nominative information, and comprehension of only complex syntactic structures is impaired (Zuriff & Caramazza, 1976).

The lesion, in most cases, involves the posterior superior temporal-parietal area. Wernicke's aphasia is also called fluent aphasia, posterior aphasia, syntactic aphasia (Head, 1926), sensory aphasia (Luria, 1964), Group II (Schuell *et al.*, 1964), and pragmatic aphasia (Wepman & Jones, 1964).

Anomia. This syndrome is a mild form of Wernicke's aphasia where the major impairment is a word-finding deficit for specific nominative information. The patient has no other obvious difficulty in his or her spontaneous speech although speech flow may be constantly interrupted because of the patient's difficulties with word retrieval. These patients are usually aware of their word-finding difficulties, although they often do not know when they have retrieved the correct word and thus continue their searching behavior. Repetition is relatively unimpaired although patients may not recognize the words, particularly nouns, that they are repeating. These patients also seem to have poor comprehension of specific nominative information (Goodglass, Gleason, & Hyde, 1970), which is not commensurate with the mild degree of their overall language impairment. Many also have marked impairments in reading and writing to a degree greater than would be expected from their oral–aural language difficulties. Alexia and agraphia may accompany this syndrome.

Although some cases exhibit this syndrome immediately following the onset of aphasia, this syndrome often occurs following significant recovery from Wernicke's or Broca's aphasia. Word-finding difficulties may result from many forms of brain injury, including diffuse lesions to the cortex. Lesions in the angular gyrus and the parietal-occipital region most frequently result in anomic aphasia. Other terms used for this syndrome include semantic aphasia (Luria, 1964; Wepman & Jones, 1964), amnestic aphasia, nominal aphasia (Head, 1926), and Group IV (Schuell *et al.*, 1964).

Conduction Aphasia. The most obvious feature of this syndrome is a repetition disorder that is not commensurate with the patient's overall degree of language impairment. This is a rare syndrome. The impairments in speech and auditory comprehension are relatively mild. Patients are fluent although not excessively so, and their main expressive problems are anomia and literal paraphasia. They are acutely aware of their errors and will frequently attempt to self-correct although not always successfully. Cuing and assistance from others is no help because of the severe repetition deficit. These patients are sometimes better at repeating digits than words. Some investigators have studied these patients in great detail because of their short-term memory impairments, which seem to be confined to the auditory modality (Saffran & Marin, 1975; Warrington & Shallice, 1969; Warrington, Logue, & Pratt, 1971).

Geschwind (1965) has attributed this syndrome to the presence of a lesion in the arcuate fasciculus, a pathway connecting the anterior expressive language area with the posterior central language region.

Transcortical Sensory Aphasia. This is also a relatively rare form of aphasia that is the converse of conduction aphasia; the patient has excellent repetition ability and retains a great deal of previously learned rote material such as the days of the week, counting, months of the year, and prayers. In contrast, propositional expressive language, confrontation naming, auditory comprehension, and reading and writing are all severely impaired. Expressive speech contains frequent neologisms due to literal and verbal paraphasias. Although patients may be aware of their errors they make few attempts to self-correct. These patients are similar to Wernicke's aphasics except they have excellent repetition ability even for long and complex sentences.

Usually there has been diffuse brain damage with a sparing of the central language areas in these patients. Some cases have occurred following carbon monoxide poisoning. Geschwind has referred to this syndrome as "Isolation of the Speech Area" (Geschwind, Segarra, & Quadfasel, 1968).

Verbal Apraxia. This is not an aphasia since the patient's language is not affected; rather it is a speech production disorder. This syndrome can occur following stroke or head injury and has also been termed aphemia, subcortical motor aphasia (Goodglass & Kaplan, 1972), phonetic disintegration (Alajouanine, Ombredane, & Durand, 1939), and cortical dysarthria (Bay, 1962). It is a specific disorder affecting oral movements during speech production. In its pure form it is rare; it occurs more often in younger patients and may appear in conjunction with an oral apraxia for nonspeech movements. Although there is no neuromotor disease as in dysarthria, and patients can read, write, and understand speech normally, in the most severe cases patients may be unable to phonate voluntarily whereas laryngeal vibration and phonation occur spontaneously during coughing. Once phonation is induced, the patient can begin to produce isolated vowel sounds, but the recovery of speech articulation is slow and awkward. Speech imitation does not assist these patients' speech production, which, following the early recovery period, remains severely distorted, slow and laborious, and sometimes sounds like a foreign accent.

The literature on this syndrome has been confused, partly because of differences between investigators' subject selection criteria. In some cases, studies have included Broca's aphasics along with verbal apractic cases, thus providing conflicting results. There is little, if any, conclusive information concerning the locus of lesions responsible for this disorder, and only cases with pure forms of the disorder who are without aphasia should be included in studies of this disorder.

To summarize the salient differences between the various syndromes, Table I is provided. Global aphasia with severe impairments is added here. These patients are close to nonverbal and often times even confuse "yes" and "no." If global aphasia persists at the end of the sixth month, the prognosis for significant language recovery even with intensive therapy is guarded (Sarno, Silverman, & Sands, 1970).

Assessment of Aphasia

Language testing with aphasic adults has the following major objectives. First, it must be determined whether the patient does indeed have a central language disorder that is affecting the various language modalities to different or equal degrees or whether the patient has a verbal dyspraxia, a dysarthria, a hearing loss, or various combinations of these. If a dementia or a psychiatric illness is suspected, other professionals should be asked to examine the patient. Second, if the patient is found to have aphasia then his or her remaining language skills must be assessed to identify the aphasic syndrome, treatment needs, and estimate prognosis for recovery.

A language assessment instrument that will be used to determine whether a patient is impaired in language for his or her age must have norms for differentiating normal from language-impaired adults at different ages. Benton (1967) and Holland (1980) have determined that aging can have a significant effect on aphasia test scores both in normal and aphasic adults. Also, Holland (1979) has

Table I. Chart of Impaired and Unimpaired Language Performance in Seven Syndromes of Acquired Speech and/or Language Disturbance

Syndrome	Speech expression				Auditory comprehension				Reading	Writing	Repetition
	Impaired fluency and syntax	Naming disorder	Impaired articulation and melody	Paraphasic errors	Normal	Mildly impaired	Moderately impaired	Severely impaired			
Broca's	⊕	+	+	−	−	+	+	+	+	+	+
Wernicke's	−	+	−	+	−	−	⊕	+	+	+	+
Anomic	−	⊕	−	−	+	+	−	−	+	+	−
Conductive	−	+	−	+	+	+	−	−	−	−	⊕
Transcortical sensory	−	+	−	+	−	−	⊕	+	+	+	−
Global aphasia	+	+	+	+	−	−	−	+	+	+	+
Verbal apraxia	−	−	⊕	−	−	−	−	−	−	−	+

+ = Impaired; − = Unimpaired; ⊕ = Primary difficulty.

demonstrated that the expected language performance of aging adults differs depending upon whether they are living at home or in an institution. Thus, if a language test is to accurately identify aging adults who are impaired in language performance, normative data for different age ranges must be provided. Two sets of norms for aging adults should also be provided; one for persons living in institutions and another for those living at home. Only two of the aphasia tests currently available have been standardized on normal adults; the *Neurosensory Center Comprehensive Examination for Aphasia* by Spreen and Benton (1969) and the test of *Communicative Abilities in Daily Living* by Holland (1978b, 1980).

The *Communicative Abilities* test uses an interview format containing various communicative tasks that are scored while the patient is involved in conversing and role playing with the examiner. The patient performs various activities involving communication such as a visit to the doctor, driving a car, buying items in a store, and using the telephone. The test is valid for predicting patients' abilities to function independently in daily living and for differentiating normally aging adults from aphasics. It can be administered reliably after reading the manual and a few practice scoring sessions. Normative data is provided for different age ranges and according to a patient's living environment. This test has also been demonstrated to be valid for measuring communicative disabilities in mentally impaired and hearing-impaired adults. The neurosensory center comprehensive examination is described in more detail below.

None of the following aphasia tests have normative data for aging adults: the *Porch Index of Communicative Ability* by Porch (1967); the *Boston Diagnostic Aphasia Examination* by Goodglass and Kaplan (1979)[1]; and the *Minnesota Test for Differential Diagnosis of Aphasias* by Schuell (1965). However, some of these tests are useful for conducting a detailed assessment of aphasia, that is, identifying a patient's syndrome type and planning a therapy program based on a patient's language rehabilitation needs. To meet this second objective, aphasia tests must have the following characteristics:

1. Performance in each of the language areas listed in Table I must be assessed by different subtests to contrast a patient's abilities across the various language modalities.

2. Scores on each of the subtests of particular language performance modalities must be referenced to the same population with similar criteria used for setting the ceiling and basal levels. For example, normal adult performance may be the ceiling while performance in severe global aphasia may be the basal.

3. Scoring intervals on each of the subtests must be equivalent.

Usually, percentile scores or z scores based on the performance of a large

[1]These authors are currently gathering data on normal aging adults on this test.

aphasic population are used. If such standardization data are not available and only raw scores for each subtest are available, performance across different language modalities cannot be compared. For example, a score of five on a ten-item subtest is not equivalent to a score of five on a subtest containing 163 possible points. Of the presently available aphasia batteries familiar to most of us today, only three provide the necessary equivalence between subtests. These are the *Porch Index of Communicative Ability*, by Porch (1967), the *Boston Diagnostic Aphasia Examination* by Goodglass and Kaplan (1972), and the *Neurosensory Center Comprehensive Examination for Aphasia* by Spreen and Benton (1969).

Each of these three tests differ in their major focus. The Porch index is a highly reliable test in the hands of an examiner who has been trained in its administration. The test content is somewhat limited in the varieties of language behaviors it samples. This examination is particularly useful for predicting degrees and rates of potential change in a patient's performance. It is most effective for assessing severely impaired patients and planning their treatment.

There are no norms for normal aging adults but percentile score equivalents are provided for raw scores on each of the subtests. Thus, a subject's performances in the different language modalities can be compared and changes in each modality can be contrasted during recovery.

The *Neurosensory Center Comprehensive Examination* follows the same form as an IQ test battery but contains 20 different subtests to assess various types of language behaviors. Benton (1967) provides a detailed account of the rationale and procedures followed during development of the test. This battery contains a shortened version of the *Token Test* (De Renzi & Vignolo, 1962), *Word Fluency Test* (Borkowski, Benton, & Spreen, 1967), and an extensive speech articulation test. Similar to the Porch index, raw scores on each of the subtests are converted to percentile ranks so that performance in different modalities as well as amount of change in each during recovery can be contrasted. This test also has normative data with adjusted scores for aphasic adults of different ages and education levels. Benton (1967) provides sample profiles for each aphasic syndrome, which can be used for identifying a patient's type of aphasia. As discussed later in this chapter, the identification of a patient's syndrome is important not only for planning treatment but also for predicting recovery. Finally, the *Neurosensory Center Comprehensive Examination* is the only aphasia battery that has normative data for determining whether aphasic patients have recovered to within the normal range for their age. Profiles are provided with percentile/score equivalents for raw test scores on each subtest based on performance in the normal adult population.

The *Boston Diagnostic Examination* is an extensive battery that samples the greatest range of language behaviors at various difficult levels. Its major feature is that it provides an objective system for classifying patients according to type of aphasia. A patient's raw test score on each of the 43 subtests is converted to a z score relative to the aphasic population on a profile sheet. Thus, z scores can be used to compare performance in different subtest areas as well as degree of recovery in each area on reexamination. This test does not have norms for normal aging adults, although the test developers are currently gathering these. It is compact and easily portable, but requires considerable practice before it can be administered reliably even by experienced clinicians.

The Porch Index is the most useful test for measuring a patient's overall degree of severity of aphasia while the *Boston Diagnostic Examination* is not as useful for this purpose (Holland, 1978b). The primary use of the *Boston Diagnostic Examination* is to provide a detailed qualitative analysis of the aphasic patient's symptoms and to identify the best modalities for tapping a patient's remaining language performance. The *Neurosensory Center Comprehensive Examination* is similar to both the *Porch Index* and the *Boston Diagnostic Examination*. Like the *Porch Index,* it discriminates well between patients on the basis of overall severity in relation to both the aphasic and normal aging populations. However, like the *Boston Diagnostic Examination,* it also provides a performance profile that can be used for identifying the patient's aphasic syndrome.

A test that is popular for assessment of auditory comprehension of spoken language is the *Token Test* (De Renzi & Vignolo, 1962). A great deal of research has been conducted using this with aphasic adults, demonstrating that it is sensitive to mild forms of aphasia in aging adults (Boller & Vignolo, 1966; Swisher & Sarno, 1969).

A more recently developed test that may be useful for assessing speech expression in a manner comparable to the *Token Test* is the *Reporter's Test,* which requires the patient to describe activities performed with tokens by the examiner (De Renzi, 1978).

Finally, there is a high incidence of hearing loss in the aphasic population and all patients' hearing should be assessed prior to initiating therapy (Karlin, Eisenson, Hirschenfang, & Miller, 1959; Miller, 1960; Terr, Goetzinger, & Konsey, 1958). Many have reported that aphasic patients have difficulties responding to sound (Street, 1957). A modified technique of using the descending approach to threshold has been found more reliable for assessing hearing in the aphasic population (Ludlow & Swisher, 1971).

Neurolinguistic Descriptions of Aphasic Language

During the last decade, with the resurgence of interest in linguistics in general, and syntax in particular, there has been a rapid growth in the number of neurolinguistic studies of aphasic patients. The rationale for these studies has been to determine the independent components of language structure and language behavior by analysis of the abilities of aphasic patients. By studying which aspects of language can be selectively impaired following brain damage, it was hoped that the structure of language and its relationship to brain organization could be determined.

An excellent review of the data gathered through neurolinguistic research that bears on this issue was published by Caramazza and Berndt (1978). The following review of some of the neurolinguistic literature highlights those findings that have relevance to the assessment and treatment of aphasic adults.

Word Finding and Anomia. As demonstrated in Table I, anomia or word-finding difficulties occur in all forms of aphasia, although to varying degrees. However, Whitehouse, Caramazza, and Zurif (1978) have found that the basis for a naming disorder may be very different in different aphasic patients. The perceptual categorization schemata for recognizing objects was found to be impaired in posterior aphasics, altering their ability to identify objects on naming tasks.

Goodglass and Baker (1976) examined aphasic and normals' knowledge of verbal associates to a target word during naming performance. They found that the patients with posterior lesions had greater difficulties in recognizing verbal associates in contrast with the normal adults and the aphasics with anterior lesions. They suggested that these subjects had a breakdown in their semantic language structure, which affected their naming behavior.

Spreen, Benton, and Van Allen (1966) found that in some patients the ability to name objects presented through the visual and tactile modalities differed. In contrast, Goodlgass, Barton, and Kaplan (1968) found that for aphasic patients as a group, there were no differences in their abilities to name objects based on whether they were presented through visual, tactile, olfactory, or auditory modalities. However, Ludlow (1977) reported that in both Broca's and fluent aphasics, tactile naming using the nondominant hand increased more rapidly during recovery from aphasia than did visual naming. These findings support the conclusion that naming behaviors must be assessed through various modalities to determine patients' full capabilities. Also, for rehabilitation, presentation of objects through a patient's best performance modality or a combination of modalities may enhance recovery of naming during therapy.

Howes (1967) and later Green (1970) made the distinction that confrontation naming is a qualitatively distinct task from noun usage in spontaneous speech. Howes (1967) notes that some patients are able to use nouns appropriately in spontaneous speech, but are extremely impaired on oral naming tasks. He interprets this as due to the specific one-word response required on naming tasks. Green (1970) indicates that naming is a significantly more complex task than noun usage in spontaneous speech. He postulates a linguistic explanation of the possible mechanisms required in naming and illustrates five distinct levels of naming tasks that must be differentiated in testing. In relation to assessment and treatment, naming to visual pictures of objects can be a more complex task than use in spontaneous speech and therefore is not an appropriate ability to be focused on early in treatment.

Several investigators have examined the frequency of use of different lexical categories in spontaneous speech. Fillenbaum, Jones, & Wepman (1961) found that the frequency of occurrence of words of different form classes in the free speech of aphasics could differentiate between two groups of aphasics and normal controls. The two groups were labeled syntactic aphasia (Broca's), with a reduction of high-frequency words in closed grammatical classes such as function words, and semantic aphasia (Wernicke's), with a reduction of low-frequency words in open lexical classes such as nouns and verbs. Thus the frequency of occurrence of words of different form classes was found to differentiate between these aphasic groups.

Howes and Geschwind (1964) and Howes (1967) found similar results when they examined the word-frequency curves contained in the free speech of aphasics and normal controls. Quantitative measures of word frequency distinguished aphasics with an anterior locus of damage from those with a posterior locus of damage in the language area. Other crucial factors that differentiated between the two groups were speaking rate and number of repetitions. Those with Broca's aphasia (anterior lesions) used a preponderance of more frequent words (lexical formatives), while those with Wernicke's aphasia deviated from normal to a lesser degree in their word-frequency curves.

The different characteristics of these two groups of aphasics indicate the need for at least two different approaches to treatment. Broca's aphasics need focus on the sentence construction in order to incorporate a higher frequency of grammatical formatives in their speech, while Wernicke's aphasics need focus on their organization of semantic knowledge: the meaning relationships between words and their visual representatives.

Syntactic Expression and Comprehension. There have been several excellent reviews of the research on syntactic breakdown in aphasia (Caramazza & Berndt, 1978; Goodglass, 1968; Goodglass, 1976; Zurif & Caramazza, 1976). Those wishing greater detail are referred to these reviews. Unlike anomia and word-finding difficulties, syntactic deficits in expressive speech seem to be less diffusely represented in the brain since they occur mostly in Broca's aphasia where the locus of the lesion is more anteriorly based. Also, in contrast with naming and word finding, syntactic breakdown does not seem to be modality specific.

Goodglass and Hunt (1958) have shown that Broca's aphasics demonstrate syntactic errors in writing the same as in their expressive speech. Caramazza and Zurif (1976) studied the ability of anterior and posterior aphasics to comprehend sentences when matching sentences to pictures, while manipulating syntactic and lexical constraints. The anterior aphasics responded correctly to lexical constraints but poorly to syntactic constraints while the posterior aphasics were poor in both cases. The posterior aphasics had poorer overall performance than the anterior aphasics. Since the agrammatic patients only had auditory comprehension difficulties with syntactic aspects, their auditory comprehension difficulties were similar to their expressive syntactic disabilities. This study also demonstrates further than the breakdown of lexical and syntactic knowledge is independent in aphasia.

The hierarchy of difficulty of various syntactic forms has been shown to be similar for all aphasics when the same type of language performance task is administered. Several studies of Goodglass (1968) and Ludlow (1973) failed to show any difference between anterior and posterior aphasics in the hierarchy of difficulty of grammatical constructions when both groups were studied on the same task. However, the hierarchies of difficulty for various syntactic structures may be altered when different methods of assessment are used such as sentence repetition, story completion, or spontaneous speech.

The results in Table II were reported by Goodglass (1968) and were found using a sentence repetition and sentence comprehension task with agrammatic and fluent aphasics. However, during both auditory comprehension tasks and repetition, memory constraints for varying lengths as well as syntactic complexity can be expected to affect performance. Thus, it is not surprising that the ability to use the same structures may not relate well to performance on repetition and comprehension tasks. This can be seen in Table III showing the hierarchy of difficulty found by Gleason, Goodglass, Green, Ackerman, and Hyde (1975) using the Story Completion Task.

Table II. Sentence Types in Ascending Order of Difficulty on Repetition and Comprehension Tasks as Reported by Goodglass (1968)

Broca's aphasia	Fluent aphasia
Imperative	WH Question
Simple Negative	Imperative
WH Question	Simple Negative
Simple Past	Simple Past
Present Progressive	Yes–No Question
Yes–No Question	Present Progressive
Conditional	Conditional
Noun + Adjectival Clause	Noun + Adjectival Clause

Table III. Syntactic Structures in Ascending Order of Difficulty from Responses on the Story Completion Test[a]

Imperative	Intransitive
Imperative	Transitive
Number + Noun	
Adjective + Noun	
WH Question	
Simple Declarative	
Comparative	
Passive	
Yes–No Question	
Direct + Indirect Object	
Embedded Sentence	
Adjective + Adjective + Noun	
Future (Use of Auxillary)	

[a] Gleason, Goodglass, Green, Ackerman, and Hyde, 1975.

Both Goodglass (1968) and Goodglass, Gleason, Bernholz, and Hyde (1975) have remarked on the noticeable differences in the hierarchy of difficulty of sentence types found on different tasks. These differences are even more striking when compared with the hierarchy of difficulty of syntactic structures found in spontaneous speech. Myerson and Goodglass (1972) described the grammatical constituents and transformational operations underlying the free speech of three Broca's aphasics representing different levels of agrammatic impairment. After deletion of verbal stereotypies, the similarities in the speech samples of the three patients were reported.

Some of the indications of the hierarchical organization of language from the Myerson and Goodglass (1972) study of free speech are summarized below.

1. The noun phrase was better preserved than the verb phrase.
2. Within the noun phrase:
   - (a) the best preserved feature was the morphological marking for plural,
   - (b) the [Integer + Noun] constituent was more preserved than indefinite quantification of the noun,
   - (c) if only one determiner was used it was the article "the" (only the least impaired patient used possessive or demonstrative articles), and
   - (d) adjectives were better preserved as predicates (following a copula) and only with reduced severity did they occur in the noun phrase.
3. Within the verb phrase:
   - (a) progressive marking of the verb was well preserved with the perfect occurring only in the less severe cases,
   - (b) "be" as an auxiliary or copula was less well preserved where used with an adjective or noun phrase in the predicate, and

(c) [verb (intransitive + particle] was well preserved in even the most severe patient.

4. Adverbial phrases containing a cardinal number plus a noun were well preserved in the speech of all patients studied.

These findings of Myerson and Goodglass (1972) indicate a hierarchy of syntax in spontaneous speech that is at variance with the hierarchies found on task performance. For example, the present progressive marking of the verb stem and simple active affirmative declarative (SAAD) were better preserved than the copula verb. Also, no use of *wh* questions or the imperative was reported. Finally, the adjective occurred mostly in the predicate form as a holophrastic phrase rather than within the noun phrase (NP). Such discrepancies may be due to the patients attempting to use structures they used most frequently premorbidly. Thus, because of premorbid usage of certain structures in spontaneous speech at a greater frequency, aphasic patients may not use those structures that are least difficult for them in their spontaneous speech. A similar conclusion was made by Ludlow (1973) when the sequence of recovery of various syntactic forms in the spontaneous speech of aphasics was found to be best predicted by the frequency of occurrence of these forms in normal adult speech.

Thus, attempts to construct a common hierarchy of syntactic difficulty in aphasia from performance on tasks such as sentence comprehension and repetition, have been quite successful. However, the structures most easily performed on repetition, comprehension, and sentence construction tasks frequently are not the same structures that aphasic patients find most useful for communication or attempt to use most often. Speech pathologists frequently note that the language behaviors most easily trained in therapy are often not used by the same patients when attempting to communicate outside of therapy (Holland, 1971). This has been referred to as the problem of carryover. If aphasic patients attempt to use syntactic structures they used most frequently premorbidly in their connected speech, then rehabilitation should concentrate on those structures even though other structures may be performed more easily. Patients may be unable to alter their premorbid tendencies to use certain structures regardless of ease of production. Thus therapy should assist patients with producing those structures they attempt to use most often. Therefore, a detailed analysis of a patient's spontaneous speech is the first step in planning a treatment program.

Recovery and Prognosis

The Time Period of Recovery from Aphasia. Following the onset of aphasia due to a cerebrovascular accident, many factors may be contributing to language recovery, such as the remission of diaschisis (see Luria, 1970a), adjustment and compensation, increased motivation, decreased depression, and language treatment. The remission of diaschisis has its greatest effect during the early postonset period. In the absence of speech and language treatment, language recovery has been found to be greatest during the first three months following the onset of aphasia due to a cerebrovascular accident (Culton, 1969; Sarno and Levita, 1971; Vignolo, 1964). When language recovery occurs early in the postonset period in the absence of treatment, it is referred to as spontaneous language recovery and is assumed to be largely due to the remission of diaschisis, adaptation, functional compensation, increased motivation, and decreased depression.

Since, during treatment, the greatest amount of recovery from aphasia also seems to occur during the early postonset period when the remission of diaschisis may still be occurring, the contribution of speech and language therapy to recovery from aphasia has been questioned. An excellent review of this question concerning the contribution of language therapy to recovery has been written by Darley (1972). At the present time, in most cases we do not know what the relative contributions are of various factors to language recovery in aphasia.

Although recovery may be most rapid between the end of the first month and during the second month following onset (Culton, 1969; Lynch, 1969), recovery has been found to continue through the third month in almost all cases. Several reports have also indicated that recovery continues to occur past the third month (Sands, Sarno, & Shankweiler, 1969). However, with increasing time past the onset of aphasia, the amount of language change is continually decreasing. These results are all based on group effects; there are cases reported in the literature of large recovery gains made on up to six years postonset (Schuell *et al.*, 1964). In all such cases there has been intensive speech and language treatment.

The various subgroups of aphasic patients differ in their patterns of recovery. The initial severity of aphasia is predictive of the amount of language recovery which can be expected. Godfrey (1959), Godfrey and Douglass (1959), Schuell *et al.* (1964), Sands *et al.* (1969), and Sarno *et al.* (1970) have found that the most severely impaired aphasics, those with a global aphasia, do not show significant recovery with or without speech therapy following the initial recovery period. In addition, although the data are not complete, it seems that the more severely impaired aphasics show the smallest gain during the recovery period (Butfield & Zangwill, 1946; Maruszewski, 1969; Schuell, 1955; Sands *et al.*, 1969).

Henri and Canter (1975) reported that the two global aphasics they studied did not show changes until later in the recovery period; that is, three months postonset, and that these changes continued throughout the remaining three months of the six-month period they were examined. These data provide a warning for clinicians. Those patients with a severe aphasia who do not show language recovery in the initial recovery period should not be discharged prematurely. Their gains may not occur until later, between three and six months postonset.

The Degree of Recovery in Different Language Modalities. The least likely ability to recover spontaneously in untreated aphasic patients is oral language production (Hagen, 1970; Kriendler & Fradis, 1968; Vignolo, 1964). This has been found to be true of patients receiving language therapy as well (Kenin & Swisher, 1972). Conversely, the area of language performance most apt to recover in both treated and untreated aphasic patients is auditory comprehension, particularly during the second and third months postonset (Hagen, 1970; Kenin & Swisher, 1972; Kriendler & Fradis, 1968; Vignolo, 1964). Henri and Canter (1975) reported the order of recovery of language modalities found in eight aphasic patients. Improvements first occurred in auditory comprehension, followed by reading, then speaking, and finally writing. Culton (1969) found the opposite to be true in his untreated aphasics whose greatest recovery was in oral production during the first and second months postonset. However, both sets of results are questionable since unstandardized tests were used in both studies. Although both authors report adequate test–retest reliability, neither equated increments on each of the tests of language performance to represent equivalent degrees of difficulty across modalities.

Kenin and Swisher (1972) used the *Neurosensory Center Comprehensive Examination for Aphasia* (which provides equivalence between different subtest scores) to study the recovery pattern of nonfluent aphasics. Each subject was first tested within two months after the onset of aphasia. However, the exact time of onset was not the same for all of the subjects studied and a retest was administered sometime between six weeks and three months following the initial examination. Since the experimental period did not cover similar time periods following the onset of aphasia in the subjects studied, the results cannot be used to indicate the time postonset at which recovery occurred in the various modalities. However, the pattern of recovery is of interest since all of the subjects were receiving intensive speech and language treatment during the experimental period.

The order of the subtests according to the amount of recovery shown from the greatest to the least is as follows. The greatest amount of recovery was on copying writing, an imitative task. Next, relatively large amounts of recovery occurred on tests of reading and auditory comprehension followed by four tests involving speech repetition and oral and written naming of single objects. Thus, the greatest amount of recovery occurred in speech and written imitation and comprehension and expression at the single word level. Those three subtests showing the least amount of recovery tested comprehension of sentences (the Token Test), use of verbs, and sentence repetition and construction. Kenin and Swisher (1972) concluded that the greatest degree of recovery occurs on simple language material, single words, and much less recovery is found at the sentence level. Also, within the different language modalities, the greatest recovery occurred in reading and auditory comprehension of simple material. Next, imitative speech production both for digits and sentences and naming showed some gains followed by written production of single nouns.

The Pattern and Prognosis for Recovery in Different Syndromes of Aphasia. Ludlow (1977) reported on the patterns of recovery of Broca's and fluent aphasics on each of the neurosensory center comprehensive examination subtests during the early recovery period while receiving treatment. Each of the subjects were tested at the end of the first, second, and third month postonset. The differences between the two groups of aphasics were as follows:

1. Throughout the three months studied the fluent aphasics had higher levels of performance than the Broca's on all tests except digit repetition.
2. The two groups were similar in their patterns of performance on the various neurosensory center comprehensive examination subtests at one month postonset and became dissimilar following recovery during the next two months.
3. The greatest recovery occurred during the second month postonset in both groups.
4. The greatest gains in the Broca's aphasics were on short-term memory speech repetition tasks while those in the fluent aphasics were on word fluency tasks.
5. The Broca's aphasics showed less recovery than the fluent aphasics in the second month but continued to show changes in the third month.
6. During the third month postonset, the fluent aphasics only had gains in speech repetition and sentence construction while the Broca's aphasics had greater gains in naming, speech repetition, and sentence construction.
7. Both the Broca's and fluent aphasics had their highest gains in auditory comprehension in the second month postonset.

The results of Ludlow (1977) indicate that the prognosis for recovery in Broca's and fluent aphasia differs significantly. Recovery was more rapid and complete in the fluent aphasics in all areas except speech repetition. The Broca's aphasics had a slower rate of recovery that continued longer.

The total recovery of the Broca's aphasics was less in all areas and particularly in word fluency, sentence comprehension, and sentence construction.

The independence of recovery found in speech comprehension from that occurring during speech expression was also found by Prins, Snow, and Wagenaar (1978). These investigators subgrouped their patients on the basis of speech expression difficulties into four groups (fluent, mixed, nonfluent, and severely nonfluent) and found the groups did not differ in their recovery patterns; all showed some recovery in speech comprehension while none showed overall improvement in speech expression.

Kertesz and McCabe (1977) grouped their patients into the five syndrome categories described in this chapter and found significant differences in the degree of recovery across the five groups during the first year postonset. These investigators had the following results:

1. Recovery rates were higher in posttraumatic than in cerebrovascular cases.

2. Anomic aphasia occurred following significant recovery of both Broca's and fluent aphasic groups and was the common end stage of evolution after recovery.

3. During the first year postonset, Broca's aphasics had the greatest amount of recovery followed by conduction aphasics.

4. However, for final outcome of resolution of the language impairment, complete recovery occurred frequently among anomic, conduction, and transcortical aphasics. Global aphasics had the least amount of recovery and Broca's and Wernicke's aphasics had intermediate levels of residual language impairment.

5. Initial level of severity was highly related to final outcome of aphasia and age was inversely related to amount of recovery seen.

6. No significant differences in recovery occurred between treated and untreated groups.

The results of these studies indicate the importance of identifying a patient's syndrome early in the recovery period. Since the pattern and speed of recovery and final outcome of aphasia differs for different syndromes of aphasia, clinicians will need to give a different prognosis to the family and physician, as well as design appropriate treatment programs based on the patient's syndrome of aphasia.

Rehabilitation

The State of the Art. Language rehabilitation in aphasia is by no means a science. Definitive texts indicating what methods are most effective are not available and most recommendations are based on unquantified and nonsystemative clinical observations of patients' responses. Aphasic patients do not seem to relearn language during recovery and rehabilitation. They frequently recover language that they have not been taught in therapy and often do not use what they have practiced in therapy sessions.

As aphasics attempt to use language, they approximate adult language structures and when they do recover expressive speech they use adult sentence forms (Goodglass, 1976). Thus, the mechanisms responsible for language recovery with or without treatment cannot be language learning or the reacquisition of language in the same way that children develop language. It seems then that we have no clear conceptual framework of the process by which aphasic patients regain language performance. Such a conceptual framework is needed to provide a basis from which various therapeutic techniques can be constructed, systematically evaluated, and the results used in further refining our understanding of the language recovery and rehabilitation process. Without such a framework, clinicans are forced to try all the methods available until they find the one that works best for a particular patient. This can be time consuming and frustrating both for the clinician and patient. Carefully designed studies contrasting the effectiveness of various treatment techniques with selected groups of aphasic patients are needed.

It was concluded in the section on recovery, that the patterns, rates and outcomes of recovery differ across the various aphasic syndromes. Therefore, investigations of the efficacy of different treatment methods must contrast treatments in groups of patients with the same type of aphasia. Two studies serve as models for this type of research. Sarno *et al.* (1970) contrasted the effectiveness of programmed instruction, traditional therapy, and no therapy on one type of aphasia: global aphasics who were past the spontaneous recovery period. Since the type of aphasia was the same for each group, the effectiveness of the two types of treatment, programmed instruction and nonprogrammed instruction, could be compared. The effects of the two types of treatment did not differ and the degree of recovery in each of the treatment groups did not differ from that found in the no treatment group. In most cases, it is unethical to withhold treatment from aphasic adults, however, this does not preclude studies comparing the results of two different types of treatment.

Helm (1978) used a different approach to determine which aphasia syndrome was most benefited by a particular treatment approach. She admitted only chronic aphasics, who had stopped making gains in traditional therapy, to a new treatment regimen using melodic intonation therapy (Albert, Sparks, & Helm, 1973; Sparks & Holland, 1976; Sparks, Helm, & Albert, 1974). Following completion of the treatment program, the amount of change found between pre- and posttreatment on the Boston Diagnostic Aphasia Examination was computed as a measure of the degree of benefit from treatment. Correlational analyses were conducted to determine which pretreatment patient characteristics were predictive of the greatest amount of benefit received from melodic intonation therapy. The syndrome of aphasia that benefited to the greatest degree from this particular treatment was labeled "phonemic articulatory disorders of speech" and included the following patient characteristics: moderately intact auditory comprehension, severe oral apraxia, poor word repetition, and restricted phonemic stereotypy (severe Broca's aphasia with meaningless verbal stereotypy, as described by Alajouanine, 1956).

Both these studies have added significant new knowledge to the field of aphasia rehabilitiation. First, patients with global aphasia past the spontaneous recovery period will not show significant gains with 40 hours of programmed or nonprogrammed language therapy. Clinicians should either not admit such patients for treatment or should provide more extensive treatment programs. Second, melodic intonation therapy is effective with patients with a severe expressive nonfluent aphasia and will result in improved propositional speech.

The second has implications for developing a conceptual framework of what may be the process responsible for language recovery in some cases. The right (non-language dominant) hemisphere has been shown to be dominant for melody perception (Spellacy, 1970) and for perception of intonation contours (Blumstein & Cooper, 1974). Thus, the recovery of expressive speech with melodic intonation therapy, in Broca's aphasia, may be the result of increased involvement of the right hemisphere in language expression with treatment.

A Conceptual Framework. Some authors have attempted to interpret what may be occurring in language recovery. It was Neilsen in 1935 who first cliamed contralateral substitution as the basis for recovery and Goldstein (1948) postulated that when language returned slowly over months and years of special training, the recovery could only be explained by a takeover of function by the nondominant hemisphere.

Such views later fell into disrepute. Penfield (1965) concluded that an adult could never establish a completely new speech center on the nondominant side. Luria (1966) has stated: "in animals differing in their degree of corticalization of functions and of cortical differentiation, the disturbed function may not recover to the same extent" (p. 30). Based on clinical observations of hundreds of patients who sustained brain damage, Luria (1966) concluded: "The cerebral cortex does not consist of separate isolated centers and that the recovery of function must not be attributed to transfer of the function to a new vicarious center but rather to a structural reorganization into a new dynamic system widely dispersed in the cerebral cortex and lower formations" (p. 29). Luria's arguments supporting the idea of restoration as a reorganization of a functional system include: "A local brain lesion may cause a disturbance of the voluntary conscious performance of a function while leaving its involuntary performance intact" (p. 30), and that "a lesion (localized) usually causes a disorganization of a behavior rather than its complete loss" (p. 30).

Thus, Luria rejected the notion of localized functions for mental capacities as well as the idea on the other extreme that various cortical regions have equipotentiality. Rather, he conceived of language behaviors as the result of complex functional systems with many interconnected components. Localized brain lesions could affect one component of several complex functional systems and particular aphasia syndromes are due to a certain component, common to several complex functional systems, being disrupted. Restoration of function is the reorganization of the functional systems underlying each of the language behaviors affected. This reorganization is a functional compensation by the tissues remaining intact that were previously (premorbidly) endowed with some language capacity but did not usually contribute to normal adult language performance. Thus, Luria did not claim that other cortical structures learned or acquired language, rather they only become more proficient with recovery.

Recently, the role of the nondominant hemisphere in language recovery following damage to the dominant hemisphere has been reemphasized. However, the role of the right hemisphere in language recovery, is now conceived quite differently than it was by Neilsen (1935).

Gazzaniga (1967) found, on testing patients following surgical section of the corpus callosum for the relief of severe epilepsy, that individuals varied in the degree of language representation found in their nondominant hemispheres.

Subsequent investigators (Gazzaniga, Le Doux, & Wilson, 1977; Levy, 1978; Levy & Trevarthen, 1977; Zaidel, 1978; Zaidel, 1980) who have studied patients who have undergone cerebral commissurotomy, have found the nondominant hemisphere is capable of the following performances without training:

1. Comprehension of spoken nouns and verbs, short sentences and phrases
2. Comprehension of written words
3. Written naming by placing letters with the left hand to name objects presented only to the right hemisphere

The deficits of the right hemisphere for language performance are:

1. Inability to produce speech
2. Inability to perform syntactic analysis
3. Poor understanding of function words
4. Inability to perform phonetic analysis, grapheme to phoneme matching, and recognition of CV syllables
5. Severely restricted short term memory

On the basis of his research on the capabilities of the right hemisphere for language function, Zaidel (1978) concluded that the deficits of aphasic patients cannot be understood unless there is a pathological interaction between the disordered language mechanism of the left hemisphere and the compensatory abilities of the right hemisphere. Kinsbourne (1971), and more recently Moscovitch (1976), suggested that in some patients poor performance may be due to the damaged left hemisphere remaining dominant for language performance and interfering with the use of the language capabilities of the right hemisphere. In this context, the beneficial effects of melodic intonation therapy for nonfluent Broca's aphasics with phonemic stereotypy could be interaction with the left hemisphere during speech expression by stimulation of its capacities for singing.

Both Gazzaniga (1967) and Luria (1970b) have stressed the variation across individuals in the capacity of the right hemisphere for language performance.

Luria (1970b) indicated that the degree of lateralization of language can be assessed by testing for latent left-handedness. Subirana (1958) concluded from studies of 161 patients with unilateral lesions that history of handedness has some bearing on the degree of recovery from aphasia and may indicate the degree of language representation in the nondominant hemisphere. Such representation may determine the degree of language reorganization possible during recovery. Subirana concluded; "the more an individual is basically right-handed, the less we shall expect to see the regression of his aphasia" (p. 424).

Although Subirana's patients were all postcerebrovascular accident and Luria's had more focal lesions due to head trauma or tumor excision, both concluded that handedness history related to recovery from aphasia. Luria (1970b) concluded:

> Thus, in the variations of hemispheric dominance we find an essential factor affecting the severity of speech disturbances that result from injury to the primary speech areas. This factor, which is, apparently genetically determined, explains both those paradoxical cases in which massive lesions of the speech areas of the left hemisphere do not produce severe or persistent aphasic symptoms and those cases in which injuries of the right hemisphere of right handers produce speech disorders—a class of cases which up till this time has been difficult to understand. (p. 57)

Thus, reorganization of a complex functional system during recovery may depend upon the utilization of all structures that were previously endowed with a partially redundant function. Patients with a bilateral redundancy in their representation of some language functions, most often reading, writing, and auditory comprehension (Gazzaniga, 1967), may undergo reorganization of their language system by incorporating the language system remaining in the left hemisphere with the language representation contained in the right hemisphere.

Wepman and Jones (1967), Alajouanine (1956), and Kertesz and McCabe (1977) have reported that patients often change from one aphasic syndrome to another during recovery. Alajouanine (1956) noted predictable sequences of syndromes evidenced in patients with good recovery. Patients with verbal stereotypy who recover, develop error awareness, become nonfluent, and subsequently exhibit Broca's aphasia and agrammatism. Those patients who initially have a jargon aphasia often develop literal paraphasias followed by verbal paraphasias during recovery.

During the first few weeks following onset, symptoms may be due to the brain or at least one hemisphere being affected as a whole by neural shock (Brain, 1965; Luria, 1963) resulting in initially severe forms of aphasia such as verbal sterotypy or jargon aphasia. As the cortical areas not directly damaged become functional with the remission of diaschisis, a less severe syndrome of aphasia emerges, which is a result of the cortical areas permanently damaged. However, aphasia patients have been observed to change their syndrome of aphasia much later, long after the period of remission of diaschisis (Kertesz & McCabe, 1977). When this occurs, the locus of permanently damaged tissue does not change and the basis of change must be due to the cortical reorganization of a system of function. Thus, although a new area may not learn a lost function, an integration of different areas of function may take place.

Some Approaches to Rehabilitation. Beyn and Shokhor-Trotskaya (1966) radically extended Luria's notions of functional reorganization as the basis of recovery. They designated primary symptoms of an aphasic syndrome as those language functions disrupted by the lesion remaining following the remission of diaschisis. They postulated that many of the symptoms described as predictable during recovery in an aphasic syndrome are due to the patient's attempts to adapt to his primary symptoms during later recovery. These authors believe that many symptoms could be prevented by early intervention with specific techniques during the second month following the completed cerebrovascular accident.

Beyn and Shokhor-Trotskaya (1966) present evidence from therapeutic intervention in an experimental group (N = 25) that emergence of the secondary symptoms of agrammatism was prevented. Their approach was to disallow nominative speech. Nouns were not used with the patients during the initial recovery period nor were the patients allowed to utter nouns at any time. Only predicate forms were used by the staff and in therapy with these patients. The major secondary symptoms of the group receiving the usual therapy methods was the high percentage of nouns and the low percentage of verbs (80% nouns—10% verbs). This was in opposition to normal speakers (36% nouns—20% verbs), and the experimental group receiving the preventive method (44% nouns—30% verbs).

Weigl (1968) and Weigl and Bierwisch (1970) have demonstrated enhancement of patients' use of their residual language performance by a method called "deblocking." In a series of studies, Weigl found that apparently lost speech functions will reappear through stimulation of other verbal functions that are preserved. This approach centers on using residual function in one language performance modality to elicit behavior in another. Weigl (1968) discovered that stimulation in a semantic "sphere of meaing" in one performance modality could elicit responses in the same sphere of meaning in another performance modality. Weigl and Bierwisch (1970) state: "there is an underlying system of linguistic competence and a rather complex set of components of performance, we presented arguments that aphasia must be considered in general as interference of these components and their subcomponents" (pp. 16–17).

Other more generalized approaches using stimulation and facilitation techniques have been provided by Wepman (1953), Schuell *et al.* (1964), and Kriendler and Fradis (1968). The reader is directed to these sources as well as reviews by Darley (1975) and Eisenson (1973) for greater detail than can be given in this chapter on specific treatment approaches.

Conclusions

This chapter has only been a cursory examination of some of the more recent findings in clinical research in aphasia and has demonstrated that a great deal more needs to be learned. The major accomplishments presented in this chapter include:

1. A relatively well-defined system for classifying the more common syndromes of aphasia
2. Methods for measuring the degree of language and communicative impairment a patient has relative to normal adults of the same age
3. Methods for identifying a patient's syndrome of aphasia
4. Description of word finding and naming impairments most common in aphasia
5. Descriptions of syntactic impairment in aphasia
6. Identification of the most rapid period of recovery following aphasia onset due to cerebrovascular accident
7. Description of the patterns and prognosis for recovery in the various syndromes of aphasia
8. A conceptual framework of what may be the process of language recovery during rehabilitation of aphasia
9. A description of some treatment approaches reported to be most helpful for certain syndromes of aphasia.

References

Alajouanine, T. Verbal realization in aphasia. *Brain,* 1956, *79,* 1–28.

Alajouanine, T., Ombredane, A., & Durand, M. *Le Syndrome de disintegration phonétique dans l'aphasie*. Paris: Masson, 1939.

Albert, M., Sparks, R. W., & Helm, N. A. Melodic intonation therapy for aphasia. *Archives of Neurology*, 1973, *29*, 130–131.

Bay, E. Aphasia and nonverbal disorders of language. *Brain*, 1962, *84*, 412–426.

Benton, A. L. Problems of test construction in the field of aphasia. *Cortex*, 1967, *3*, 32–58.

Beyn, E. S., & Shokhor-Trotskaya, M. K. The preventive method of speech rehabilitation in aphasia. *Cortex*, 1966, *2*, 96–108.

Blumstein, S., & Cooper. W. Hemispheric processing of intonational contours. *Cortex*, 1974, *10*, 146–158.

Boller, F., & Vignolo, L. A. Latent sensory aphasia in hemisphere-damaged patients: An experimental study with the Token Test. *Brain*, 1966, *89*, 830.

Borkowski, J. G., Benton, A. L., & Spreen, O. Word fluency and brain damage. *Neuropsychologia*, 1967, *5*, 135–140.

Brain, R. *Speech disorders: Aphasia, apraxia, and agnosia* (2nd ed.). London: Butterworths, 1965.

Butfield, E., & Zangwill, O. L. Reeducation in aphasia: A review of 70 cases. *Journal of Neurology, Neurosurgery, and Psychiatry*, 1946, *9*, 75–79.

Caramazza, A., & Berndt, R. A. Semantic and syntactic processes in aphasia: A review of the literature. *Psychological Bulletin*, 1978, *85*, 898–918.

Caramazza, A., & Zurif, E. B. Dissociation of algorithmic and heuristic processes in language comprehension: Evidence from aphasia. *Brain and Language*, 1976, *3*, 16–27.

Culton, G. L. Spontaneous recovery from aphasia. *Journal of Speech and Hearing Research*, 1969, *12*, 825–832.

Darley, F. L. The efficacy of language rehabilitation in aphasia. *Journal of Speech and Hearing Disorders*, 1972, *37*, 3–21.

Darley, F. L. Treatment of acquired aphasia. In W. J. Friedlander (Ed.), *Advances in Neurology: Volume 7*. New York: Raven Press, 1975.

De Renzi, E., & Vignola, L. The Token Test: A sensitive test to detect receptive disturbances in aphasics. *Brain*, 1962, *85*, 665–678.

De Renzi, E. The Reporter's Test: A sensitive test to detect expressive disturbances in aphasics. *Cortex*, 1978, *14*, 279–293.

Eisenson, J. *Aphasia in adults*. New York: Harper and Row, 1973.

Fillenbaum, S., Jones, L. V., & Wepman, J. M. Some linguistic features of speech from aphasic patients. *Language and Speech*, 1961, *4*, 91–108.

Gazzaniga, M. S. The split brain in man. *Scientific American*, 1967, *217*, 24–29.

Gazzaniga, M. S., LeDoux, J. E., & Wilson, D. H. Language, praxis, and the right hemisphere: Clues to some mechanisms of consciousness. *Neurology*, 1977, *27*, 1144–1147.

Geschwind, N. Disconnexion syndromes in animals and man. *Brain*, 1965, *88*, 237–294.

Geschwind, N. Language and the brain. *Scientific American*, 1972, *226*, 76–83.

Geschwind, N., Segarra, J., & Quadfasel, F. A. Isolation of the speech area. *Neuropsychologia*, 1968, *7*, 327–340.

Gleason, J. B., Goodglass, H., Green, E., Ackerman, N., & Hyde, M. R. The retrieval of syntax in Broca's aphasia. *Brain and Language*, 1975, *7*, 451–471.

Godfrey, C. M. A dysphagia rehabilitation clinic. *Canadian Medical Association Journal*, 1959, *80*, 616–618.

Godrey, C. M., & Douglas, E. The recovery process in aphasia. *Canadian Medical Association Journal*, 1959, *80*, 618–624.

Goldstein, K. *Language and language disturbance*. New York: Grune and Stratton, 1948.

Goodglass, H. Studies on the grammar of aphasics. In S. Rosenberg and J. Kopkin (Eds.), *Developments in applied psycholinguistics research*. New York: Macmillan, 1968.

Goodglass, H. Agrammatism. In H. Whitaker and H. A. Whitaker (Eds.), *Studies in neurolinguistics* (Vol. 1). New York: Academic Press, 1976.

Goodglass, H., & Baker, E. Semantic field, naming, and auditory comprehension in aphasia. *Brain and Language*, 1976, *3*, 359–374.

Goodglass, H., & Hunt, J. Grammatical complexity and aphasic speech. *Word*, 1958, *14*, 197–207.

Goodglass, H., & Kaplan, E. *The assessment of aphasia and related disorders*. Philadelphia: Lea & Febiger, 1972.

Goodglass, H., Barton, M. E., & Kaplan, E. Sensory modality and object naming in aphasia. *Journal of Speech and Hearing Research*, 1968, *11*, 488–496.

Goodglass, H., Gleason, J., & Hyde, M. Some dimensions of auditory language comprehension in aphasia. *Journal of Speech and Hearing Research*, 1970, *13*, 595–606.

Goodglass, H., Gleason, J. B., Bernholz, N. A., & Hyde, M. R. Some linguistic structures in the speech of a Broca's aphasic. *Cortex*, 1972, *7*, 191–212.

Green, E. On the contribution of studies in aphasia to psycholinguistics. *Cortex*, 1970, *6*, 216–235.

Hagen, C. *The effect of speech and language therapy on communication disordered stroke patients*. Paper presented at the convention of the American Speech and Hearing Association, New York, November, 1970.

Head, H. *Aphasia and kindred disorders of speech*. New York: Macmillan, 1926.

Helm, N. *Criteria for selecting aphasic patients for melodic intonation therapy*. Presented at the Annual Meeting of the American Association for the Advancement of Science, Washington, D.C., 1978.

Henri, B. P., & Canter, G. J. *A longitudinal investigation of patterns of language recovery in eight recent aphasics*. Paper presented at the American Speech and Hearing Association Convention, Washington, D.C., November 1975.

Holland, A. L. *Psycholinguistic and behavioral variables underlying recovery from aphasia* (Project No. RD-3116-SH-69). Final Report submitted to Social Rehabilitation Service, June 1971.

Holland, A. *Estimators of aphasia patients' communicative performance in daily life*. Presented at the annual meeting of the American Association for the Advancement of Science, Washington, D.C., 1978. (a)

Holland, A. *Estimators of aphasic patients' communicative performances in daily life* (Contract No. NO1-NS-5-2317). Final Progress Report to the Communicative Disorders Program, the National Institute of Neurological and Communicative Disorders and Stroke, National Institutes of Health, Department of Health, Education and Welfare, 1978. (b)

Holland, A. *A measure of communicative abilities in daily living*. Takoma Park, Md.: University Park Press, 1980.

Howes, D. Hypotheses concerning the functions of the language mechanism. In K. Salzinger & S. Salzinger (Eds.), *Research in verbal behavior and some neurophysical implications*. New York: Academic Press, 1967.

Howes, D., & Geschwind, N. Quantitative studies of aphasic language. In D. Rioch & E. Weinstein (Eds.), *Disorders of communication*. Baltimore: Williams & Wilkins, 1964.

Jakobson, R. Aphasia as a linguistic problem. In S. Saporta & J. R. Bastian (Eds.), *Psycholinguistics: A book of readings*. New York: Holt, Rinehart, & Winston, 1961.

Karlin, I. W., Eisenson, J., Hirschenfang, S., & Miller, M. H. A mutlievaluational study of aphasic and non-aphasic right hemiplegic patients. *Journal of Speech and Hearing Disorders*, 1959, *24*, 369–379.

Kenin, M., & Swisher, L. P. A study of pattern of recovery in aphasia. *Cortex*, 1972, *8*, 56–68.

Kerschensteiner, M., Poeck, K., & Brunner, E. The fluency–non-fluency dimension in the classification of aphasic speech. *Cortex*, 1972, *8*, 233–247.

Kertesz, A., & McCabe, P. Recovery patterns and prognosis in aphasia. *Brain*. 1977, *100*, 1–18.

Kinsbourne, M. The minor hemisphere as a source of aphasic speech. *Archives of Neurology*, 1971, *25*, 302–306.

Kriendler, A., & Fradis, A. *Performance in aphasia: A neurodynamical diagnostic and psychological study*. Paris: Gauthier-Villars, 1968.

Levy, J. Lateral differences in the human brain in cognition and behavioral control. In Buser & Rougeul-Buser (Eds.), *Cerebral correlates of conscious experience INSERM Symposium No. 6*. New York: Elsevier/North-Holland Biomedical Press, 1978.

Levy, J., & Trevarthen, C. Perceptual, semantic and phonetic aspects of elementary language processes in split-brain patients. *Brain*, 1977, *100*, 105–118.

Ludlow, C. L. *The recovery of syntax in aphasia*. Paper read at the Academy of Aphasia, Albuquerque, New Mexico, October 1973.

Ludlow, C. L. Recovery from aphasia: A foundation for treatment. In M. Sullivan & M. S. Kommers (Eds.), *Rationale for adult aphasia Therapy*. Omaha: University of Nebraska Press, 1977.

Ludlow, C. L., & Swisher, L. P. The audiometric evaluation of adult aphasics. *Journal of Speech and Hearing Research*, 1971, *14*, 535–543.

Luria, A. R. *Restoration of function after brain injury*. New York: Macmillan, 1963.

Luria, A. R. Factors and forms of aphasia. In A. V. De Reuck & M. O'Connor (Eds.), *Ciba foundation symposium in disorders of language*. Boston: Little, Brown, 1964.

Luria, A. R. *Higher cortical functions in man*. New York: Basic Books, 1966.

Luria, A. R. The functional organization of the brain. *Scientific American*, 1970, *222*, 66–78. (a)

Luria, A. R. *Traumatic aphasia*. The Hague: Mouton, 1970. (b)

Maruszewski, M. Neuropsychology in neurological rehabilitation. In *Proceedings of the XVIth international congress of applied psychology*, Amsterdam: Swets and Zeitlinger, 1969.

Miller, M. H. Audiologic evaluation of aphasic patients. *Journal of Speech and Hearing Disorders*, 1960, *25*, 333–339.

Moscovitch, M. On the representation of language in the right hemisphere of right-handed people. *Brain and Language*, 1976, *3*, 47–71.

Myerson, R., & Goodglass, H. Transformational grammars of three agrammatic patients. *Language and Speech*, 1972, *15*, 40–50.

Neilsen, J. M. Spontaneous recovery from aphasia: Autopsy (Report of a case). *Bulletin of the Los Angeles Neurological Society*, 1935, *18*, 147–148.

Penfield, W. Conditioning the uncommitted cortex for language learning. *Brain*, 1965, *88*, 787–798.

Porch, B. E. *Porch index of communicative ability*. Palo Alto, Calif.: Consulting Psychologists Press, 1967.

Prins, R. S., Snow, C. E., & Wagenaar. Recovery from aphasia: Spontaneous speech versus language comprehension. *Brain and Language*, 1978, *6*, 192–211.

Saffran, E. M., & Marin, O. S. Immediate memory for word lists and sentences in a patient with deficient auditory short-term memory. *Brain and Language*, 1975, *2*, 420–433.

Sands, E., Sarno, M. T., & Shankweiler, D. Long-term assessment of language function in aphasia due to stroke. *Archives of Physical Medicine and Rehabilitation*, 1969, *50*, 202–207.

Sarno, M. T., & Levita, E. Natural course of recovery in severe aphasia. *Archives of Physical Medicine and Rehabilitation*, 1971, *52*, 175–179.

Sarno, M. T., Silverman, M., & Sands, E. Speech therapy and language recovery in severe aphasia. *Journal of Speech and Hearing Research*, 1970, *13*, 609–623.

Schuell, H. M. Diagnosis and prognosis in aphasia. *AMA Archives of Neurological Psychiatry*, 1955, *44*, 308–315.

Schuell, H. *Differential diagnosis of aphasia with the Minnesota Test*. Minneapolis: University of Minnesota Press, 1965.

Schuell, H., Jenkins, J. J., & Jimenez-Pabon, E. *Aphasia in adults*. New York: Harper and Row, 1964.

Sparks, R. W., & Holland, A. L. Method: melodic intonation therapy for aphasia. *Journal of Speech and Hearing Disorders*, 1976, *41*, 287–297.

Sparks, R., Helm, N., & Albert, M. Aphasia rehabilitation resulting from melodic intonation therapy. *Cortex*, 1974, *10*, 303–316.

Spellacy, F. Lateral preferences in the identification of patterned stimuli. *Journal of the Acoustical Society of America*, 1970, *47*, 574–578.

Spreen, O., & Benton, A. L. *Neurosensory center comprehensive examination for aphasia* (Rev. ed.). Victoria, B.C.: University of Victoria, 1969.

Spreen, O., Benton, A. L., & Van Allen, M. W. Dissociation of visual and tactile naming in amnesic aphasia. *Neurology*, 1966, *16*, 807–814.

Street, B. Hearing loss in aphasia. *Journal of Speech and Hearing Disorders*, 1957, *22*, 60–67.

Subirana, A. The prognosis in aphasia in relation to cerebral dominance and handedness. *Brain*, 1958, *81*, 415–425.

Swisher, L. P., & Sarno, M. T. Token test scores of three matched patient groups: Left brain-damaged with aphasia; right brain-damaged without aphasia; non-brain-damaged. *Cortex*, 1969, *5*, 264–273.

Terr, M., Goetzinger, C., & Konsey, C. A study of hearing acuity in adult aphasic and cerebral palsied subjects. *Archives of Otolaryngology*, 1958, *67*, 447–455.

Vignolo, L. A. Evolution of aphasia and language rehabilitation: A retrospective exploratory study. *Cortex*, 1964, *1*, 344–367.

Warrington, E. K., & Shallice, T. The selective impairment of auditory verbal short-term memory. *Brain*, 1969, *92*, 885–896.

Warrington, E. K., Logue, V., & Pratt, R. T. The anatomical localization of selective impairment of auditory verbal short-term memory. *Neuropsychologica*, 1971, *9*, 377–387.

Weigl, E. On the problem of cortical syndromes: Experimental studies. In M. L. Simmel (Ed.), *The reach of the mind: Essays in memory of Kurt Goldstein*. New York: Springer, 1968.

Weigl, E., & Bierwisch, M. Neuropsychology and linguistics: Topics of common research. *Foundations of Language*, 1970, *6*, 1–18.

Weisenberg, T. H., & McBride, K. E. *Aphasia*. New York: Commonwealth Fund, 1935.

Wepman, J. M. A conceptual model for the processes involved in recovery from aphasia. *Journal of Speech and Hearing Disorders*, 1953, *18*, 4–13.

Wepman, J. M. & Jones, L. Five aphasias: A commentary on aphasia as a regressive linguistic phenomenon. In D. Rioch & E. Wienstein (Eds.), *Disorders of communication*. Baltimore: Williams & Wilkins, 1964.

Wepman, J. M. & Jones, L. V. Aphasia: Diagnostic description and therapy. In W. S. Fields & W. A. Spencer (Eds.), *Stroke rehabilitation: Basic concepts and research needs*. St. Louis: Warren Green, 1967.

Whitehouse, P., Caramazza, A., & Zurif, E. Naming in aphasia: Interacting effects of form and function. *Brain and Language*, 1978, *6*, 63–74.

Zaidel, E. Lexical organization in the right hemisphere. In Buser & Rougeul-Buser (Eds.), *Cerebral correlates of conscious experience, INSERM Symposium No. 6*. New York: Elsevier/North-Holland Biomedical Press, 1978.

Zaidel, E. The split and half brains as models of congenital language disability. In C. L. Ludlow & M. E. Doran-Quine (Eds.), *Neurological bases of langauge disorders in children: Methods and directions for research*. Washington, D.C.: Government Printing Office, 1980.

Zurif, E. B., & Caramazza, A. Psycholinguistic structures in aphasia. In H. Whitaker & H. A. Whitaker (Eds.), *Studies in neurolinguistics* (Vol. 1). New York: Academic Press, 1976.

6

Sheldon M. Frank and R. W. Rieber

LANGUAGE DEVELOPMENT AND LANGUAGE DISORDERS IN CHILDREN AND ADOLESCENTS

Introduction

The first part of this chapter will provide an overview of the authors' orientation; this will be followed by a second part on normal language development and then by a third part on language disorders. Thus the discussion of language disorders is the ultimate goal of this chapter. "Language-affecting disorders" would be a better, though more cumbersome term—it connotes the disturbance of the language function while avoiding the issue of primacy of affected modality. The emphasis is meant to be on the factor "disorders," as they appear to a professional involving the patients he or she sees. The disorders discussed will include "mental" and "physical" language disorders; the focus will be on the kind of background knowledge, information, and viewpoint that would benefit a clinician in trying to understand a patient's language as a given part of behavior and as a clue to many aspects of functioning and of organization. Any clinician—speech pathologist, special educator, internist—must know of the many factors (biological, psychological, and social) that may combine to produce a symptom or syndrome. The language of the patient's presentation may be a central focus or may be something on the periphery of the patient's main problem, much as high blood pressure may be an idiopathic disease entity or a sign of nephritis, or learning disability may be a syndrome or one sign of schizophrenia.

Sheldon M. Frank • Departments of Psychiatry and Pediatrics, University of Miami, School of Medicine, Miami, Florida 33101.
R. W. Rieber • Department of Psychology, John Jay College, CUNY, and Columbia University, College of Physicians and Surgeons, New York, New York 10019.

The authors' account of normal language development is not meant to be a final word on that subject, which is of such intense study and preoccupation among linguists, communication scholars, and psychologists at the current time. Undoubtedly, their own selectivity of reading and influence of professional activities and education, as well as the peculiarities of the subject population to which they have been exposed, will all contribute to possible skewing of views. Perhaps if enough of these admittedly somewhat skewed views are superimposed onto each other, a more definitive picture of this complex development might be possible. The authors' schema is based on a developmental view of normal and abnormal psychology[1]: six stages of language acquisition are presented. It should be noted that these stages are not brought forth as Piagetian stages in the sense that one must traverse a first stage to get to the second stage and so forth. This may well turn out to be the case, but it has not been established. An important problem in investigating order and progress from one stage to another lies in the very nature of language. The competence–performance dichotomy makes such evidence very hard to come by.[2] Indeed, in a more clinically oriented approach to language it is very common that the surface-perceived linguistic behavior (performance) may be either below or more advanced than the true language competence of a child. A final complicating issue is that language encompasses at least four major, independent systems that have individual patterns of development and of deviance from the norm:

1. Semantics
2. Syntax
3. Phonology
4. Communicativeness (includes speaker–listener interaction and speech act aspects)

Normal Language Development in Children and Adolescents[3]

A schema for delineation of language stages is given in Table I. There appear to be minimal requirements for a child to progress from stage to stage in areas of psychology, linguistics and (neuro-) biology:

1. *Biological requirements:* Adequate brain growth and maturation in the

[1]With elements of Freudian and Piagetan as well as more additive views: viz. Freud (1905/1953, 1939/1953); Flavell (1963), Inhelder and Piaget (1958) and McCarthy (1954).

[2]Wherein N. Chomsky (1957) and subsequent authors have distinguished surface productive phenomena (talking, sign-language motions) from inner, relationship/language structure sense, i.e., performance from competence.

[3]This section is an expanded and revised version of Frank (1973).

first 2 years. Intactness of certain cortical and subcortical areas of one hemisphere. Intact auditory apparatus or substitute receptive apparatus. Intact speech apparatus or substitute productive apparatus. In addition, some or all of the particular neurobiological maturation features of Stage III (1½–3 years) may qualify it as a "critical period" in language acquisition (see below).

2. *Linguistic requirements:* Exposure to appropriate quantity and quality of language within the critical language-learning period. The mother usually has the role of this exposure, in most of contemporary society. Too much or too little language is not optimum: a trilingual, confused and stuttering three-year-old encountered socially (cured by elimination of one language) shows the former; children with language lag based on cultural deprivation show the latter. Substitution or addition of television language is beginning to be looked at in this respect. It may indeed aid in the quantity of environmental input (Clarke-Stewart, 1973); effects on language quality are unclear, apart from the obvious effect on surface features of some abnormal children, e.g., autistic and schizophrenic children whose imitated commercials are a significant portion of their language production.

3. *Psychological requirements:* A relationship with a nurturing person with adequate initial closeness, normal progress through separation–individuation and absence of other severe psychopathology in mother or child at early developmental periods.

I. Prelanguage (0–7 months)

An autonomously unfolding series of vocal, kinesic and sensory—perceptual—prelinguistic competences/performances is one view of this phase (see Table I). Just as in related fields of attention and object relations, one sees the framework, even in the first days of life, of abilities and phenomena that will be subsumed within the more general rubric of language e.g., the initially reflexive *smile* develops into part of object relations. Brazleton (1975) has revealed neonates able to follow moving objects quite readily with their eye muscles, to smile responsively to another person, and to begin participation in back-and-forth gestural interaction. These behaviors are present given only minimum requirements of adequate warmth, nutrition, and environmental homeostasis. Even in later infancy (from approximately 6 weeks), when the smile becomes more a responsive one to environmental stimuli (a forehead and two eyes in the work of Spitz, 1946), the evidence and impression are of innately determined behavior patterns. Similarly, recent work by Condon and Sander (1974) shows similarly determined psycholinguistic behavior; the neonate has the beginnings of linguistically oriented, caretaker-tuned body movements. Furthermore, work done by

Table I. Stages of Normal Language Development

Psycholinguistic phenomena	Psychological stage	Neurobiological correlates	Conditions affecting language
I. *Prelanguage (0–7 months)* Perception of phonemes Smiling; crying: response to internal state/reflex Cooing: typology/order autonomous Discourse rules begin	*Symbiosis* Mother–infant unity	Rapid growth and maturation of brain Equipotentiality of cerebral loci and hemispheres Development of articulatory apparatus Reflex basis of nervous system function	Central nervous system pathology Congenital, trauma, infection, vascular, electrophysiological, etc. includes central nervous system manifestations of systemic disorders or abnormal chemical and/or hormonal state) Psychosis in caretaker or child
II. *From baby talk to words (7–18 months)* Babbling: intonation, consonants of mother language Semantics: signal—sign—symbol Syntax: one-word sentences Imitation strategy/feedback	*Separation–Individuation* Object constancy (Freud, Piaget) Beginning of self as separate—locomotion	Brain maturation less rapid; 60% complete Equipotentiality continues Motor development rapid, control over skeletal muscles and sphincters increases. Learning increases as basis of nervous system function; reflexes decrease	Central and/or peripheral nervous system pathology a. brain / with or without mental retardation b. sense organ deficit: blindness, deafness c. articulatory deficit Parental deficit a. deafness/psychosis in parent b. deprivation: maternal, cultural, language Psychosis (autism, atypical psychosis)
III. *Mother tongue 1: Toddlerese (1½–3 years)* Semantics: Refinement; differentiation; higher level specificization Increased vocabulary (750)	*Separation–Individuation* ''Independence''/negativism (concern over control—anal)	Slower rate of brain matuation Cortical hemisphere specialization begins but equipotentiality holds	Above, plus: Developmental conflict/reaction Neurosis

Syntax: 2-word sentences; child (phrase-) grammars Language of mother: maternal modification; expansion: imitation; "naming"/ostensive questions			
IV. *Mother tongue 2: Peer accent (3–6 years)* Adult grammar prototype attained Peer influence; i.e. regional accents Rule acquisition rather than imitation Vocabulary increases	*Preschool/"oedipal"* Triangular relationships Sexual identity	Left hemisphere localization of linguistic mode Secondary aphasis after trauma; full recovery rapid, recrossing stages	Above, plus: Specific Disorders of speech/language a. stuttering b. cluttering c. other articulation disorder d. developmental aphasia
V. *Mother tongue 3: Literacy (6–13 years)* Grammar refined; transformations embedding added Written language influence	*Middle childhood/latency* operational thinking	Completion of brain maturation Slow but complete recovery from secondary aphasia	Above, plus: Minimal brain dysfunction and associated disorders Movement disorder (tics, Tourette syndrome) Specific learning disabilities (dyslexia, Gerspun, etc.) Psychosis: schizophrenia (adult type)
VI. *Mother tongue 4: Teen idiolect (13–18 years)* Accents; new mother tongue impossible by original mechanism Peer: subculture/dialect; relation to identity issue True abstract idea expression	*Adolescence* Identity formation higher level Resurgent conflicts of above stages	Permanent residual from secondary aphasias	Above, plus: Psychosis: manic-depressive disorder Altered chemical state—especially substance abuse

Eimas (1971) and others reveal the ability of neonates to respond, in a similarly rudimentary form, differently to linguistically differentiated stimuli, such as phoneme and word boundaries. R. Miller (1979), in a recent review, summarizes much of the work on development in Stages I–III.

As reviewed a decade ago by Wolff (1969), who synthesized relevant data and theories, the infant's *cry* can be understood similarly. Swedish research, which was summarized there revealed four acoustically differentiated cries in the neonatal and early infantile period. These cries were interpreted to communicate existence of internal tension states, namely hunger, pain, cold, and wet states. It would not be surprising if acoustic study of animals determined similarly differentiated behaviors in vocal or other kinds of apparatus. It is questionnable to what extent this constitutes communication in terms of a social exchange of meaning—the cries seem to take place regardless of the presence and/or responding pattern of the caretaker. These patterns, however, comprising both surface vocal behavior and a rudimentary association of mental central nervous system events with the behavior, could constitute a framework for a later development of such communication and/or for the supersegmental and paralinguistic music, which is the accompaniment of language production.[4]

Cooing involves the production of vowel sounds, performed with a fairly constant intonation pattern. Studies by Jakobson (1956) and Irwin (1947, 1948) showed that the progression of individual vowel sounds produced in the development of this phase of prelanguage is fairly constant from infant to infant. These findings state that the back vowels (produced in the throat area) are produced first, with more frontal vowels produced later. The sounds produced, together with that of the later babbling stage, include all known sounds of all languages; the specificity of the child's—or parent's—language development does not affect this language stage.

Lenneberg (1967) epitomizes these views of the innateness of language development in this stage. He stresses that the phenomena of this stage appear identically in children of deaf mute parents and of congenitally deaf children, neither of whom have an optimal linguistic environment—none at all in the latter. From this and other far-reaching data, including work with retardates[5] and comparison of language and motor development, he postulates a close correlation between central nervous system maturation and language acquisition. In his view,

[4]Work in progress by Corwin and Golub (1979) has apparently extended and computerized the relationship of crying patterns to abnormal as well as normal states, including disease entities such as phenylketonuria and meningitis.

[5]The same stages are traversed at a slower rate by retardates.

rapid central nervous system maturation[6] is characteristic of this stage. There is not yet any lateralization of language function. Severe trauma to one hemisphere, or even its removal, will not drastically affect language development; the critical period does not seem to have been reached.

Work that is more interaction-oriented has begun in the past few years, casting shadows on the pure nativist view presented above. For example, Stern (1976) and his colleagues have shown that series of vocal-oriented interactions are common in most families during this period of the infant's life and involve the beginnings of back-and-forth vocal interchange with certain rules already in force. Many of these early vocal interactions are initiated by the infant, who is capable of relatively long communicative volleys (1½ minutes and longer) by age 4 months (shown also by Bateson, 1975). Gaze activity and kinesis are incorporated into this early communicative behavior, which on the mother's side, contains very special, rule-governed linguistic input. This work is consonant with an interactionist reading of the early "tuning" ability of the infant described above and with similar biologic rhythm turning to the principal caretaker referred by Brazleton (1975) and others. Clinical opinions that early maternal deprivation specifically affects language, and does so beyond the extent expected by consideration of biological/nutritive or cultural components have not been proven, though some evidence exists. (See Frank, 1980.)

II. From Baby Talk to Words (7–18 months)

In Stage II, the child increases his or her repetoire of consonants in the same anatomically determined sequence as he did vowels. He or she combines them in sundry ways and the result is *babbling*, or "baby talk." What gives this stage its charm is that the intonation patterns of environmental speakers—parents, siblings—are rapidly built into the pattern. The combination of nonsense sounds and sensible intonation patterns is striking. Here linguistic differences appear—a babbling young Japanse *is* distinguishable from a babbling young Turk. Furthermore, offspring of the deaf mute and those with hearing problems are also deviant by this stage. Interestingly, as selection of the wide spectrum of sounds produced in Stage I and early Stage II are incorporated, an order which is the reverse of the original order holds, namely, from front to back of the vocal apparatus. "Mamma" and "dada," e.g., are produced at the rostral extreme, and occur early. The word "groggy" would occur rather late (viz., Irwin, 1948; Jakobson, 1956; Lenneberg, 1967).

[6]Measured by Lenneberg as the composite of increase in nervous cell mass, proliferation of dendrites, myelinization and biochemical and electrophysiologic properties.

In terms of *semantic* development, Werner and Kaplan (1967) gave a major impetus to our understanding. With help from the mother at forming distinctions, "words" emerge. For example, the reinforcement of the child's "mama" helps give him or her an utterance that becomes a *signal* for food, for older sister or mother. This later evolves into a *sign,* then *symbol* for the mother. Nelson (1973) and other more recent investigators have found individual variations in semantic strategies in that some children are more referential and others more expressive in their approach. The first group develop, during Stage II and early Stage III, with emphasis on nouns and pronouns in their language production; the second group has more emphasis on function words.

Certain features have been found to correlate with superior language development during this stage: birth order (eldest), gender (female), quality and quantity of linguistic input (mother and others—even T.V.), and intelligence (Clarke-Stewart, 1973; Lenneberg, 1967; Nelson, 1973).

By the end of this stage, a variety of words are uttered with a presyntactic structure described as *one-word sentences*. Namely, the child can use language in context, with self and others, with a one-word limit of length that training cannot overcome. This holophrastic output combines with gestures and elements of the context for meaing (Bloom, 1975, 1976; Bloom & Lahey, 1978). Neurobiological trends of Stage I continue, with moderate decrease in slope of the brain maturation curve.

III. Mother Tongue 1. Toddlerese (1½–3 years)

At the age where the child's growing sense of *separation* and *individuation* from the principal caretaker is a psychological hallmark,[7] the beginnings of linguistic reflection of the processes appear.[8] For the first time word combinations appear, quickly constituting a system capable of forming novel grammatical constructions. Indeed the acquisition, practice and novelty-forming of rule-oriented language behavior is characteristic. Weir's (1962) taped nighttime monologues read like a foreign language student's practicing with a Berlitz record, with the toddler trying different words in grammatical slots.

A current conjecture is that this specific pattern may play a role in language learning, although documentation has been equivocal. The authors cited, and others—Moerk, (1975) after Snow (1972)—have documented a finely tuned mother-to-child modulation of adult language, which has been correlated with the

[7]As described by Mahler (1975).

[8]The use of "no," "yes," and the vocative may be a direct expression of the toddler's increased sense of separateness.

child's progress in language attainment. It has been further shown by Shipley, Smith, and Gleitman (1969) that the child at this stage is more responsive to language structure slightly beyond his (i.e., "expanded") rather than adult language or his own two- or three-word level.

The net result of these trends is a language whose structure, accent, and idiosyncracies are, in normal circumstances, those of the mother. This is obviously a simplification and a generalization, but it is the common pattern, and the authors have used the concept of mother tongue as a central theme in organizing this and subsequent stages of language development. Stage II is cited as a critical period in language acquisition by Lenneberg (1967), depending heavily on neurosurgical evidence: as much as the entire hemisphere can be removed through this age period without subsequent secondary aphasia. If one stretches (by a year) the further limits, this is consistent with the quantum decrease in positive prognoses of autistic children who have not attained language by that age.

There is a further element in this process of the beginnings of maternal speech/language, described more fully in different aspects by Piagetian writers on the one hand, and by Freudians on the other as constituting egocentric speech and/or the *Anlage* of superego internalization of the parents. These views demand a separate exposition and discussion; suffice it to say that they may—together or separately—help integrate the many psycholinguistic events and trends mentioned in Stages III and IV, as various psycholinguistic features of the parents' speech and language are incorporated.

Syntactically, throughout this stage, the child evolves through a number of sets of rules describable as serial discrete grammers, though ones with obvious relationships to each other and to adult language. For example, in the two-word sentence phase, Braine (1963) conceived of "open" and "pivot" categories and rules of usage in those categories that govern the child's language production; noun, verb and other adult categories cannot. E.g. "allgone milk" is a grammatical sentence at this stage consisting of pivot + open words; the pivot "allgone" can easily be recombined with many other open words whereas the latter has a more fixed and limited use. More recent work by Bloom (1975) has used other categorizations. She shows that in this stage, too, the child is using elements of the environmental context in addition to linugistic elements to convey meaning. Thus while more independent of the environment (including mother), he or she hasn't yet a complete ability to communicate by linguistic elements alone; gestures too are incorporated into the communicative stream as signs and symbols, not only as paralinguistic features.

Semantically, the number of vocabulary items increases greatly, often to over 750 words. Differentiation from earlier, more global terms occurs, with words attaining much more closely adult qualities of abstractness and degree of generalization. "Ginger" may now refer to one particular dog, for example, instead of the class "dog." For the latter concept, indeed, the word "dog" might be employed, but may well apply to wolves, bears, or perhaps even more remotely related four-legged creatures (viz., Werner & Kaplan, 1967).

In a speech act view, certain functions predominate in this stage:

1. "*Naming*" and associated ostensive question-asking (viz., Allen, 1973; Frank, Allen, Stein, & Myers, 1976) are very characteristic behaviors, as a child saying "That a chair. What's *that?* (pointing to table.)" They constitute a large proportion of mother–child linguistic behavior, as seen by Baldwin and Baldwin (1973). High instances of questions and answers by mothers are characteristic. Close to half their language is comprised of questions and answers at age 2½ (Frank *et al.*, 1976).

2. *Imitation* is a typical feature in many children's production in Stage III, both immediate and delayed (viz., in Weir's monologues, 1962). There is some doubt as to the universality of its importance in language acquisition as few have found the pattern stressed by Brown and Bellugi (1964) to a great degree (Utterance by child—"expansion" by Mother—imitation of Mother's expansion, Bellugi, 1964, e.g., "allgone milk"—"Oh, your milk's all gone now, eh?"— "Milk allgone now").

In Stage III, the rate of brain maturation slows greatly, handedness begins to emerge and there is evidence that hemisphere specialization is just beginning. From this point on, injury to language areas of one hemisphere will result in greater and longer lasting aphasias. The locus of language in the overwhelming majority of people is in the left hemisphere. Over 95% of right-handed individuals and over two-thirds of left-handed persons have their language function localized there—a mysterious asymmetry of nature (Kimura, 1976). Hand-dominance appears to be necessary but not sufficient for language acquisition; clinically one sees language lags in children with or without motor dominance problems, whereas a large percentage of those with motor dominance difficulties do have language delays.

IV–VI. Further Elaboration of the Mother Tongue

From Stage IV on, the rules of (adult) grammar and syntax can be used to describe the child's language (see Table I). The child at 3 years has few transformational rules (Frank & Osser, 1970). He or she adds them over this period, including embedding[9] among the more complex rules (C. Chomsky, 1969; In-

[9]Ingram points out the Piagetan prediction that only by the age range of 6–12 can a child perform rules on rules (Ingram, 1975).

gram, 1975). The *peer group's* is the language that is now emphasized, whether similar to or different from the Mother's language. This may mean becoming bilingual or changing a regional or foreign accent. This development is obviously culture bound and dependent on psychological and social issues; in contemporary society it is the norm, however, and its nonoccurrence calls for explanation, e.g. fear of peers and/or overattachment to mother.

Overt imitation is observed less in these years and has been shown to be a less effective manner of incorporating new language at this stage (Blank & Frank, 1971); autistic children, who persist in imitation, are deviant. Conscious rule-learning, by contrast, useless at earlier stages, now expedites the learning of language (Krashen, 1975).

The major development of Stage V is the addition of *written* features of language to the established vocal–auditory system. This too is a culture-bound process, untrue for many individuals otherwise not different in development up to this point. Quite different neuromuscular, perceptual, and cognitive skills are now called upon, including left-right orientation, sequencing in time and space, visual-motor and cross-modal corrdination of sensory input. Spoken langauge effects, positive and negative, may stem from those. The cognitive changes beginning in Stage V—towards concrete, then formal operations—build on the linguistic groundwork. (Just as neurological maturation, growth, and psychological development is the groundwork for the language development of Stage IV.) A main quantitative change in syntax, for example, is the incorporation of more and more transformational rules. The main feature of the cognitive state is the attainment of abstract thought as defined by Piaget. Some questions exist, however, as to how complete such thinking is of an average—or even bright—adolescent (viz., Dulit, 1972).

Stage VI is a culmination of the above stages, and perhaps of a critical period in some respects. Many, with Lenneberg (1967), feel true new-language learning can no longer take place in this stage. Results of neurological lesions, experience with "wolf boys," and other evidence point to the impossibility of a "virgin" right hemisphere fully learning language. However, experiments by Gazzaniga (1970) in patients with a severed corpus callosum suggest some languagelike capabilities of the nonlanguage hemisphere as does other work with humans (viz., Kimura, 1976) and with apes (several authors beginning with Gardner & Gardner, 1969). Experience shows that many individuals can acquire, at least, accented abilities in foriegn languages past this age, with the unexpected preeminence of later-learned languages in cases of neurological or psychiatric illness. One could explain this as rapid translation, where the thought is represented by language 1 word, then language 1 word by language 2 word. There is some evidence that this mechanism, admittedly existing in many cases, is not always at work in late-life multilinguals, but that this representation may be from thought directly to a second language. (See Marcos & Alpert, (1977) and Krashen's (1975) comparison of errors in second-language learning.) Clearly, these later stages have not been written about as extensively as the earlier ones. Once again, the fewer overt or dramatic changes in language production may hide developments in overall language competence that are latent: a 15-year-old's sentence on a term paper may differ little from a 10-year-old's. The trends in language production in nonschool activities may even seem to present regression of syntactic competence by the great, complex effect of peer and identity issues discussed in Shapiro (1979).

Disordered Language in Children and Adolescents

A schema for classification of language-affecting disorders is presented in Table II. It is organized on the basis of etiology and primary system of the organism affected, with neurological disorders in category I, emotional/mental disorders in category II. Prior trends in understanding and classification of neurological and mental disorders relate to both problems in ordering and definition of language aspects of those disorders:

1. It was hoped, aimed for, and/or assumed for many years that commonly recognized illness states—in this case, schizophrenia, depression, hysteria, stroke—would ultimately be definable in unitary terms with clearly specified pathognomonic features.

2. With the advent of biochemical advances, even more specificity was hoped for—"one gene, one enzyme (abnormality), one disease." On the model of sickle-cell anemia, one different protein would cause a clearly spelled out disease entity, through a chain of events on different levels of organization.

3. In linguistics over the same years, the tradition was equally one of focus on discrete entities—language or dialect—each with its own system of rules in semantics, syntax, and phonology. The entity might realte to others in time (historical change) or space (dialectology), but was a discrete unit. Even child-language study in the 1960s fell under this influence when investigators stressed the independent (from adult) nature of the rule system in "child grammars."

4. With these preconceptions, many investigators attempting to study language in various abnormal states hoped for a grammar of autism, for example, to the point where one psycholingustic talks about "autists" in the same vein as she may speak of Francophones or Xulu speakers. This corresponded to neurosurgical findings of Penfield and others that many parts of the brain wave have quite

Table II. Taxonomy of Conditions with Disordered Language

I. *Neurological disorders*
  A. Peripheral and/or localized central nervous system disorders
    1. Central speech–Language Area—Socialized Disorder
      a. Acquired aphasias
      b. Acquired speech (articulation) disorders

    2. Sensory disorders
      a. Deafness
      b. Blindness

    3. Motor disorders
      a. Congenital articulation disorders, stuttering
      b. Cerebellar disorders
      c. Extrapyramidal motor system disorders: Tics, Tourette syndrome, Parkinson's disease, Huntington's chorea

  B. Generalized (or uncertainly localized) central nervous system disorders
    1. Mental retardation (see pediatric or psychiatric text for subdivisions) Includes deprivation syndromes—maternal cultural—sometimes referred to as "psychosocial retardation"

    2. Developmental aphasia

    3. Learning disability—with or without minimal brain dysfunction, dyslexia

II. Emotional/mental disorders

  A. Psychosis

    1. Autism
    2. Atypical childhood psychosis
    3. Schizophrenia
    4. Manic-depressive psychosis

  B. Neurosis/behavior disorder

  C. Other

specific functions; a brain locus for noun phrases was also hoped for. Clinically, there was hope to use a language phenomenon as one might a laboratory test, as a clear evidence of presence of a disease entity.

More recent trends (see A.P.A., 1978) have the following characteristics:

5. Appreciation of the multiple dimensions (axes in the A.P.A., 1978, DSM-III) of human functioning with the possibility of differential assessment of each axis, for example: intelligence, neurodevelopmental maturity, psychopathology, and social circumstances.

6. Syndromes, rather than individualized disease entities, were recognized

as composing many if not most disorders. A syndrome includes some of a list of several "typical" abnormalities. The implications are of quantitative rather than qualitative distinctions and of existence of partial and combined syndrome states.

7. Better delineation of childhood psychoses, partly in light of points 4 and 5 above. The earlier occurring syndrome, now labeled autism, is characterized by

a. Onset prior to 30 months
b. Pervasive lack of responsiveness to other human beings
c. Gross deficits in language development
d. Peculiar speech patterns such as echolalia, metaphorical language, pronoun reversal
e. Bizarre responses to various aspects of the environment.

Schizophrenia in its active phase has at least one of the following:

a. Delusions of characteristic types
b. Hallucinations of characteristic types
c. Disturbance of semantic aspect of language; may include incoherence, loose associations, illogicality, improverished content
d. Flattened affect/intonation
e. Disorganized or catatonic behavior.

Schizophrenia is described as a chronic profound disturbance in function, cognition, relating to others, and at times, perception. Its manifestations occur rarely in young children, but often arise at or shortly after puberty.
Atypical psychosis is characterized by onset of psychosis between 2½ and 12 years with some features of autism (yet less marked an effect on development) and some of schizophrenia. Such children often appear normal or mildly disturbed except when under stress, when their psychosis becomes evident.

Manic-depressive psychosis has only recently been seen as a distinct, affective disorder in children and adolescents, and it is rare. Philosophical trends underlying this reclassification may be summarized.

8. All diseases appear to have both "organic" and "other" components, the latter including "functional" and "environmental" factors. Feedback systems among the components are the rule.

In the balance of this section, the authors will review findings on a major condition from each of the categories of schema of Table II. With the points above (especially 8) in mind, it must be stressed that one aspect of many classifications, and of the confusion and/or disagreement in diagnosis of clinical as well as of language features, is that of the specialist who is primarily interested in the disorder. Thus, the diagnosing professional in disorders I.A. would be a neurologist, of I.A.2. an audiologist, and in those of II.A. a psychiatrist. Few see many patients in the other's domain. Many tend to assign patients with borderline or overlapping symptoms to their personal pigeonhole. It is to be hoped that all collaborate with language, hearing, and/or speech specialists in treatment.

The entities discussed are developmental aphasia and childhood psychosis. Please note that the other conditions are discussed elsewhere in this volume. A cautionary note in the area of mental retardation will be mentioned first. Intellectual development is an important axis in itself and one which overlaps with that of language development. By definition, intellectual development is defined by scores on standard I.Q. tests (A.P.A., 1978), and there may be great overlap with other conditions. Just as everyone has a height, everyone has an I.Q., and by the Gaussian standard distribution curve, a certain percentage will be in the "retarded" range. Furthermore, the conditions themselves may well hinder ability to do well on an I.Q. test and thereby foster misdiagnosis or secondary mental retardation. Thus, in assessing language characteristics of either aphasia or childhood psychosis, language features of mental retardation must be taken into account. Lenneberg (1967) reviewed these and found essentially that the parameters (such as sentence length) of language developed in similar sequence to that in normally intelligent children but at a slower rate. The rate was equal to that of other aspects of development, i.e., there was an across-the-board slowing. Depending on the I.Q., the final level achieved can well be lower than normal, and will vary from child to child.

Developmental Aphasia[10]

The diagnosis and treatment of childhood aphasia has been and will probably continue to be one of the most challenging clinical problems in the field of communication disorders.

Myklebust (1954, 1957a,b) distinguishes between symbolic and nonsymbolic language disorders. After pointing out that "aphasia is traditionally and typically viewed as a language disorder which results from damage to the brain" (1957a, p. 507; cf. Berry and Eisensen, 1956, p. 420), he goes on to suggest that the term aphasia (or dysphasia) should be used with children only to refer to disturbances of auditory symbolic language functioning (cf. Wood, 1960). In his definition of aphasia, he includes all degrees of disturbance of use of the verbal symbol, though he makes a distinction between receptive and expressive aphasia as diagnostic categories (Myklebust 1957b). Under the heading of nonsymbolic language disorders, Myklebust (1957a) includes the language disorders of the congenitally deaf, the congenitally blind, the mentally defective, those with

[10]Portions of this section were taken from Kendall (1966).

disorders of auditory perception, those with auditory agnosia, and those with psychological disorders such as psychoses or severe anxieties. Although this classification is of value in distinguishing between the sort of language disorder theoretically to be expected as a result of sensory deprivation or imperception and that to be expected as a result of what is presumably a specific central, integrative defect, the distinction in practice is not always easy to maintain. As other writers have indicated (Davis & Goldstein, 1960; McGinnis, 1963; Miller, 1951), symbolic language disorders in children rarely occur in a pure form, and they are often accompanied by (or expressive of) sensory and perceptual disorders or global mental or personality defects. It is, for example, difficult to regard the language problems of some psychotic children (Goldfarb, 1961), or severely retarded children (O'Connor & Hermelin, 1963) as being nonsymbolic, or necessarily categorically different from those of aphasic children described by Myklebust. Furthermore, input transmission defects such as hearing loss act as a definite barrier to the development of symbolic language, and at the expressive level, there are often similarities between the errors in production of deaf and aphasic individuals.

McGinnis (1963), like Myklebust, emphasizes the distinction between the expressive and receptive types of aphasia in children. In contrast with the theoretical formulation of Wepman, Jones, Bock, and Van Pelt (1960), she regards nonperipheral transmission defects, the agnosias and apraxias, as "forming the distinctive features of what we are calling childhood aphasia" (McGinnis, 1963, p. xviii), and bases her classification system on two main groups, sensory (receptive) aphasia and motor (expressive) aphasia. McGinnis describes her classification as being "oriented towards symptoms of disordered communication; pointing to the requirements for instruction" (McGinnis, 1963, p. xix), rather than as having a speculative anatomical orientation (cf. Davis, 1962).

The differences in terminology between those writers are relatively unimportant. What is of more pressing concern is whether the concept of developmental aphasia can stand on its own, as a specific unitary disorder, capable of being differentiated from other types of language disorder, or whether the degree of overlap in language pathologies (as these have so far been conceptualized) is sufficient to render such as distinction unrealistic or impractical. It is clear from the work of McGinnis (1963) and of other writers (DiCarlo, 1960; Monsee, 1961; Myklebust, 1954; Wood, 1960, 1964) that many children categorized as aphasic demonstrate symptomatology that seems to go far beyond the strict limits of verbal symbolic dysfunction. In particular, coexisting hearing loss, motor disorders, mental retardation, and emotional or personality disorders are reported rather frequently for this group (McGinnis, 1963). However, as Goldstein, Landau, and

Kleffner (1960) and Rapin (1978) have shown, it is not easy to identify from neurological testing a cluster of symptoms that differentiates from hearing-impaired children those whose difficulties in learning language lead to a classification of aphasia. Neuhaus (1959) has reported a similar difficulty for differentiating between mentally retarded and aphasic children on the basis of clinical observations of behavior. Wilson *et al.* (1960), working with small groups of aphasic and hearing-impaired children on an auditory perceptual task, found that although some of the aphasic children failed to learn the task within the allotted number of trials, others performed as well as the hearing-impaired controls, and that it was not possible to identify characteristics within the aphasic subgroup that differentiated between those who learned the task and those who did not. Doehring (1960) also found overlapping between his deaf, aphasic, and normal subjects on visual perceptual tasks though the aphasic group tended to make more errors than the other groups. Apart from children with obvious motor defects, it would certainly appear to be difficult to apply strictly the requirement that the term aphasia be reserved for those cases where there is demonstrable brain damage (Berry & Eisensen, 1956; Karlin, 1954).

Stark (1966) studied aphasic children including the administration of the ITPA, and found that for most of these children the auditory–verbal subtests were much too sophisticated and the children tended to have highly individual patterns even in visual areas. The language characteristics of the aphasic child "are often similar to those of a young normal child. Thus, he may omit plural endings, confuse verb tenses and personal pronouns. . . . One seven-year-old boy said 'No home Jean (his sister)' and 'No cartoons on . . . on Rawhide.' He tended to reverse the orders of words, omit the verbs and articles, and produce utterances which were reduced in length" (Stark, Foster, Giddan, Gottsleben, & Wright, 1968, p. 149).

N. Chomsky (1957), Brown and Bellugi (1964), Smith and G. Miller (1966), and the work of others stimulated research in language disorder. Menyuk (1963), Lee (1966), and Carrow (1968) were among the first to describe the specific nature of the language-impaired child's deviant syntax. Experimental editions of new tests such as the Northwestern Syntax Screening Test (Lee, 1969), Test for Auditory Comprehension of Language (Carrow, 1968), and the Assessment of Children's Language Comprehension (Foster, Giddan, & Stark, 1969) emerged. Tests such as these and a revised ITPA were provided for the many clinicians who were interested in childhood language disorders.

Morehead and Ingram (1973) studied the language of fifteen normal and fifteen aphasic children based upon a corpus elicited during free play and interaction with the clinician or the parents. Their results demonstrated that these children did not have qualitatively different language than normally developing children. They postulated a generalized representational deficit, *à la* Piaget, whose work was also beginning to ride the crest of the wave. They also noted that the aphasic child failed to use language as creatively as normal children. This kind of research led to the development of more formal language-sampling procedures (Tyack & Gottsleben, 1974) and had a significant effect on the assessment and management of the language-impaired child.

Authorities are not at all in agreement on classification of this disorder, nor are they in any more agreement on the etiology of it. One of the major questions in this field that has stimulated a great deal of controversy is the problem of how much of the syndrome of developmental dysphasia can be accounted for outside of the cognitive and/or linguistic domain. Lashley's (1951) seminal paper on the problem of serial order in behavior laid the foundation for many of the studies that stressed the importance of auditory–temporal sequencing. Efran (1963) developed Lashley's position and postulated that temporal order is essential to the processing and production of speech. Monsee (1961) took the position that the core of language disability in childhood aphasia is a disorder of the perception of temporal sequencing, especially auditory and perhaps visual. There have been a number of studies that have provided evidence for this position; see Furth (1964) and Tallal and Piercy (1973, 1974, 1975).

Rees (1973) challenged the auditory–temporal sequencing position and claimed that there was little evidence to support it. In her paper she suggests an alternative explanation in terms of a deficit in cognitive function. Benton (1978) tends to support the auditory perception theory, whereas Cromer (1978) is not convinced of the specificity of the problem in terms of auditory perception. Cromer prefers to describe the problem of the dysphasic child primarily in terms of a language deficit. His research, however, is primarily based upon written language ability of dysphasic children as compared with deaf children.

Menuyk (1978) takes still a different position and suggests the possibility that "this difficulty may be due to the inability of these children to keep in mind more than the basic portions of the sentence (predicate or subject and predicate) to allow for an in-depth analysis of the sentences heard. This possibility is indicated by their more sophisticated ability to generate utterances spontaneously than to recall utterances immediately." She goes on to state that the possibility exists that dysphasic children's characteristic is a difficulty in categorizing linguistic relations within the normal processing time.

In summary, some of the problems surrounding the diagnosis of childhood aphasia are due to:

1. The difficulty in excluding other causal conditions that may have contributed to the delay in language

2. The overriding importance of the auditory channel to initial language learning, with the result that the early appearance of verbal symbolic behavior is almost invariably dependent upon hearing

3. The lack of knowledge about, and uncertain relationship between, anatomical or structural defects and disordered cognitive-language function

4. The changes in symptoms that may occur as a result of maturation and development

5. The lack of before-and-after contrast, as in acquired aphasia

6. The two-way interaction between language and other aspects of development, especially in the cognitive and social–emotional areas

7. The comparative lack of instruments able to measure reliably processes and functions contributing to language learning in young children.

For many clinicians, the key feature in the diagnosis is not the identification of an unequivocal pattern of symptoms, but the recognition that the delay in language is out of proportion to deviations in auditory sensitivity, mental development, or personality development, and cannot be explained by these. The realization that something else must be involved lends an immediate attraction to any plausible explanatory principle involving hypothetical central mechanisms. Yet there is a danger that the term aphasia may introduce a ghost into the machine: That a vague, ill-defined label will be applied to any unexplained language-learning difficulty, without a proper appreciation of the underlying mechanisms, or the variety of processsing, organizing, and formulating defects that may be present. Despite the arguments that have been advanced for the use of aphasia as a differentiating term for symbolic disorders in children, there may be still greater advantages in requiring the clinician to specify more precisely what kinds of behavioral dysfunction are involved.

We have seen a rather general agreement that there are at least three important aspects of language behavior and learning: the system's reception and decoding of the incoming signal, the integration and elaboration of the response, and the output transmission. Rather than categorize defects in this complex series by a single static and potentially overworked term, it would contribute more to the understanding of the dynamics of the system if it were possible to measure more specifically the integrity and functioning of its main components. Here, the approach outlined by McCarthy and Kirk (1961) in the ITPA appears to be potentially fruitful, both for the operational specification of factors contributing to language dysfunction and for therapeutic and educational treatment. Finally, the increasing use of diagnostic therapy and diagnostic teaching (Barry, 1961; Kleffner, 1962; McGinnis, 1963; Monsee, 1961) seems to reflect a growing appreciation of the difficulty of reaching an adequate diagnosis during a brief clinical examination, and an understanding of the need for longer term study of the variety of perceptual, learning, and adjustment problems associated with language disorders.

Zangwill (1978) ends his review of the concept of development aphasia with the following positive conclusion: "In the light of these considerations, it appears likely that issues of cerebral dominance will come to throw light on the neurology of developmental aphasia as they have in the past on that of acquired aphasia. This would appear an obvious target for future research."

Child and Adolescent Psychosis

Kanner (1943), in his classic description of autism (which crystallized the dramatic discovery of psychosis in children and adolescents in many centers in the 1930s), branded "inability to relate" and "anxiously obsessive desire for the maintenance of sameness" as pathognomonic features of autism.[11] The unusual features are emphasized in his clinical description: "the children did not use (language) for the purpose of communication," "parrot-like repetitions of heard word combinations . . . immediate (and) . . . *delayed echolalia,*" "*personal pronouns are repeated just as heard.*"

A recent review and investigation by Baker, Cantwell, Rutter, and Bartak (1976) elaborates a spectrum of language abnormalities (in syntax, semantics, and phonology as well as communicativeness) found in autistic children. This included: echolalia, lack of questions, lack of proper use of personal pronouns, shorter sentences despite occasional employment of rotely performed automatic phrases, atypical/mechanical prosody, and metaphorical (personalized) language use. The authors found, in comparing a group of autistic children with a control group of aphasic children matched for language developmental stage and I.Q.,[12] that the suggestion of Rutter (one of the authors) was supported: there is a set of

[11]The same features are emphasized in the current nomenclature; see above and A.P.A. (1978). However, the main etiological factors found recently include pre- peri- and postnatal physical trauma or infection to the central nervous system and the role of professional class "refrigerator-parents" is not thought to be causative. See Frank, Kraus, and Oberfield (1979a) and Ritvo (1976).

[12]They stressed that some "autistic" language findings, viz., shorter sentence length, are a function of the I.Q. of the patient, i.e., a cooccurring or secondary mental retardation—see previous discussion.

"autistic language characteristics" of which different children may exhibit different subsets. Thus there was no specific autistic language pattern, and the differences between the two groups was a quantitative one, with different subsets of features tending to be found in each group.

Shapiro's work in this area (Shapiro & Fish, 1969; Shapiro & Heubner, 1976; Shapiro & Lucy, 1977) focuses on (1) echoing and (2) communicativeness—more recently on the speech act aspect of the latter. To him, diagnostic categorizations are less important than the age/stage at which deviant development began, i.e., when normal development was blocked (schizophrenia simply represents a later block of the developmental path—qualitatively no different from that in autism). Within these two categories he formulates substages of development/pathology. Assessing patients in both frameworks gives a language level that can be well correlated with future development.

Shapiro's recent chronometric study of echoing (Shapiro & Lucy, 1977) confirmed his hypothesis that ending of the communicative sequence was the intention in exact echoing, which thus appears as an anticommunicative speech act.

Frank's most recent work, in collaboration with Kraus and Oberfield, (Frank *et al.*, 1979a) reveals in yet a further dimension the defect in the communication aspects of language in autistic children by exposing (in family settings) the lack of back-and-forth dialogue and its correlates. This work reviews prior work concerning parental linguistic influence on autistic children's language development, and concludes—as did Frank *et al.* (1976) and Cantwell, Baker, and Rutter (1977)—that the language environment of autistic children is if anything richer than that of children with other developmental abnormalities. A further finding confirms others' recent findings, including reports by Rutter (1976) and by Campbell (1978), that siblings of autistic children have a greater incidence of language disorder. A speculation stemming from these and other data is that autism represents a final common path for early deviation in object relations with a closely associated deviation in language; both genetic predisposition and pre- peri-, and/or postnatal physical trauma and/or infection seem to combine in many or most cases.

Schizophrenia language studies have been done largely in adults, or on mixed groups of psychotic children containing autisitic children and those now called atypical childhood psychosis. Semantics has been the area where most abnormalities have been found (see Maher, 1972; Tucker, 1975). Clusters of semantic trends toward certain cognitive–emotional concepts, e.g., the self, power, war–peace, have been identified. They have not yet been looked for in younger schizophrenic patients. Difficulties in gauging and/or using the appropriate level

Table III. Communication Findings in Autism and Schizophrenia

	Speech pathology	Phonetic or intonation abnormality	Syntactic abnormality	Semantic abnormality	Communicativeness
Autistic child	Dysfluencies common (occasionally are mute)	Exact echoing (e.g., adult or TV pattern of intonation) (immediate and delayed echoing)	Imperfect control with special difficulties with pronouns, negative, compound sentences, 2nd person, short (spontaneous) sentences	Poor vocabulary But may rotely echo nonunderstood words or phrases Extreme concreteness, difficulty with function words connoting emotion and many other complexities	Poor to absent, avoids (similar approach to human and nonhuman objects) Vocatives, even for mother, often omitted
Schizophrenic child (Active Phase)	Dysfluencies not typical	Phonemes normal May be flat intonation or bizarre pattern	Normal (occasionally rigidly so) except in rare 'word salad' seen acutely.	Good vocabulary Often with distorted and/or idiosyncratic elements. Imagery can be inappropriate Level of abstraction often misjudged Confusion of word and thing	Often poor or distorted (e.g., regarding one person as another or as an aspect of self, or vice versa) Less redundant? (Evidence in adults)
Nonpsychotic normal child	(Fluent speech by 6–7)	(most adult-like phonemes by 5–6)	(See Table I.) Basic grammar by 5. Complex transformations by 10–12.	(Adult-like semantics and vocabulary by 6–7. True abstraction by 13–15)	(Some is present as neonate; importance of object related, vs. egocentric language by third year and increases through adolescence)
Mentally retarded child (nonpsychotic)	Dysfluencies very common	Individual phonemes (consonants especially) slowly or never. Intonation pattern behind normals	Syntactic progress evenly behind that of normal I.Q. children. Shorter, simpler sentences for age.	Slowness in increase of vocabulary often limited Special difficulty with function words	Similar progression, with chronological lag

of abstractness is a further finding, but this is difficult to separate from developmentally expected stages in children and adolescents. Adult schizophrenic language has been found more difficult to understand by others (viz., Salzinger, 1972), but this may be due to a combination of deviances in semantics and in desire to communicate with others. Seekers of deviant syntax (e.g., Chaika, 1974) have been disappointed, except when confronted with the acute schizophrenic patient in a decompensated "word-salad" state and/or in hallucinosis (see review in Frank, Rendon, & Siamopoulous, 1979b). Indeed, there is evidence in work of some investigators that the syntactic systems may be overdeveloped and/or rigidified in schizophrenia (Carpenter, 1976; Rochester, 1977).

The principle language findings in childhood psychosis have been summarized in Table III (viz., DeHirsch, 1967, for further comparison to aphasia).

There is no literature yet on language disorders in atypical childhood psychosis or in juvenile manic-depressive psychosis; both conditions have only recently been described. The former has no counterpart in adults and literature on language in manic-depressive psychosis in adulthood is sparse. A study by Andreasen (1976) may be the first of further work in this area. From clinical data and preliminary work, one might expect principal language findings in depression to be paucity of complex syntax, depressed intonation pattern and word rate, with communicativeness low and semantics skewed toward sad, colorless vocabulary. Mania might show the opposite.

References

Allen, D. *The development of predication in child language*. Unpublished doctoral dissertation, Columbia University Teacher's College, 1973.

American Psychiatric Association. *DSM-III, 1-15-78 Draft*. Washington, D.C.: A.P.A., 1978.

Andreasen, N. Linguistic analysis of speech in affective disorders. *Archives of General Psychiatry*, 1976, *33*, 1361–1367.

Baker, L., Cantwell, D., Rutter, M., & Bartak, L. Language and Autism. In S. Ritvo (Ed.), *Autism*. New York: Spectrum, 1976, pp. 121–149.

Baldwin, A., & Baldwin, C. Study of mother–child interaction. *American Scientist*, 1973, *61*, 714–721.

Barry, H. *The young aphasic child: Evaluation and training*. Washington, D.C.: Alexander Graham Bell Association, 1961.

Bateson, M. Mother–infant exchanges: The epigeneis of conversational interaction. In D. Aaronson & R. Rieber (Eds.), *Developmental psycholinguistics and communication disorders* (Vol. 263). Annals of the New York Academy of Sciences, 1975.

Benton, A. Cognitive functioning. In M. Wykie (Ed.), *Developmental dysphasia*. London: Academic Press, 1978.

Berry, M. F., & Eisensen, J. *Speech disorders*. New York: Appleton-Century-Crofts, 1956.

Blank, M., & Frank, S. Story recall in kindergarten children: Effect of presentation method. *Child Development*, 1971, *42*, 1.

Bloom, L. *Language development, form and function in emerging grammars*. Cambridge, Mass.: M.I.T. Press, 1976.

Bloom, L. *One word at a time: The use of single word utterances before syntax*. The Hague: Mouton, 1975.

Bloom, L., & Lahey, M. *Language development and language disorders*. New York: John Wiley, 1978.

Braine, M. The ontogeny of English phrase structures: The first phase. *Language*, 1963, *39*, 1–14.

Brazleton, T. *et al*. *Early mother–infant reciprocity in parent–infant interaction*. Ciba Foundation Symposium No. 3, 1975.

Brown, R., & Bellugi, U. Three processes in the child's acquisition of syntax. *Harvard Educational Review*, 1964, *34*, 133–151.

Cantwell, D., Baker, K., & Rutter, M. Families of autistic and dysphasic children. II. Mothers' speech to the children. *Journal of Autism and Childhood Schizophrenia*, 1977, *7*, 313–327.

Carpenter, M. D. Sensitivity to syntactic structure: good vs. poor premorbid schizophrenics. *Journal of Abnormal Psychology*, 1976, *85*, 41–50.

Carrow, M. A. The development of auditory comprehension of language structure in children. *Journal of Speech and Hearing Disorders*, 1968, *33*, 99–111.

Chaika, E. A linguist looks at "schizophrenic" language. *Brain and Language*, 1974, *1*, 257–276.

Chomsky, C. *The child's acquisition of syntax between five and ten*. Cambridge, Mass.: M.I.T. Press, 1969.

Chomsky, N. *Syntactic structures*. The Hague: Mouton, 1957.

Chomsky, N. *Aspects of the theory of syntax*. Cambridge, Mass.: M.I.T. Press, 1965.

Clarke-Stewart, K. A. Interactions between mothers and their young children: Characteristics and consequences. *Society for Research in Child Development Monograph 153*. Chicago: Society for Research in Child Development, 1973.

Condon, W., & Sander, L. Synchrony demonstrated between movements of the neonate and adult speech. *Child Development*, 1974, *45*, 456–462.

Corwin, R. The basis of childhood dysphasia: A linguistic approach. In M. Wykiel (Ed.), *Developmental Dysphasia*. London: Academic Press, 1978.

Corwin, M., & Golub, H. Reported on in *N.Y. Times*, June 13, 1979, p. A21.

Davis, H. Occam's razor and congenital aphasia. *Psychosomatic Medicine*, 1962, *24*, 1.

Davis, H., & Goldstein, R. Special auditory tests. In H. Davis & S. R. Silverman (Eds.), *Hearing and deafness*. New York: Holt, Rinehart, 1960.

DeHirsch, K. Differential diagnosis between aphasic and schizophrenic language in children. *Journal of Speech and Hearing Disorders*, 1967, *32*, 3–10.

DiCarlo, L. M. Differential diagnosis of congenital aphasia. *Volta Review*, 1960, *62*, 361.

Doehring, D. Visual spatial memory in aphasic children. *Journal of Speech and Hearing Research*, 1960, *3*, 138.

Dulit, E. Adolescent Thinking à la Piaget: The Formal Stage. *Journal of Youth and Adolescence*, 1972, *1*.

Efran, R. Temporal perception, aphasia and déjà-vu. *Brain*, 1963, *86*, 403–424.

Eimas, P. *et al*. Speech perception in infants. *Science*, 1971, *171*, 303–306.

Erikson, E. C. *Childhood and society*. New York: Norton, 1950.

Flavell, J. *The developmental psychology of Jean Piaget*. Princeton: Van Nostrand, 1963.

Foster, C., Giddan, J. J., & Stark, J. *Assessment of children's language comprehension*. Palo Alto, Calif.: Consulting Psychologists Press, 1969.

Frank, S. *Language acquisition*. Paper presented to staff conference, New York University Medical Center, March 7, 1973.

Frank, S. Review of language effects of maternal deprivation. Unpublished manuscript, 1980.

Frank, S., & Osser, H. A psycholinguistic model of syntactic complexity. *Language and Speech*, 1970, *13*, 38–53.

Frank, S., & Seegmiller, M. *Children's language environment in free play situation*. Paper presented to Society for Research in Child Development, Philadelphia, 1973.

Frank, S., Allen, D., Stein, L., & Myers, B. Linguistic performance in vulnerable and autistic children and their mothers. *American Journal of Psychiatry*, 1976, *133*, 909–915.

Frank, S., Kraus, M., & Oberfield, R. Family communication patterns in autism. In M. Simpson (Ed.), *Clinical psycholinguistics*, New York: Elsevier–North Holland, 1979. (a)

Frank, S., Rendon, M., & Siamopoulous, G. Language in hallucinations of adolescent schizophrenics. In R. Rieber (Ed.), *Childhood language disorders and mental health*. New York: Plenum, 1979. (b)

Freud, S. Three essays on the theory of sexuality. In J. Strachey (Ed.), *Standard Edition, Complete psychological works*. London: Hogarth, 1953. (Originally published, 1905.)

Freud, S. Outline of psychoanalysis, In J. Strachey (Ed.), *Standard Edition, Complete psychological works*. London: Hogarth, 1953. (Originally published, 1939.)

Furth, H. Sequence learning in aphasic and deaf children. *Journal of Speech and Hearing Research*, 1964, *9*, 441–449.

Gardner, R., & Gardner, B. Teaching sign language to a chimpanzee, *Science*, 1969, *165*, 664–672.

Gazzaniga, M. *The bisected brain*. New York: Appleton-Century-Crofts, 1970.

Goldfarb, W. *Childhood schizophrenia*. Cambridge: Harvard University Press, 1961.

Goldstein, R., Landau, W., & Kleffner, F. Neurologic observations in populations of deaf and aphasic children. *Annals of Otology, Rhinology, & Laryngology*, 1960, *67*, 468.

Ingram, D. If and when transformations are acquired by children. Paper presented to Georgetown (Linguistics) Round Table, 1975.

Inhelder, B., & Piaget, J. *The growth of logical thinking from childhood to adolescence*. New York: Basic Books, 1958.

Irwin, O. Infant speech: Consonantal sounds according to place of articulation. *Journal of Speech and Hearing Disorders*, 1947, *12*, 397–401.

Irwin, O. Infant speech: Development of vowel sounds. *Journal of Speech and Hearing Disorders*, 1948, *13*, 31–34.

Jakobson, R., & Halle, M. *Fundamentals of language*. The Hague: Mouton, 1956.

Kanner, L. Autistic disturbance of affective contact. *Nervous Child*, 1943, *2*, 242–250.

Karlin, I. Aphasias in children. *American Journal of Dis. Child.*, 1954, *87*, 752.

Kendall, D. C. Language and communication problems in children. In R. Rieber & R. Brabaker (Eds.), *Speech pathology*, Amsterdam: North Holland, 1966.

Kimura, D. The neural basis of language qua gesture. In H. Whitaker & H. Whitaker (eds.), *Studies in neurolinguistics* (Vol. 2). New York: Academic Press, 1976, pp. 145–156.

Kleffner, R. Aphasic and other language deficiencies in children. In W. Daley (Ed.), *Speech and language therapy with the brain-damaged child*. Washington, D.C.: Catholic University of America, 1962.

Krashen, S. The critical period of language acquisition and its possible bases. In D. Aaronson & R. Rieber, *Developmental psycholinguistics and communication disorders* (Vol. 263). New York: Annals of the New York Academy of Sciences, 1975.

Lashley, K. The problem of serial order in behavior. In L. Jeffress (Ed.), *Cerebral mechanisms in behavior*. New York: Wiley, 1951.

Lee, L. Developmental sentence types: A method for comparing normal and deviant syntactic development. *Journal of Speech and Hearing Disorders*, 1966, *31*, 311–320.

Lee, L. *Northwestern Syntax Screening Test*. Evanston, Ill.: Northwestern University, 1969.

Lenneberg, E. *Biological foundations of language*. New York: Wiley, 1967.

Lowe, A., & Campbell, R. Temporal discrimination in aphasoid and normal children. *Journal of Speech and Hearing Research*, 1965, *8*, 313–315.

Maher, B. The language of schizophrenia: A review and interpretation. *British Journal of Psychology*, 1972, *120*, 3–17.

Mahler, M. *The psychological birth of the human infant*. New York: Basic Books, 1975.
Manschreck, T., Maher, B., Rucklos, M., & White, M. The predictability of thought disordered speech in schizophrenic patients. *British Journal of Psychology*, 1979, *134*, 595–601.
Marcos, L., & Alpert, M. Bilingualism. In R. Rieber (Ed.), *Childhood language disorders and mental health*. New York: Plenum, 1977.
McCarthy, D. Language development in children. In P. Mussen (Ed.), *Carmichael's manual of child psychology*. New York: Wiley, 1954.
McCarthy, J., & Kirk, S. *The Illinois Test of Psycholinguistic Ability*. Urbana: University of Illinois, 1961.
McGinnis, M. *Aphasic children: Identification and education by the association method*. Washington, D.C.: Alexander Graham Bell Association, 1963.
McReynolds, L. Operant conditioning for investigation speech sound discrimination in aphasic children. *Journal of Speech and Hearing Research*, 1966, *9*, 519–528.
Menyuk, P. Syntactic structures in the language of children. *Child Development*, 1963, *34*, 407–422.
Menyuk, P. Linguistic problems in children with developmental dysphasia. In M. Wykie (Ed.), *Developmental Dysphasia*. London: Academic Press, 1978.
Miller, G. *Language and communication*. New York: McGraw-Hill, 1951.
Miller, R. Development from 1 to 2 years: Language acquisition. In J. Noshpitz (Ed.), *Basic handbook of child psychiatry* (Vol. 1). New York: Basic Books, 1979, pp. 127–144.
Moerk, E. L. Verbal interactions between children and their mothers during the preschool years. *Developmental Psychology*, 1975, *11*, 788–794.
Monsee, E. Aphasia in children. *Journal of Speech and Hearing Disorders*, 1961, *26*, 83.
Morehead, D., & Ingram, D. The development of base syntax in normal and linguistically deviant children. *Journal of Speech and Hearing Research*, 1973, *16*, 330–352.
Myklebust, H. *Auditory disorders in children*. New York: Grune and Stratton, 1954.
Myklebust, H. Aphasia in children—Language development and language pathology. In L. Travis (Ed.), *Handbook of speech pathology*. New York: Appleton-Century-Crofts, 1957. (a)
Myklebust, H. Aphasia in children—diagnosis and training. In L. Travis (Ed.), *Handbook of speech pathology*. New York: Appleton-Century-Crofts, 1957. (b)
Nelson, K. Structure and strategy in learning to talk. *Society for Research in Child Development*, Monograph 161, 1973, *38*, 1–2.
Neuhaus, E. Psychological aspects. In *The concept of congenital aphasia from the standpoint of dynamic differential diagnosis*. Washington, D.C.: American Speech and Hearing Association, 1959.
O'Connor, N., & Hermelin, B. *Speech and thought in severe subnormality*. New York: Macmillan, 1963.
Rapin, I., & Wilson, B. Children with developmental language disability: Neurological aspects and assessments. In M. Wykie (Ed.), *Developmental dysphasia*. London: Academic Press, 1978.
Rees, N. Auditory processing factors in language disorders: The view from Procrustes' Bed, *Journal of Speech and Hearing Disorders*, 1973, *38*, 304–315.
Rees, N., & Shulman, M. I don't understand what you mean by comprehension. *Journal of Speech and Hearing Disorders*, 1978, *43*, 208–219.
Rochester, S. R., *et al*. Sentence processing in schizophrenic listeners. *Journal of Abnormal Psychology*, 1973, *3*, 350–356.
Rosenthal, W. Auditory and linguistic interactions in developmental aphasia: Evidence from two studies of auditory processing. *Stanford University Papers and Reports on Child Language Development*, 1972, *4*, 19–35.
Shapiro, T. Language in adolescence. In R. Rieber (Ed.), *Childhood language disorders and mental health*. New York: Plenum, 1979.
Shapiro, T., & Fish, B. A method to study language deviations as an aspect of ego organization in young schizophrenic children. *Journal of Child Psychiatry*, 1969, *8*, 36–56.

Shapiro, T., & Huebner, H. Speech patterns of five psychotic children now in adolescence. *Journal of Child Psychiatry*, 1976, *15*, 278–293.

Shapiro, T., & Lucy, P. Echoing in autistic children: A chronometric study of semantic processing. *Journal of Child Psychology & Child Psychiatry*, 1977, *19*, 373–378.

Shipley, E., Smith, C., & Gleitman, L. A study in the acquisition of language: Free responses to commands. *Language*, 1969, *45*, 322–342.

Smith, F., & Miller, G. (Eds.). *The genesis of language*. Cambridge, Mass.: M.I.T. Press, 1966.

Snow, C. Mothers' speech to children learning language. *Child Development*, 1972, *43*, 549–565.

Spitz, R. A. The smiling response: A contribution to the ontogenesis of social relations (with assistance of K. M. Wolf, Ph.D.). *Genetic Psychology Monographs*, 1946, pp. 57–125.

Stark, J. Performance of aphasic children on the ITPA. *Exceptional Child*, 1966, *33*, 153–161.

Stark, J. A comparison of the performance of aphasic children on three sequencing tests. *Journal of Communication Disorders*, 1967, *1*, 31–34.

Stark, J., Poppen, R., & May, M. Effects of alterations of prosodic features on the sequencing performance of aphasic children. *Journal of Speech and Hearing Research*, 1967, *10*, 849–855.

Stark, J., Foster, C., Giddan, J., Gottsleben, R., & Wright, T. Teaching the aphasic child. *Exceptional Child*, 1968, *35*, 149–154.

Stern, D. A microanalysis of mother-infant interaction: Behavior regulating social contact between a mother and her 3½ month old twins. In E. Rexford, L. Sanders, & T. Shapiro (Eds.), *Infant psychiatry*. New Haven: Yale University Press, 1976.

Strauss, A. A., & Kephart, N. C. *Psychopathology and education of the brain injured child (Vol. 2)*. New York: Grune and Stratton, 1955.

Strauss, A. A., & Lentinen, L. E. *Psychopathology and education of the brain injured child (Vol. 1)*. New York: Grune and Stratton, 1947.

Tallal, P. Rapid auditory processing in normal and disordered language development. *Journal of Speech and Hearing Research*, 1976, *19*, 561–571.

Tallal, P., & Piercy, M. Developmental aphasia: Impaired rate of nonverbal processing as a function of sensory modality. *Neuropsychologia*, 1973, *11*, 389–398.

Tallal, P., & Piercy, M. Developmental aphasia: Rate of auditory processing and selective impairment of consonant perception. *Neuropsychologia*, 1974, *12*, 1–11.

Tallal, P., & Piercy, M. Developmental aphasia: The perception of brief vowels and extended stop consonants. *Neuropsychologia*, 1975, *13*, 69–74.

Tucker, G. Senorimotor functions and cognitive disturbance in psychiatric patients. *American Journal of Psychiatry*, 1975, *132*, 17–21.

Tyack, D., & Gottsleben, R. *Language sampling, analysis and training: A handbook for teachers and clinicians*. Palo Alto, Calif.: Consulting Psychologists Press, 1974.

Weir, R. *Language in the Crib*, The Hague: Mouton, 1962.

Wepman, J. M., Jones, L. V., Bock, R. D., & VanPelt, D. Studies in aphasia. *Journal of Speech and Hearing Disorders*, 1960, *25*, 323.

Werner, H., & Kaplan, B. *Symbol formation*. New York: Wiley, 1963.

Whitaker, H., & Whitaker, H. *Studies in neurolinguistics* (2 Vols.). New York: Academic Press, 1976.

Wolff, P. H. The natural history of crying and other vocalizations in early infancy. In B. Foss (Ed.), *Determinants of infant behavior* (Vol. 4). London: Methuen, 1969, pp. 81–109.

Wood, N. E. Language development and language disorders. *Society for Research in Child Development*, Monograph, 1960, *77*, 25, 3.

Wood, N. E. *Delayed speech and language development*. Englewood Cliffs, N.J.: Prentice-Hall, 1964.

Zangwill, O. The concept of developmental dysphasia. In M. Wykie (Ed.), *Developmental dysphasia*. London: Academic Press, 1978.

PART III

PSYCHOLOGICAL–INTERPERSONAL DISTURBANCES

7

A. C. Nichols

ARTICULATION DISORDERS

PSYCHOLOGICAL FACTORS

While organic factors underlying articulation problems have been confirmed as etiologically significant (see Nichols' earlier Chapter 1), perception, learning and social–emotional factors have long been known to affect articulatory disintegration and to be important in therapy. Important theoretical studies have been designed to study the *developmental learning* aspects of phonological competence. Many studies have dealt with the *discrimination of speech sounds* among speakers with articulation problems, and some have dealt with the contributions of *social and emotional factors* to such disorders. As required in science, the theoretical canon dealing with articulation disorders has been demonstrated to have descriptive (diagnostic) power and predictive (therapeutic) power. The present chapter will treat these psychological topics with comments on their powers of description and prediction.

Developmental Learning

Learning Theories

Articulation problems are due to faulty learning, postulated by Milisen (1954). The normal language-learning pattern has been discussed in an earlier chapter. The reader will note that such a definition excludes certain articulatory

A. C. Nichols • Department of Speech Pathology, San Diego State University, San Diego, California 92182.

deviations from the standard phonatory pattern from the category of articulation defects, notably those due to dialectal or foreign language background. A discussion of dialectal articulation properly belongs in a chapter on normal language development and will thus not be treated in the present chapter.

Several learning theories have been used by speech pathologists to explain how articulation is acquired and how defects may occur. The most common base for articulation therapy is some form of stimulus–response theory.

Two-Factor Theory. One complex stimulus–response model that has been proposed is Mowrer's (1953, 1958) two-factor theory. Its relevance to articulation problems and their remediation may be explained (admittedly simplistically) as follows:

> Learning is motivated by admiration or affection. The child wants to be like the "nice" or "strong" person who cares for him (or her). One aspect of that person that may be retained by the child when absent is his (or her) speech pattern. Thereby the child's own speech can evoke the presence of the beloved person. The more the resemblance between the child's speech and that of the "nice person," the more effective the imitation and the more pleasant to the child's ear; the child's speech is self-reinforcing in proportion to his success in imitation. The imitated speech is the stimulus; the satisfaction of successful imitation is the reinforcement.

This model goes far toward explaining why speakers may articulate like their favorite people. It also implies that a child with an articulation defect may be quite satisfied that he has come "close enough" to his speech model to reinvoke that person's presence, and may indeed be unaware that he is making an error.

Mowrer's theory (1953, 1958) has the implication that reception, not production, is the fundamental process in articulation learning or rehabilitation. The child learns production later, when the "nice person" is absent. There is no need to evoke his or her presence by spoken imitation when he or she is present. Therapy based on this model (the child was stimulated by speech, but response was not required) proved equally as effective with mentally retarded children as therapy in which production was required of the child, according to Rigrodsky and Steer (1961). Although this is a weak demonstration of therapeutic power, it may be noted that this is one of the rare experiments dealing with comparisons of therapeutic theories in the literature of speech pathology. Many therapeutic theories are proposed, but few are tested against other theories; a major requirement of the philosophy of science is thus neglected.

It will be noted that imitation—as the child imitates his model—is also crucial to Mowrer's two-factor theory. Several studies have shown that children may improve their articulation (in comparison to their spontaneous productions) by imitating the speech patterns of an examiner (e.g., Carter & Buck, 1958; Moore, Burke, & Adams, 1976; Snow & Milisen, 1954). Several investigations have also emphasized the predictive power of measures of improved articulation in imitative procedures, which have come to be called *stimulability* measures. Children who display high stimulability tend, over time, to improve their articulation without therapy (Carter & Buck, 1958; Snow & Milisen, 1954). Elbert and McReynolds (1978) demonstrated the therapeutic power of such abilities, noting that generalization of correct articulation followed close upon the child's learning to imitate correct articulation. They concluded, "Articulatory concept should be operationally defined as the point at which a subject can imitate a sound correctly and generalizes this production to new exemplars" (p. 149).

Programmed Instruction. As has been noted, stimulus–response theory has dominated articulation therapeutics. A major stream of stimulus–response theory, operant-conditioning (or, variously, programmed instruction, behavior modification, etc.) procedure, has had a powerful impact on speech pathology during the past decade. Costello (1977) provided an explanation of such procedures in a review of the literature. She identified the following as fundamental to this approach:

> (1) *successive approximations*—moving from the child's closest approximation to the correct sound by "small, carefully graduated steps" to successful production. (2) *active participation*—the learner is required "to make frequent, continuous overt responses to the stimulus material" allowing (indeed, requiring) the observer to confirm learning when it has occurred. (3) *immediate knowledge of results*—each of the client's correct responses is immediately confirmed. (4) *mastery learning and self-pacing*—the client must demonstrate mastery of one program step before proceeding to another and is allowed to determine his own pace for accomplishing the steps. (5) *fading stimulus support*—various "models, prompts and cues" that guide early learning steps are gradually eliminated until only natural stimuli remain. (6) *concept learning through varied repetition*—the desired response is "evoked by a variety of similar stimuli," to facilitate acquisition (and presumably generalization and retention).

Examples from the literature that conformed to these principles, and some that did not, make up much of Costello's (1977) presentation.

In an accompanying article, Gerber (1977) criticized some of the commercially available programs that have appeared on the market. Many were critically variant from programmed instruction principles. For example, provision for individual adjustments (*successive approximation*) is rarely discussed in such presentations, leading to inefficiencies in application.

It may be added that Gerber's criticism was limited to programmed instruction criteria; other critical orientations were neglected, e.g., a program that recommended clenched teeth as a basic step in eliminating a lingual protrusion lisp, which from the standpoint of psychological phonetics is incompetent. Gerber mentioned this step as an example of entrance behavior applying the principle of successive approximation. The program in question does not do this, nor is it written so as to achieve successive progressions. Successive approximations must be guided by the perceptions of a skilled articulation therapist; no program can specify such perceptions.

One of Gerber's major criticisms of the commercially available programs dealt with a lack of intrapersonal evaluation; the client is not asked to monitor or evaluate his own production. This is due to the operant-conditioning ancestry of programmed instruction. Cognitive processing is not considered in operant theory, although it appears to be a powerful factor in articulation therapy. Gerber (1977) noted that therapy involving interpersonal—therapist to client—feedback about articulatory success was much less successful (0 to 16 met criterion learning levels) than intrapersonal—client self-evaluation—feedback (8 of 16 met criterion) in an experiment she carried out a decade before (Gerber, 1966). She also criticized the variance in terminal steps in the programs. We may note that the neglect of cognitive processing makes carry-over a difficult though not impossible step for programmed instruction; people think for themselves when they learn and when they talk, and the program must take this into account.

In spite of such weaknesses, clients do process cognitively even in the absence of programmed provisions, and therapists with skills do apply the programs. Therefore, the therapeutic power of prediction of programmed instruction for articulation has been confirmed by many studies (e.g., Clark, 1974; Costello & Onstine, 1976; Evans & Potter, 1974; Gray, 1974). Certain elaborations of the technique have also been put to therapeutic test. Costello and Ferrer (1976), for example, have demonstrated that positive reinforcements for correct production of an error sound and "punishment" for incorrect productions were more effective than positive reinforcement alone. Van Hattum, Page, Baskervill, Dugay, Conway, and Davis (1974) provided a taped therapy program that they demonstrated to be effective with mild to moderate articulation problems in the schools. It is in such tests of the therapeutic powers of prediction of articulation modification procedures that the most substantial progress in the field of articulation therapy is being made.

Distinctive Features

Chomsky's (Chomsky & Halle, 1968) propositions have had a profound impact upon language-learning theory. From his point of view, the speaker has developed his phonological competence by actively formulating and testing hypotheses about the forms and sequences that are permitted in the language he is speaking. In articulation theory this might be restated: the child makes up his sounds to be like those he perceives as correct. If so, how do errors occur? One explanation may be based on distinctive feature theory. A distinctive feature analysis of phonology provided by Chomsky and Halle (1968) has been widely applied to articulation research. This research notes that speech sounds are complex. The child's hypothesis about how a sound should be made may be partly correct, attending to one feature, but neglecting another. For example, he may articulate the unvoiced and coronal (alveolar) features of a /s/, but fail to consider or produce it as a continuant; instead he may regard it and speak it as a stop, /t/. Thus, an error has been learned. Singh's (1976) review of the literature in the application of distinctive feature theory to articulation deviation provides substantial support for the validity of this concept.

In recent years, the descriptive (diagnostic) power of the distinctive features to delineate the errors of articulation made by children has been demonstrated by such investigators as Oller (1973), McReynolds, Engmann, and Dimmitt (1974), and Lewis (1974). The point has often been made in such descriptions that a greater economy of notation is possible (see Compton, 1970; Oller, 1973); that is, systems of, let us say 7 to 13, distinctive features can efficiently explain the errors of the some 26 consonant sounds of English.

The power of remedial hypotheses generated by distinctive feature analysis has been demonstrated by such investigators as Compton (1970, 1975), McLean (1970), McReynolds and Bennett (1972), Pollack and Rees (1972), and Costello (1975). A common and important result was that learning a feature for one phoneme may generalize to other phonemes. Transfer across morphological contexts has also been demonstrated (Holdgrafer, Kohn, & Williams, 1976). Such results are powerful evidence that distinctive feature theory has high validity.

The elaboration of distinctive feature theory developed by Compton (1970, 1975) shows promise of more refined therapeutic hypotheses. He proposed that phonology is learned as a system or network. For example, the systematic relations /θ/—/s/—t/ may be found to describe one portion of a child's phonological network. Compton (1972) has proposed the specified relationship as part of a general universal model of sound change for 23 consonants of the English consonants. (The /h/ and /hw/ were not included.) Underlying such a proposition is the hypothesis that articulation disorders in which substitutions that are made across intervening phonemes in the network, e.g., /t/ for /θ/ in the sequence above, should be approached in therapy by dealing with the whole pattern beginning with the relations of the substituted and target sounds to the intervening /s/. As noted above, the therapeutic predictive power of this theory has been demonstrated in articulation therapy (Compton, 1970, 1975).

Developmental Sequence

The sequence of development has for many years been considered in terms of a maturation pattern. It has been acknowledged that growth is fundamental to articulation maturity and that children grow at different rates. The term "maturation," however, has a technical sense in learning theory that has been employed, but not acknowledged, in the speech pathology literature. That is, by definition, maturation is an *unlearned* change in behavior due to growth and integrative factors, which is not due to drugs or temporary physiological states such as fatigue (Hilgard, 1956). Corollary to this definition is the argument that such behavior changes are not learnable; that teaching behaviors (e.g., articulation therapy) would be futile until growth has occurred. This position has been proposed as support for delay for therapy until the child is over eight years of age, when classical studies of speech sound development show that no further articulation gains are made by children without therapeutic intervention (e.g., Poole, 1934; Roe & Milisen, 1942; Templin, 1957).

One weakness in the speech pathology literature dealing with maturation is that no study has attempted to separate the relative contributions of growth and learning. The assumptions that speech sounds are unlearnable at any age or that growth will produce articulatory development are thus untenable. Indeed, much evidence and most theory would argue that articulation is learned.

Interpreting the maturation literature also involves the arbitrary nature of the criteria by which maturation of the speech sounds have been specified by the investigators who have done the classical studies. This has usually been done by percentile–age measures, e.g., an investigator may find that 75% of normally developing children have achieved the articulation of the /s/ sound by age eight. The age of maturation is then specified as eight. Such estimates run the risk of specifying maturation at an age, when the learning readiness period for speech sounds—the age range when learning may proceed with the least interference from prior learning—may have passed for most children. This reverses the sequence proposed by educational learning theorists. In their perspective, maturation precedes behavior acquisition.

Recent studies of development by Sander (1972) and Prather, Hedrick, and Kern (1975) have provided data estimating the age at which the 50th percentile for correct production has been reached. If the assumption were made that maturation had probably occurred by the 50% percentile–age points for the "average" normally developing child, then sound acquisition periods of the 50th to 75th percentile range would provide an estimate of the "articulation readiness period" for such children. Failures to develop articulation by the beginning of this period could be tentatively attributed to faulty learning. From this perspective it is of interest to note that groups of randomly selected preschoolers reach the 50th percentile point for all sounds by the age of four. There is other evidence, notably the (statistically detectable) plateau of development from ages 4.5 to 6 shown in Templin's (1957) articulation data that tends to support the hypothesis that maturation has been completed by age four. If so, delay of therapy for four or five more years (until the child is eight or nine) seems risky; habituation of faulty learning over this period of time may result in strong interference with therapeutic intervention (Nichols, 1966). This is not to say that there is no evidence that maturation does not occur earlier than age four. Indeed, Strayer (1930) reported that one of a pair of identical twins, who was kept away from speech stimulation for several weeks starting at one year eight months, had not developed articulation as well as his twin at the end of the vocabulary-training experiment in which they were participating. (Vocabulary was equal at the end.) No other studies of normally developing twins has been done, according to Sloan (1967). We may note that co-twin study is the only methodology that controls for maturation and learning effects.

All developmental research on articulation has been undermined by the lack of a test that measures phonological competence directly. It is possible that recent studies of the dichotic listening of children with speech problems may produce such a measure. Dichotic listening involves presenting words to the right and left ears simultaneously. Because the left side of the brain is the center for language, and because the nerves of the right ear lead to the left side of the brain, words presented to the right ear are heard, processed, and reported somewhat more frequently than words presented to the left ear in normal listeners. Slorach and Noehr (1973) and Sommers, Moore, Brady, and Jackson (1976) have shown that children with severe articulation problems have less well-established left cerebral dominance for language than "minimally brain-damaged" children or normal controls. The question that arises is: may left cerebral dominance be equated with language and articulatory maturation? Kimura's (1963, 1967; also see Ingram, 1975; Lenneberg, 1967) work on the development of left cerebral dominance for speech suggested that this effect increased with age. Kinsbourne (1975; also see Hiscock & Kinsbourne, 1977) disputed the validity of such findings, however, hypothesizing that left cerebral dominance is strongest in the very young child and that learning language serves to reduce left cerebral dominance. Kinsbourne explained Kimura's and Ingram's results as artifacts of procedure. Celano (1978) confirmed previous findings to the extent that her results showed that with an appropriate procedure certain three- and four-year-old children respond to minimally contrasted dichotic stimuli (e.g., /Ed—said/) and tend to demonstrate left cerebral dominance for syllable initial consonants. They did not, however, at this age range, demonstrate left cerebral dominance for syllable terminal consonants (e.g., B—beet), although adults do (Unger, 1978), providing potential support for Kimura's developmental hypothesis. Should this line of inquiry lead to a valid measure of maturation and readiness, a major issue in articulation therapy would be resolved.

Studies of articulatory development have also shown that the "blends" (or consonant "clusters") may be better articulated than the same sounds in vowel context (Templin, 1957; Wellman, Case, Mengert, & Bradbury, 1931; also see Curtis & Hardy, 1959; Spriestersbach & Curtis, 1951). This observation has led to a diagnostic (Dorsey, 1959) and therapeutic (McDonald, 1964) theory and procedures based on the premise that beneficial consonant context effects may provide clues to a child's articulatory competencies that may be strengthened and generalized. Articulation therapy based upon this assumption has been reported to be highly effective (McDonald, 1964). Phonological theory has not dealt with the blends and their implications for learning models, though studies of juncture effects may provide the basis for such a synthesis. However, a therapeutic study based on juncture theory did not show juncture effects to be powerful in a predictive sense (Elbert & McReynolds, 1978). In the absence of individualized information dealing with prior learned patterns of articulation (such as that utilized by the Dorsey–McDonald procedures), juncture seems a weak, rather than a strong effect. However, where juncture may be assessed for the individual child by real-time sonographic techniques, Kopp (1979) has shown that very powerful diagnostic and therapeutic procedures may be developed. This appears to be consistent with competence theories; it is the individual child's errors that must be detected and modified. No amount of group data will efficiently substitute for such evidence in the clinical setting.

This section may be summarized by noting that stimulus–response theories have been the primary models for articulation therapy and have generated procedures of substantial descriptive and therapeutic power. Competence theories have also been brought into play and strong theories involving distinctive features have contributed to the speech pathologist's diagnostic and therapeutic techniques. Maturation has been discredited as a model for articulatory development, but a learning readiness period remains a useful hypothesis that may be detectable by dichotic-testing techniques. Juncture effects work diagnostically and therapeutically for individual cases but not groups. It is in this individualizing direction that the most promising and potentially resolving research seems to be tending.

Discrimination

The perceptual skills of the client are clearly involved in articulation acquisition. Among such skills, auditory perception—called *discrimination*—is the most studied by speech pathologists, though many perceptual feedback processes for speech have been studied (e.g., see *oral stereognosis* in Chapter 1 on Articulation—Organic Factors). Speech scientists such as Fairbanks (1954) have demonstrated the descriptive power of a model of the speech mechanism that specifies self-guidance, accomplished by the speaker's monitoring his own sensory systems. For remediation, however, auditory processing remains the primary system. Alternative models such as the motor theory of speech learning have yet to contribute significantly to articulation therapy.

The standard auditory discrimination test involves asking the listener whether two contrasted syllables or words, e.g., one—run or run—run, are the "same" or "different." Standardized and commercially available versions of this kind of measure are widely used in the diagnosis of articulation disorders (e.g., Goldman, Fristoe, & Woodcock, 1970; Templin, 1943; Wepman, 1958).

It has been demonstrated that children with poor discrimination tend to have poor articulation (Sherman & Geith, 1967). Further, Allison (1975) showed that children with self-monitoring discrimination problems for a set of sounds have more articulation problems than children with good self-monitoring. Some studies have supported the inverse hypothesis, i.e., that children with articulation problems have poor discrimination (e.g., Kronvall & Diehl, 1954; Monnin & Huntington, 1974; Travis & Rasmus, 1931), while others have not found children with articulation problems to have discrimination deficits (e.g., Hall, 1938; Hansen, 1944; Reid, 1947; Sommers, 1962). This latter result may be related to the finding reported by McReynolds, Kohn, and Williams (1975) that (some) children with articulation defects discriminated features and phonemes that they did not produce. Another issue was raised by Aungst and Frick (1964), who showed that the severity of children's /r/ defects was significantly related to their monitoring of their own speech, but not to their discrimination of the articulatory productions of others.

The designs of discrimination experiments probably account for much of the variance in their results. The many factors that may be involved in articulation problems allow samplings that may include children with discrimination deficits underlying their articulatory defects, while including those who have good discrimination but poor articulation due to other causes. We may hypothesize that the Sherman and Geith (1967) and Allison (1975) studies avoided this sampling problem by testing for discrimination first to insure an inclusion of all levels of discrimination ability and thus demonstrated that there is a relationship between poor discrimination and poor articulation. The Aungst and Frick (1964) results, however, showed that a child's Templin Test discrimination score did not predict /r/ articulation deficits, whereas his self-monitoring discrimination score for the /r/ did relate strongly to /r/ articulation scores. Such a result is consistent with distinctive feature-competence theory. That is, the child may develop a partial learning deficit for a specific phoneme, which is unrelated to his general discriminating and articulating competence. Some evidence confirming this hypothesis (for the /r/ phoneme) was found by Monnin and Huntington (1974). As has been noted, Gerber (1966) found intrapersonal feedback to have predictive therapeutic power, thus providing more evidence of the importance of self-monitoring.

The therapeutic power of discrimination training is in question. Some studies have shown that such therapy improves articulation (e.g., Mann & Baer, 1971; Winitz & Preisler, 1965). Holland and Matthews (1963), Guess and Baer (1973), and Williams and McReynolds (1975) did not find this effect. Still others have shown that articulation training improves discrimination (Winitz & Bellerose, 1963; Williams & McReynolds, 1975), and some other studies have indicated no changes may occur with such training (Winitz & Bellerose, 1962; Guess & Baer, 1973). Few of these studies had sufficiently large numbers of subjects to justify generalized conclusions. In the absence of a statistically sound study, however, the weight of positive effects must take theoretical precedence. We may conclude that training discrimination may improve articulation and improving articulation may improve discrimination.

Gains in the design of discrimination procedures have been made by the study of the 4IAX method (Pisoni & Lazarus, 1974). The method deals with phonological differentiation on a precompetence basis. The procedure is straightforward, involving four (4I) items spoken as two word-pairs. Three of the items are the same word, the fourth another word that differs in some feature or features, e.g., "one—one; one—run." The listener responds to the question, "Which two words are *more* different?" (*which* "A" is not equal to "X"?). In theory he attends to the phonetic difference, even if the phonemic distinction is not made in his linguistic system, that is, if his system says: /w/ = /r/. Locke (1976) has adapted this procedure to discrimination testing with children with articulation disorders, and found they were able to "hear" distinctions between word pairs that they did not differentiate on a standard discrimination test. Hamrick (1978) confirmed that children's 4IAX scores are higher than (standard) word-pair test scores. Standard word test scores are higher after exposure to the 4IAX test, an indication of possible predictive therapeutic power for the procedure. That is, we may hypothesize that once a child has "tuned in" to the phonetic distinction between word pairs, he can learn to hear the phonemic distinction that makes the words different in the language, without further learning. Again the self-structured competence of the speaker is emphasized in emerging articulation theory.

We may summarize the discrimination literature by saying that self-structured competence leading to improved self-monitoring discrimination seems to be a powerful factor in articulation development and rehabilitation. Coupled with other aspects of distinctive-feature competence theory, self-monitoring promises to generate improved models for articulation therapy in this developing area of speech pathology.

Emotional–Social Factors

Many, perhaps most, self-aware speakers have noted that emotional or social stress may bring about articulatory disturbances in their own speech. Rousey and Moriarty (1965) proposed that since consonants carry the intelligibility load of communication, that errors in consonant production are evidence of disruption of interpersonal interaction and may be viewed as defense mechanisms. The authors presented 17 premises, seven of which dealt with particular articulation problems, e.g., *6*. The /θ/ for /s/ substitution is an indication that the speaker is psychologically fixed at a pregenital level of personality development; *10*. The /f/ for /θ/ is an indication that the speaker has a disturbed relationship with his/her father; *11*. The /d/ for /ð/ is an expression of anger.

These hypotheses were then applied to articulation test results, and compared to independent psychological assessments done by a psychiatrist. With the population of 11- to 14-year-old children studied, the hypotheses were found to have strong descriptive power. Several cautions must be pronounced regarding this study. It was, at best, a preliminary effort. The depth of its validation was not great. No other populations were studied; most importantly, a "normal" control group was not included. Moreover, the authors made no claim that psychiatric diagnoses could be made by articulation test results. It is unfortunate that they did not clearly disclaim such applications of their findings, for there would be no basis to that extension of their work. At best, the psychiatrist may find hints in speech errors to be followed up by more adequate procedures. Reports of predictive therapeutic power were not provided, nor have they appeared in the literature in the decade since the publication of this monograph.

Another source of evidence that phonological–emotional factors might play a part in articulatory disturbances is the literature that deals with the speech sounds that are stuttered. The emotional, anxiety-based factors in stuttering have been well established. Johnson and Brown (1935) reached the conclusion that "each individual stutterer" had his own pattern of blocking on particular sounds, though for the group of stutterers they studied it was possible to rank order the phonemes of English (/z/ is stuttered more than /l/, which is stuttered more than /j/, etc.). The wide variations in individual patterns Johnson and Brown attributed to learning, that is, to anticipatory anxiety reactions. Fairbanks (1937), however, showed that the Johnson and Brown results were correlated significantly with the defect pattern of two-year-old children, with speech sound intensity, with speech sound duration, and perhaps with fundamental pitch. All of these he noted were consistent with the interpretation that the more difficult or effortful the speech sound, the more likely that it may be stuttered. This research has not been adequately followed up. A relationship between articulation and stuttering is, however, clear from these studies.

The studies reviewed in the preceding paragraphs have the implication that social interactions are among the most important factors in the emotional stability of the speaker. The relevance of this hypothesis to articulation disorders may explain the results of Perrin (1954) and Woods and Carrow (1959) that showed that children with articulation problems were rated lower than their peers with normal articulation on various measures of social esteem. In another study, Mowrer, Wahl, and Doolan (1978) have shown that speakers with lisps were rated lower by listeners in the areas of speaking ability, intelligence, education, masculinity, and friendship, than speakers with normally developed /s/ sounds. The impact of such opinions upon self-esteem may underlie the results of Barrett and Hoops (1974), who found that children who correctly developed their /s/ and /r/ phonology between the first and third grades had higher self-concept scores than those who did not. Some studies showed no serious social penalties were inflicted upon children with articulation problems. Freeman and Sonnega (1956) found that persons noticed speech problems, but did not rate them down on "friendship" or "social acceptability traits." Marge (1966), too, found little social handicap among the speech handicapped. Young's (1958) account of her own painful childhood memories of her articulation problem may explain to the reader the personal impact of the foregoing findings if the experimental data will not.

There are, of course, studies that did not find relationships between psychological factors and articulation. Spriestersbach (1956) reviewed studies of the relationship between personality and articulation problems. Five studies reporting relationships were found, and four that found no relationships balanced them. His criticism concluded that well-controlled studies using appropriate tools had yet to be done.

In summary, the literature includes substantial evidence that phonological disturbance may be attributed to emotional factors. Personality patterns have been proposed as the basis for some articulation disorders of a more generalized nature, while the anxieties of stuttering have been implicated in individual patterns of articulatory disruptions identified by other investigators. Finally, social evaluations and self-concept measures seem to show that interpersonal and intrapersonal relationships may be disrupted by articulation problems.

References

Allison, G. The relationship between articulation disorders and self-monitoring auditory discrimination. *Journal of Communication Disorders*, 1975, *8*, 349–356.

Aungst, L., & Frick, J. Auditory discrimination ability and consistency of articulation of /r/. *Journal of Speech and Hearing Disorders*, 1964, *29*, 76–85.

Barrett, C. M., & Hoops, H. R. The relationship between self-concept and the remission of articulatory errors. *Language, Speech and Hearing Service Schools*, 1974, *5*, 67–70.

Carter, E. T., & Buck, M. Prognostic testing for functional disorders among children in the first grade. *Journal of Speech and Hearing Disorders*, 1958, *23*, 124–133.

Celano, K. M. *Consonant discrimination in a dichotic listening task for young children*. Unpublished master's thesis, San Diego State University, 1978.

Chomsky, N., & Halle, M. *The sound pattern of English*. New York: Harper and Row, 1968.

Clark, B. J. Using a short-term lisp correction program for more effective distribution of clinician's time. *Language, Speech and Hearing Service Schools*, 1974, *5*, 152–155.

Compton, A. J. Generative studies of children's phonological disorders. *Journal of Speech and Hearing Disorders*, 1970, *35*, 315–339.

Compton, A. J. Personal correspondence, 1972.

Compton, A. J. Generative studies of children's phonological disorders: A strategy of therapy. In S. Singh (Ed.), *Measurement procedures in speech, hearing and language*. Baltimore: University Park Press, 1975, pp. 55–90.

Costello, J. Articulation instruction based on distinctive features theory. *Language, Speech and Hearing Service Schools*, 1975, *6*, 61–71.

Costello, J. M. Programmed instruction. *Journal of Speech and Hearing Disorders*, 1977, *42*, 3–28.

Costello, J., & Ferrer, J. Punishment contingencies for the reduction of incorrect responses during articulation instruction. *Journal of Communication Disorders*, 1976, *9*, 45–61.

Costello, J., & Onstine, J. M. The modification of multiple articulation errors based on distinctive feature theory. *Journal of Speech and Hearing Disorders*, 1976, *41*, 199–215.

Curtis, J. F., & Hardy, J. C. A phonetic study of misarticulation of /r/. *Journal of Speech and Hearing Research*, 1959, *2*, 244–257.

Dorsey, H. A. *The relationship between performances of kindergarten children on a three-position test and a deep test of articulation*. Unpublished doctoral dissertation, The Pennsylvania State University, 1959.

Elbert, M., & McReynolds, L. V. An experimental analysis of misarticulating children's generalization. *Journal of Speech and Hearing Research*, 1978, *21*, 136–150.

Evans, C. M., & Potter, R. E. The effectiveness of the S-Pack when administered by sixth-grade children to primary-grade children. *Language, Speech and Hearing Service Schools*, 1974, *5*, 85–90.

Fairbanks, G. Some correlates of sound difficulty in stuttering. *Quarterly Journal of Speech*, 1937, *23*, 67–69.

Fairbanks, G. Systematic research in experimental phonetics: 1. A theory of the speech mechanism as a servosystem. *Journal of Speech and Hearing Disorders*, 1954, *19*, 133–139.

Freeman, G. G., & Sonnega, J. A. Peer evaluation of children in speech correction class. *Journal of Speech and Hearing Disorders*, 1956, *21*, 174–182.

Gerber, A. J. The achievement of /r/ carryover feedback. *Pennsylvania Speech and Hearing Association Newsletter*, 1966, *7*.

Gerber, A. Programming for articulation modification. *Journal of Speech and Hearing Disorders*, 1977, *42*, 29–43.

Goldman, R., Fristoe, M., & Woodcock, R. W. *Goldman-Fristoe-Woodcock Test of Auditory Discrimination*. Circle Pines, Minn.: American Guidance Service, 1970.

Gray, B. B. A field study on programmed articulation therapy. *Language, Speech and Hearing Service Schools*, 1974, *5*, 119–131.

Guess, D., & Baer, D. M. An analysis of individual differences in generalization between receptive and productive language in retarded children. *Journal of Applied Behavioral Analysis*, 1973, *6*, 311–329.

Hall, M. Auditory factors in functional articulatory speech defects. *Journal of Experimental Education*, 1938, *7*, 110–132.

Hamrick, K. S. *Comparison of the 41AX and the same/different methods of testing auditory discrimination*. Unpublished master's thesis, San Diego State University, 1978.

Hansen, G. The application of sound discrimination tests to functional articulatory defectives. *Journal of Speech Disorders*, 1944, *9*, 347–355.

Hilgard, Ernest, R. *Theories of learning*. (2nd Ed.). Appleton-Century-Crofts, 1956.

Hiscock, M., & Kinsbourne, M. Selective listening asymmetry in preschool children. *Developmental Psychology*, 1977, *13*, 217–224.

Holdgrafer, G., Kohn, J., & Williams, G. Transfer of articulation training across morphological contexts: A brief report. *Human Communication*, 1976, *1*, 9–15.

Holland, A. L., & Matthews, J. Application of teaching machine concepts to speech pathology and audiology. *Journal of the American Speech and Hearing Association*, 1963, *5*, 474–482.

Ingram, D. Cerebral speech localization in young children. *Neuropsychologica*, 1975, *13*, 103–105.

Johnson, W., & Brown, S. F. Stuttering in relation to various speech sounds. *Quarterly Journal of Speech*, 1935, *21*, 481–496.

Kimura, D. Speech lateralization in young children as determined by an auditory test. *Journal of Comparative Psychology*, 1963, *56*, 899–902.

Kimura, D. Functional asymmetry of the brain in dichotic listening. *Cortex*, 1967, *3*, 163–178.

Kinsbourne, M. The ontogeny of cerebral dominance. In D. Aaronson & R. Rieber, *Developmental psycholinguistics and communication disorders* (Vol. 263). New York: Annals of the New York Academy of Sciences, 1975, pp. 244–250.

Kopp, H. G. *Some Functional Applications of Phonetic Principles*. (Rev. Ed.). New York: Neyenesch Press, 1979.

Kronvall, E., & Diehl, C. The relationship of auditory discrimination to functional articulatory defectives. *Journal of Speech and Hearing Disorders*, 1954, *19*, 335–338.

Lenneberg, E. *Biological foundations of langauge*. New York: Wiley, 1967.

Lewis, F. C., Jr. Distinctive feature confusions in production and discrimination of selected consonants. *Language and Speech*, 1974, *17*, 60–67.

Locke, J. *Miniseminar*. ASHA Regional Conference, Portland, Oregon, 1976.

McDonald, E. T. *Articulation testing and treatment*. Pittsburgh: Stanwix House, 1964.

McLean, J. E. Extending stimulus controle of phoneme articulation by operant techniques. *ASHA Monographs*, 1970, No. 14, 24–47.

McReynolds, L. V., & Bennett, S. Distinctive feature generalization in articulation training. *Journal of Speech and Hearing Disorders,* 1972, *37,* 462–470.

McReynolds, L. V., Engmann, D., & Dimmitt, K. Markedness theory and articulation errors. *Journal of Speech and Hearing Disorders,* 1974, *39,* 93–103.

McReynolds, L. V., Kohn, J., & Williams, G. C. Articulatory-defective children's discrimination of their production errors. *Journal of Speech and Hearing Disorders,* 1975, *40,* 327–338.

Marge, D. K. The social status of speech handicapped children. *Journal of Speech and Hearing Research,* 1966, *9,* 165–177.

Milisen, R. A rationale for articulation disorders. *Journal of Speech and Hearing Disorders,* 1954, Mongr. Suppl. *4,* 5–18.

Monnin, L. M., & Huntington, D. A. Relationship of articulatory defects to speech-sound identification. *Journal of Speech and Hearing Research,* 1974, *17,* 352–366.

Moore, W. H., Burke, J., & Adams, C. The effects of stimulability on the articulation of /s/ relative to cluster and word frequency of occurrence. *Journal of Speech and Hearing Research,* 1976, *19,* 458–466.

Mowrer, D. E., Wahl, P., & Doolan, S. J. Effect of lisping on audience evaluation of male speakers. *Journal of Speech and Hearing Disorders,* 1978, *43,* 140–148.

Mowrer, O. H. Speech development in the young child: 1. The autism theory of speech development and some clinical applications. *Journal of Speech and Hearing Disorders,* 1953, *17,* 263–268.

Mowrer, O. H. Hearing and speaking: An analysis of language learning. *Journal of Speech and Hearing Disorders,* 1958, *23,* 143–152.

Nichols, A. C. Public school speech and hearing therapy. In R. W. Rieber and R. S. Brubaker (Eds.), *Speech pathology.* Amsterdam: North Holland, 1966, pp. 111–134.

Oller, D. K. Regularities in abnormal child phonology. *Journal of Speech and Hearing Disorders,* 1973, *38,* 36–47.

Perrin, E. H. The social position of the speech defective child. *Journal of Speech and Hearing Disorders,* 1954, *19,* 250–252.

Pisoni, D. B., & Lazarus, J. H. Categorical and noncategorical modes of speech perception along the voicing continuum. *Journal of the Acoustical Society of America,* 1974, *55,* 328–333.

Pollack, E., & Rees, N. S. Disorders of articulation: some clinical applications of distinctive feature theory. *Journal of Speech and Hearing Disorders,* 1972, *37,* 451–461.

Poole, I. Genetic development of the articulation of consonant sounds in speech. *Elementary English Review,* 1934, *11,* 159–161.

Prather, E. M., Hedrick, D. L., & Kern, C. A. Articulation development in children aged two to four years. *Journal of Speech and Hearing Disorders,* 1975, *40,* 179–191.

Reid, G. The etiology and nature of functional articulatory defects in elementary school children. *Journal of Speech Disorders,* 1947, *12,* 143–150.

Rigrodsky, S., & Steer, M. D. Mowrer's theory applied to speech habilitation of the mentally retarded. *Journal of Speech and Hearing Disorders,* 1961, *26,* 237–243.

Roe, V., & Milisen, R. The effect of maturation upon defective articulation in elementary grades. *Journal of Speech and Hearing Disorders,* 1942, *7,* 37–50.

Rousey, C. L., & Moriarty, A. E. *Diagnostic implications of speech sounds.* Springfield, Ill.: Charles C Thomas.

Sander, E. K. When are speech sounds learned? *Journal of Speech and Hearing Disorders,* 1972, *37,* 55–63.

Sherman, D., & Geith, A. Speech sound discrimination and articulation skill. *Journal of Speech and Hearing Research,* 1967, *10,* 277–280.

Singh, S. *Distinctive features: Theory and validation.* Baltimore: University Park Press, 1976.

Sloan, R. F. Neuronal histogenesis, maturation and organization related to speech development. *Journal of Speech and Hearing Disorders,* 1967, *1,* 1–15.

Slorach, N., & Noehr, B. Dichotic listening in stuttering and dislalic children. *Cortex*, 1973, *9*, 295–300.

Snow, K., & Milisen, R. Spontaneous improvement in articulation as related to differential responses to oral and picture articulation tests. *Journal of Speech and Hearing Disorders*, 1954, Mongr. Suppl. *4*, 45–49.

Sommers, R. K. Factors in the effectiveness of mothers trained to aid in speech correction. *Journal of Speech and Hearing Disorders*, 1962, *27*, 178–186.

Sommers, R. K., Moore, W. H., Jr., Brady, W., & Jackson, P. Performances of articulatory defective, minimal brain dysfunctioning and normal children on dichotic ear preference, laterality, and fine-motor skills tasks. *Journal of Special Education*, 1976, *10*, 5–14.

Spriestersbach, D. C. Research in articulation disorders and personality. *Journal of Speech and Hearing Disorders*, 1956, *21*, 329–336.

Spriestersbach, D. C., & Curtis, J. Misarticulation and discrimination of speech sounds. *Quarterly Journal of Speech*, 1951, *37*, 483–490.

Strayer, L. C. Language and growth: The relative efficacy of early and deferred vocabulary training. Studies by the method of co-twin control. *Genetic Psychology Monographs*, 1930, *8*, 209–319.

Templin, M. C. A study of speech sound discrimination ability of elementary school pupils. *Journal of Speech Disorders*, 1943, *8*, 127–132.

Templin, M. C. *Certain language skills in children* (Institute of Child Welfare Monograph 26) Minneapolis: University of Minnesota Press, 1957.

Travis, L., & Rasmus, B. The speech-sound discrimination ability of cases with functional disorders of articulation. *Quarterly Journal of Speech*, 1931, *17*, 217–226.

Unger, T. K. J. *Equipotentiality of hemisphere function consequent to conditioning in a dichotic listening experiment*. Unpublished master's thesis, San Diego State University, 1978.

Van Hattum, R. J., Page, J., Baskervill, R. D., Dugay, M., Conway, L. S., & Davis T. R. The speech improvement system (sis) taped program for remediation of articulation problems in the schools. *Language, Speech and Hearing Service Schools*, 1974, *5*, 91–97.

Wellman, B., Case, I., Mengert, I., & Bradbury, D. Speech sounds of young children. *University of Iowa Studies in Child Welfare*, 1931, *5*, 1–82.

Wepman, J. M. *Auditory discrimination test: manual of directions*. Chicago: Language Research Associates, 1958.

Williams, G. C., & McReynolds, L. V. The relationship between discrimination and articulation training in children with misarticulations. *Journal of Speech and Hearing Research*, 1975, *18*, 401–412.

Winitz, H., & Bellerose, B. Sound discrimination as a function of pretraining conditions. *Journal of Speech and Hearing Research*, 1962, *5*, 340–348.

Winitz, H., & Bellerose, B. Effects of pretraining on sound discrimination learning. *Journal of Speech and Hearing Research*, 1963, *6*, 171–180.

Winitz, H., & Priesler, L. Discrimination pretraining and sound learning. *Perceptual and Motor Skills*, 1965, *20*, 905–916.

Woods, F. J., & Carrow, M. A. The choice rejection status of speech defective children. *Exceptional Child*, 1959, *25*, 279–283.

Young, E. H. A personal experience with speech. *Journal of Speech and Hearing Disorders*, 1958, *23*, 136–142.

8

Harris Winitz

CONSIDERATIONS IN THE TREATMENT OF ARTICULATION DISORDERS

Currently there is no general consensus regarding the treatment of articulatory errors. Rather, a variety of technical procedures are available some of which are intuitively based, while others seem to rest on a firm theoretical foundation. The primary focus of this chapter is to explore treatment considerations. The discussion will be restricted to procedures of learning, omitting treatments specific to medically linked articulation errors. Additionally, the special role of auditory knowledge will be considered.

In the classical treatments of articulation errors and in the more modern treatments, psychologies of learning, as well as specific routines of learning, were advanced. A psychology of learning refers to a set or format of organized practices for training, whereas a routine is a specific practice, often narrow in scope, which is designed to teach a sound or remediate an error.

An early format for training was provided by Van Riper (1939/1978), who suggested that articulation training be guided by the following four stages of discrimination training prior to articulation instruction. They are: (a) isolation—detection of the presence of a sound, (b) stimulation—repetitive hearing of the correct sound, (c) identification—verbal description of the sound, and (d) discrimination—hearing the difference between two sounds.

Van Riper (1939/1978) also provided a number of specific routines to teach

Harris Winitz • Division of Speech and Hearing Science, University of Missouri–Kansas City, Kansas City, Missouri 64110.

new articulatory gestures. For example, he recommended that the /r/ sound may be taught by mechanically manipulating the tongue while it was poised to make the /l/ sound. A tongue depressor was used to push the tongue tip back to provide a space between the alevolar ridge and the tongue tip. If an /r/ sound was not produced, another technique was offered: produce a /z/ and drop the lower jaw.

It is interesting to observe that although many routines for teaching sounds have been presented, systematic investigation of their efficiency has not been tested. This issue will be returned to later when programming procedures are discussed.

Explicit Instruction

Explicit training is a generally accepted clinical practice (Ryan, 1971; Van Riper, 1939/1978). Explicit training refers to the various procedures used to instruct a client to gain conscious control of an articulatory pattern. Explicit training usually involves the following: (1) the client, if a child, is told he has an articulation problem that he is to correct, and (2) the client is given explicit instructions as to how to produce a sound.

A multitude of training procedures and devices have been used to impart explicit knowledge of articulation. Mirror practice is usually accompanied by explicit verbal instructions as to the placement of the articulators. The client might be told to "bite down hard," "move the tongue forward until it can be felt behind the upper teeth," or "put the lips together and blow out." The underlying assumption of explicit training is that conscious knowledge is essential for articulation improvement. This conclusion may be true, but it is a curious fact that native speakers apparently do without explicit knowledge when they generate sentences (Chomsky, 1965) and perform articulatory gestures. The native speaker is really unaware of the movement of the tongue or the target it forms for the many and varied sounds of a particular natural language.

In some cases this knowledge can be made explicit, but this issue is not to the point. If an individual is asked to describe tongue placement, for example, he will first make the movement and then try his best to describe it. However, explicit recovery of articulatory knowledge does not prove that explicit knowledge is critically important for the execution of articulatory patterns. Many years ago Wendell Johnson (1946) suggested that explicit knowledge would probably impede the smooth flow of speech. For him speech production was an implicit process.

There is no evidence to suggest that explicit instruction plays a role in language development. Until recently, the focal point of child language investigations, as might be expected, was the child. The development of sentence structure and phonology was charted in great detail. Large-scale investigations (Poole, 1934; Templin, 1957) provided an age-by-age description of the development of sounds. For example, Templin's study reported consonantal development as follows:

Age (year)	Sound
3.0	m, n, ŋ, p, f, h, w
3.5	j
4.0	k, b, d, g, r
4.5	s, ʃ, tʃ
6.0	t, θ, v, l
7.0	ð, z, ʒ, dʒ

A most recent developmental profile was provided by Prather, Hedrick and Kern (1975) for children up to 48 months:

Age (year)	Sound
2.0	m, n, h, p, ŋ
2.3	f, j, k, d
2.7	w, b, t
3.0	s, g
3.3	r, l
3.8	ʃ, tʃ
4.0	ð, ʒ

The above profiles, however interpreted, are merely descriptions of end points. They reflect the age at which most children show correct usage of a particular sound. However, these profiles say nothing of the essential input that led to the development of these sounds.

Parental style is apparently the critical ingredient that makes phonetic growth possible. A number of studies, summarized in Snow and Ferguson (1977), indicates that parents use a different style of language when addressing young children. Their style of syntax is simple. They use short sentences, noncomplex grammar, and repeat and rephrase often. Simplification at the phonological level is probably restricted to an increase in word duration, a decrease in the rate of speech, and the use of single word utterances.

In addition to simplification of input, parents often ask for repetition when utterances are unintelligible. Such requests no doubt encourage children to alter their articulatory patterns. Other than simplification of input and request for clarification, there is little else that parents do to teach their children to articulate correctly. There is no evidence or reported observation that explicit training in the mechanics of articulation takes place. Parents clearly do not give their young children instructions on the dimensions of place, manner, and voicing.

Although explicit instruction in the mechanics of articulation is not a teaching routine practiced by parents, it does not necessarily follow that explicit instruction should not be used as a clinical routine. It is also true that phoneticians often study acoustic and physiological data to help them master nonnative productions (Ladefoged, 1964). However, use of explicit instruction represents a process that seems not to take place in native language acquisition. The use of explicit instructions to remediate articulation errors raises some additional questions, which will be turned to shortly.

Auditory Input

For the moment, however, the learning process as it applies to articulation acquisition should be reconsidered. It was remarked that early phonetic learning is largely an implicit process. Young children, for example, might seem to know that phonological contrasts mark words (Weir, 1962), but not at a level of consciousness that is regarded as explicit.

Auditory input is central to articulation learning (Winitz, in press). Without auditory input, acquisition of phonological patterns and articulatory gestures is difficult (Ingram, 1974; Winitz, 1969). However, phonological mastery can take place when medical deficits prevent the acquisition of articulation (Lenneberg, 1962; MacNeilage, Rootes, & Chase, 1967). That is, there can be understanding of the phonology of a language without previous mastery of the articulatory movements of the language.

However, in order to produce the motor units of speech, the speaker's knowledge must include more than phonology. The speaker must know (tacitly, that is) how to produce an articulatory response. He or she need not know what a speech motor response unit is or the nature of the command unit that governs a motor response (Kent, 1976).

The fact that phonetic discriminations can often be made independently of success in articulation will be carefully considered in this chapter. We will attempt to present the point of view that auditory comprehension of the phonological system of a language, including distinctions among phonetic productions, requires, initially, massive amounts of auditory input.

Although implicit knowledge is stored in some form, the precise nature of this stored information is unknown. Of particular interest are the processes that lead to the accumulation of this stored knowledge. Conceivably, not all the information about how to produce an articulatory gesture is gained through auditory experience alone. It may well be that practice in the production of speech sounds is an essential prerequisite for acquiring knowledge about how to produce speech sounds. That is, articulation knowledge may require experience in both listening and producing. This issue will be considered shortly.

Varieties of Articulation Responses

For purposes of treatment, articulatory responses have been categorized in a variety of ways. Milisen (1954), among others, proposed several categories of articulatory proficiency. Below are several that have received considerable attention:

(1) *Stimulability*—ability to articulate a sound given auditory input. Presumably, for some children, the stored articulatory information is not sufficient to produce the target response and recent auditory information is required (Milisen, 1954). According to observations by Winitz and Bellerose (1972), sounds which are stimulable are those for which the appropriate motor gesture has been stored, but for which there is difficulty in its retrieval. Failure to retrieve a sound can occur at any of several linguistic levels: isolation, word, sentence, etc.

(2) *Isolation*—production of continuant consonants in nonvocalic contexts, and noncontinuants in consonant–vowel contexts in which the vowel is usually central (/ə/-like), short, and whispered. The claim is often made (Van Riper, 1939/1978) that sounds should first be taught in isolation prior to their placement in words and sentences.

There is disagreement as to whether sounds should initially be taught in isolation. Critics have commented that isolated sounds are citation forms, unrepresentative of sounds in connected speech.

Research in speech science (Daniloff & Moll, 1968; Kent, 1976; Liberman, Cooper, Shankweiler, & Studdert-Kennedy, 1967) has demonstrated that speech sounds are best described as sequences of coarticulated segments. Segments, such as /s/ or /t/, will have varying articulatory shapes, depending upon their phonetic and syntactic context. Advanced programming, timing of the innervations to the speech articulators, and inertial lag are some of the reasons given for the great variety observed in running speech.

The fact that many clinicians report success with isolation training does not necessarily mean that it is a useful procedure. Usually isolation training is one of several steps of a programmed package (Costello, 1977) and, therefore, its effectivness is not usually assessed. In these programs the first stage of training is isolation, followed by other verbal frames, especially word and sentence training. The usefulness of isolation training cannot be assessed in these programs because acquisition and transfer are not independently evaluated. For example, if /s/ is initially trained in isolation, and then trained in the syllable [sa] or word [sæt], it is unknown whether the /s/ sound could have been acquired without isolation training. The program developers usually assume that it cannot be.

However, the major issue at hand pertains to transfer and not to acquisition. The purpose of isolation training is to teach a response that can be transferred to a verbal context. Isolation training in and of itself is of no clinical interest.

One investigation that bears on this issue was conducted by McReynolds (1972). The findings showed that transfer of correct articulatory productions to words was far more effective when the sound was trained in syllables than in isolation. It appears, then, that isolation training needs to be considered relative to its overall effectiveness in transfer as well as in acquisition.

(3) *Phonetic feature training*—when the phonetic feature or distinctive feature of a sound is the focal point of training rather than the individual sound, the training procedure is called phonetic feature training. Here articulatory elements are independently emphasized.

A common error among children is the substitution of stops for fricatives (e.g., /p/ for /f/, and /t/ for /s/). One clinical procedure that might be employed is to teach "frication" as an abstract activity. For example, the bilabial fricative (lips slighly parted so that outgoing air will produce local turbulence) is a sound that can be taught prior to the correction of particular sound errors. Perhaps such training will lead to the development of frication as a concept. A second approach is to select one of a child's several stop/fricative substitutions for training. The procedure here is to teach a particular feature without concern for any of the remaining features of the training sound. Thus, if /t/ is substituted for /s/, and training results in changing /t/ to /f/, the fricative feature would be regarded as successfully learned. At the same time, the fricative feature in the form of /f/ may have been generalized to the remaining stop/fricative substitution errors.

There have been several training studies in which feature acquisition, as well as feature generalization, has been observed (Costello & Onstine, 1976; McReynolds & Bennett, 1972).

Focusing on features would seem to have two important clinical uses: (1) the error sound is described and treated in terms of the articulatory feature, and (2) a feature can be selected for training that will maximize generalization across a class of sounds.

(4) *Conversational speech*—this mode of responding refers to the use of articulatory responses freely and easily in conversation. Wright, Shelton, and Arndt (1969) have used the term *automaticity* to refer to speech production in the absence of conscious or deliberate effort. Deliberate or conscious effort, these researchers reason, would impair the smooth flow of speech. Since the final aim of articulation treatment is automatization of speech motor routines, would it not be advisable to avoid deliberateness in speech patterning during all phases of articulatory instruction?

Three Types of Articulation Disorders

Articulatory Control

Traditionally only one type of articulation disorder was emphasized. Errors in articulation, it was believed, were caused by inadequate control of the speech motor system. An articulation error was generally described as an inability to perform a skilled motor movement. Remediation, therefore, centered on the teaching of new articulatory productions.

What was generally emphasized in teaching, however, were target positions. At /t/, for example, was taught as a voiceless, apico-alveolar sound, and a student was expected to master this sound by learning this static, target position. There was little emphasis on movement and target variability until the mid-1960s when McDonald (1964) emphasized that articulation responses should be taught as a series of overlapping movements.

Phonological Considerations

In 1951, Spriestersbach and Curtis demonstrated that articulation defectiveness is characterized more by inconsistency than by consistency. They found that children identified as misarticulating the /s/ could articulate /s/ correctly in some phonetic environments. Later research confirmed this observation, as summarized in Winitz (1969).

Students of articulation explored this issue further, and concluded that articulation errors are not really inconsistent. There is apparently an underlying consistency, which can be described well by using phonological rules. Ingram (1976) has provided an excellent reivew of studies devoted to this topic.

When articulation errors of individual children are examined in detail, a rule or set of rules can be developed to describe the variance between the child's articulation performance and that regarded as the community norm. A conclusion that is generally made is that articulation defectiveness is an instance of an incorrect, or at least an inappropriate, phonological representation.

There are two ways in which defectiveness in phonological development may be defined (Ingram, 1976). A phonological system is delayed in development when it is similar in form to the phonological system of children much younger in age. Phonological development is usually complete at about eight years of age. A phonological system is deviant when it contains forms commonly not observed in normally developing children. Presently these two types of phonological differences are not well defined.

One consideration apart from the use of phonological classifications is that of prediction. Regardless of the label applied to the phonological system—deviant, defective, delayed, or different (Leonard, 1973)—without predictive validity in the correlational sense, the classifications are not operationally useful.

Two types of predictions can be made: (1) improvement with treatment, and (2) improvement without treatment (Winitz, 1969). The second category is often defined as "spontaneous change." Further study is required to determine whether the categories of "delay" and "deviant" are operationally useful for predicting spontaneous improvement or for developing a program of instruction.

Conversational Patterns

A particularly perplexing problem in the treatment of articulation disorders relates to the mastery of sounds in connected discourse. It is a commonplace observation among speech clinicians that articulatory errors continue to occur in conversational speech long after they have been eliminated in other more deliberate forms of speech (Van Riper, 1939/1978; Winitz, 1975). One explanation for their persistence is that the underlying phonological system has not been altered. Possibly, articulation training has produced changes in articulatory control without a corresponding shift in the underlying phonological representation.

Failure to use corrected articulatory responses in conversational speech has been called the "problem of carry-over." It has been observed that although articulation responses are used correctly in imitation and in picture-naming tasks, often these responses fail to carry over into conversational speech (Bankson & Byrne, 1972; Johnston & Johnston, 1972; Shelton, Johnson, & Arndt, 1972). I once questioned a native-born German as to why he failed to pronounce /θ/ and /ð/ in conversational speech when it was clear that he could capably produce these sounds in isolated words. I received the same reply as I later did from a native-born French woman: "It's not that I can't say 'th,' it's just that I must put too much effort into it while I am talking." In these instances we cannot regard the errors as phonological deviations. Rather, failure to obtain carry-over into conversational speech would seem to be a matter of articulatory control, which is more demanding in conversational speech than in isolated phrases or words.

Auditory Considerations and Articulation Errors

In this section auditory knowledge and its relationship to articulation development and treatment will be briefly reviewed. The areas of discussion will involve auditory knowledge and:

a. Articulation errors
b. Articulation development
c. Phonological disorders
d. Articulatory control
e. Conversational speech

Auditory knowledge refers to the set of cognitive operations and rules essential for the production of speech. Although thorough understanding of the form and content of these rules is not currently available, there is an important theory in speech perception called "analysis by synthesis," which provides a general framework for studying the process of speech perception (Halle & Stevens, 1962). According to this theory, speech sounds are analyzed (or perceived) by an internal generating process. Very simply, here is how it is presumed to work. The listener hears the stimulus item and performs a preliminary analysis on the speech signal. The analyzer takes note of this preliminary information and feeds it into a generator. The generator refers to a corpus of language rules (in this case, phonetic rules) and synthesizes a speech signal. A comparator mechanism matches the generated signal and compares it with a stored version of the original signal. If the match is reasonably close, the received signal is given the same interpretation as the generated signal. For example, if /p/ is heard, the acoustic representation of /p/ will be stored momentarily; certain acoustic segments will be analyzed such as the burst frequency, the transition, and perhaps the vowel steady state. From this preliminary analysis an idealized acoustic wave form of the category /p/ will be generated. If there is a close match between the major components (e.g., features) of these two signals, the received signal will be declared a /p/. Note that a complex set of algorithms would need to be used to make the final match.

One may ask why so much machinery is required to perceive a /p/. Can't it simply be assumed that /p/ will be identified by its distinguishing acoustic characteristics? The answer to these two questions cannot be given easily. In 1967, an important summary paper on this issue was published by Liberman *et al.* They emphasized that individual sounds often have very different acoustic cues. Furthermore, phonetic context and speaker differences contribute thousands of different patterns, all of which listeners can assign to a phonetic category with relative ease. If only auditory comparisons were involved, listeners would be required to sort through thousands, or possibly millions, of patterns to achieve a phonetic percept.

Evidence in support of an analysis-by-synthesis speech perceptual process is the finding that most phonological perceptions are categorical (Liberman *et al.*, 1967). Categorical perception means that listeners make discriminations *at* phoneme boundaries, but not *within* phoneme boundaries. Identifications are categorical in that subjects make absolute phonetic judgments between sounds (e.g., either a /p/ or /b/, but not something in between).

The analysis-by-synthesis processor uses acoustic information to access abstract phonetic forms. Acoustic information is fed into the processor which produces an abstract neural form. Almost simultaneously, a second abstract phonetic form is generated from the acoustic features of the sound. At this abstract neural level, these two forms are compared. If an acceptable match is made, the two abstract forms are regarded as the same or different, but not as something in between. A categorical decision has been made.

An auditory decoder would not involve reference to the processes of speech production. Rather, the incoming signal would be compared with the stored auditory signal in some complex way. Auditory signals would be matched, but no perceptual categories would be involved. For example, the determination that A is an *A* would be made by matching it with *A* by a set of complex mathematical algorithms. In effect, the *A*'s would be superimposed on each other, and a goodness-of-fit test would be made.

The superimposition of the received *A* on the ideal *A*, equated for size by a special algorithm, would have the following appearance:

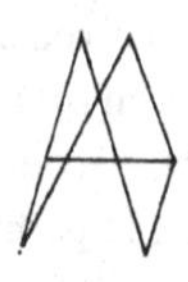

A determination of whether A is in fact an exemplar of *A* would depend upon the match between these two symbols. In the same way, every *A* would need to be matched. The formula would have to be capable of testing millions of letter patterns.

In contrast, in an analysis by synthesis model, A suggests a certain neural impression. At the same time, an examination is made of the features of A (that is its lines are analyzed as to size, shape, direction and intersection). A candidate letter is proposed by this analysis, possibly *A*. It also has a certain neural impression. These two neural impressions are compared, and if the match is regarded as "close," the written letter A is taken to be an A.

A special case of the generalized analysis-by-synthesis model is the "motor theory." According to the motor theory, "speech is perceived by processes that are also involved in its production" (Liberman *et al.*, 1967, p. 452). Exactly what these processes of articulatory coding are is unknown. For the time being we can think of them as abstract neuromotor mechanisms common to perception and production.

The motor theory of speech perception has the advantage of simplicity in that a relatively small number of neuromotor correlates of phonological categories would be involved in contrast to the infinitude of potential wave forms required of a strictly auditory decoder.

The motor theory can be misinterpreted when applied to the treatment of articulation disorders. One possible source of confusion centers around the word "production." It may be recalled that the basic postulate of the motor theory is that perception involves reference to the processes of speech production. The implication is not that articulatory movements are involved, only that a set of neuromotor rules are common to perception and production.

A neuromotor rule provides an input to the speech motor system. The function of a rule is to activate a "motor command unit," which in turn sends neural signals to the speech musculature. Arbitrarily we might say that auditory knowledge ends at the point at which the motor command system is activated.

The motor theory, in its current form (Liberman *et al.*, 1967), leaves open the question of how the processes of articulatory coding are acquired. In an early presentation of the motor theory, Liberman, Cooper, Harris, and MacNeilage (1963, p. 177) speculated that, "in time, these articulatory movements and their sensory feedback (or, more likely, the corresponding neurological processes) become part of the perceiving process mediating between the acoustic stimulus and its ultimate perception."

Of particular clinical interest is the role of production practice in perception. Possibly, alternate training trials of production and perception are more advantageous for acquiring phonetic percepts than training involving only perceptual practice. Clearly, there is a need to test a variety of training routines before conclusive statements can be made.

Auditory Knowledge and Articulation Errors

Auditory factors continue to be a primary source of concern in the search for etiological contributors to articulation disorders. There is general agreement, although not universally held, that an impairment in auditory discrimination will cause articulation development to be delayed. Comparative studies, summarized by Winitz (1969), generally indicate that children with articulation errors evidence reduced acuity on tests of speech sound discrimination. The results of some of the investigations proved to be negative, largely, I believe, because of two factors: (a) subject differences, and (b) test differences. The subjects of the various investigations differ in severity. In some experiments the subjects had relatively few articulation errors, and in other studies the opposite was the case. Furthermore, the discrimination and articulation assessments are substantially different from study to study.

Differences in population samples and composition of tests can substantially alter results. For example, a short discrimination test will not include a sufficiently large number of phonological contrasts to assess the speech sound discrimination of children with mild articulation errors.

It has also been demonstrated that discrimination errors and articulation errors are related functions when specific speech sounds are involved (Aungst & Frick, 1964; Monnin & Huntington, 1974; Prins, 1962; Spriestersbach & Curtis, 1951). There are two recent studies that seem to argue against error specificity and the importance of discrimination training in teaching production. These studies have been cited as counterevidence to the belief that discrimination is importantly related to the acquisition of articulatory responses, and for this reason are reviewed in detail.

In the first investigation (McReyolds, Kohn, & Williams, 1975), the topic of error specificity was examined. It was concluded that children with articulation errors evidence little difficulty in speech sound discrimination. As the design of this study is examined, it will be seen that the concept of error specificity in discrimination and production cannot be rejected, primarily because an adequate test of error specificity was not made.

The auditory discrimination and articulation of a small group of articulatory-defective and normal-speaking children were studied in a rather interesting way. The auditory discrimination test was developed so that six Chomsky–Halle features (continuancy, nasality, voicing, stridency, anterior, and coronal) could be assessed. Mean correct scores in percentages for three of the six features for the subjects was as follows:

	Continuancy		Voicing		Stridency	
	Production	Discrimination	Production	Discrimination	Production	Discrimination
Defective	17	83	41	79	6	84
Normal	98	94	98	88	95	94

It may be observed from the above that the discrimination skills of the articulatory-defective children were somewhat depressed relative to the normal-speaking children. The differences are not as great as would be expected, but the reasons accounting for this result have been outlined above. In particular, the 80 minimal-pair test contained only 20 types of same–different pairs (e.g., /l/—/r/) replicated by altering order (e.g., /r/—/l/) and 40 same pairs (e.g., /l/—/l/ and /r/—/r/). These contrasts included a few pairs composed of target sounds and common substitutions (e.g., /w/—/r/, /b/—/v/, /w/—/l/, lateral lisp, and /s/) (Snow, 1964). Also, the test contained contrasts that reflect rare occurrences of substitutions (/l/—/r/, /d/—/n/, /z/—/v/) found in the speech of young children (Snow, 1964).

The discrimination test was not necessarily appropriate for each child, because each child had a different set of articulation errors. For example, the error profile of the first child tested revealed ten articulation errors of which 7 pairs of the 20 pairs were composed of contrasts in which neither sound was misarticulated. Therfore, at best, only 65 percent of the pairs were relevant. Most importantly, however, was the fact that there is no way of knowing whether individual substitution errors were reflected by the 13 putatively relevant contrasts, as McReynolds, *et al.* (1975) do not report substitution errors.

The fact that substitution errors were not considered in this investigation makes the authors' second analysis and conclusion inappropriate, as we shall see. They initially categorized a child's production of a test sound as correct if it was produced correctly on 40% or more of the instances tested on the McDonald Deep Test of Articulation, and incorrect if this criterion was not achieved. Subsequent to this analysis, discrimination performance was examined in the following way: when a sound was produced *correctly* the number of pairs in which this sound appeared and was incorrect was tallied, and when a sound was produced *incorrectly* the number of pairs in which this sound appeared and was incorrect was tallied. There were no doubt some cross pairings because the sounds of a pair could easily involve one sound that was articulated correctly and one that was articulated incorrectly.

The results of this subsequent analysis indicated that the discrimination scores for the sounds that were correctly articulated were only slightly above that for the incorrectly articulated sounds. The authors' conclusion was: "Certainly it does not appear to be a large enough difference to suggest that the children were not able to discriminate their error phonemes as well as they discriminated phonemes they could articulate" (McReynolds, *et al.*, 1975, p. 336). This conclusion is unwarranted because the contrasts were not necessarily relevant ones, and the criterion of correctness was arbitrarily selected and would not appear to distinguish the two groups radically, on the basis of articulatory performance. The criterion of correct production is further inappropriate because it indicates that there was considerable variability in the production of each sound. This result is to be expected with the McDonald deep test, a research tool particularly useful in assessing articulatory performance in a variety of phonetic contexts. Since articulation and discrimination were not assessed with regard to context, a strong relationship would not be expected between each individual target sound and its respective error.

In an investigation by Winitz and Preisler (1965), it was demonstrated that discrimination training effectively resulted in self-corrected production responses for contrasts involving the target sound and the error sound, but not for unrelated contrasts. This investigation involved normal-speaking children and the training sound was a non-English cluster. In a subsequent study (Winitz & Bellerose, 1967), a small number of children with /r/ errors were trained to discriminate between /r/ and typical /r/ error substitutions. They did not show improvement in the production of /r/ following intensive speech sound discrimination training in which their discrimination of /r/ was significantly improved. One interpretation of these apparent conflicting results is that discrimination training will produce changes in articulation if the relevant articulatory features can be articulated correctly, as was the case in the Winitz and Preisler (1965) study.

When the relevant features are not present, a certain degree of production practice is probably important. This latter conclusion receives support from several studies (Guess, 1969; Ruder, Smith, & Hermann, 1974; Ruder, Hermann, & Schiefelbusch, 1977), in which a small number of imitation trials following auditory training were sufficient to develop correct production responses.

In summary, there is no reason to reject the value of discrimination training. It seems most plausible that a degree of articulatory practice is essential before articulatory control is established, but the role of auditory discrimination cannot be discounted.

In a second investigation, McReynolds and her associates (Williams & McReynolds, 1975) tackled again the issue of production and discrimination. However, this time they introduced a different experimental design. Rather than comparing these two functions as to performance levels, they experimentally manipulated these two functions. Unfortunately this investigation was inappropriately designed to inquire about the relationship between discrimination and production.

Both production and discrimination training were individually tailored for four children, who misarticulated, at least, four or more phonemes. The design was as follows with two children in each group:

Group I	Group II
Production training Discrimination probe	Discrimination training Production probe

The subjects in Group I were trained to articulate several sounds correctly, followed by a test of discrimination, and the subjects in Group II were trained to discriminate between several sounds and production was subsequently tested. This design might appear to be straightforward if production and discrimination can be regarded as independent training functions. However production training involved auditory input as part of the training routine. Additionally, however, an unusual design was employed to teach the sounds.

In both production and discrimination training, a picture association task was used. For example, the picture–sound association task for one subject in production training and one subject in discrimination training was as follows:

Production training	Discrimination training
Picture A } /ʃa/ Picture B } Picture C — /dʒa/ Picture D — /θa/	/va/ — Picture A /ga/ — Picture B /θa/ { Picture C /θa/ { Picture D
Assessment of discrimination	Assessment of production

After the children learned to acquire the production task (naming the pictures correctly) or the discrimination task (selecting the correct picture or pictures), discrimination and production were respectively assessed.

In the discrimination probe or assessment, the instructions were to listen to the taped presentation of each of the three syllables and select the appropriate picture or pictures. The two subjects performed well on the discrimination test, a result that indicates that discrimination was a factor contributing to or related to successful articulatory production, although this interpretation was not offered by the authors.

The production probes following discrimination training revealed that articulatory performance did not change at all. The authors drew the conclusion that "discrimination training had no direct influence on production" (Williams & McReynolds, 1975, p. 411), realizing that this finding could be interpreted to imply that discrimination practice should be omitted in clinical training.

Discrimination pretraining cannot be easily dismissed as an unimportant clinical routine. Most importantly, discrimination practice provides experience in learning to make phonetic discriminations as well as experience in learning to know about the phonological system.

One reason for rejecting the findings and interpretations of the Williams and McReynolds (1975) investigation is: The authors did not correctly program the discrimination contrasts. An evaluation of the discrimination contrasts can be illustrated best by considering the subject who had errors of /ʃ/, /dʒ/, and /θ/. The training, as mentioned above, involved discrimination among these three sounds. Not considered here is the well-accepted fact that for discrimination training to be effective it should involve training between the target sound and the error sound (Winitz, 1975). In this case the child might easily have learned to discriminate between the three target sounds /ʃ/, /dʒ/ and /θ/ without being able to discriminate between each target sound and its respective error. In summary, this investigation by Williams and McReynolds (1975) provides no substantive support for the conclusion that discrimination performance and articulation errors are unrelated. The value of discrimination training cannot be discounted. The evidence is strong that discrimination training is an essential prerequisite for acquiring articulatory control. However, with regard to training regimens, further research on discrimination training and articulation learning is needed.

Auditory Considerations and Articulatory Development

The capacity for making phonetic distinctions appears early in the life of the newborn (Eimas, 1974). Over time these discriminations are refined to fit the phonological properties of the community language (Edwards, 1974; Templin, 1957).

Several writers (Eilers & Oller, 1976; Menyuk, 1969; Winitz, 1969) have taken the position that discrimination skills precede articulation skills. Three stages posited by Menyuk (1969), follow:

$$\begin{bmatrix} -\text{discrimination} \\ -\text{production} \end{bmatrix} \quad (1)$$

$$\begin{bmatrix} +\text{discrimination} \\ -\text{production} \end{bmatrix} \quad (2)$$

$$\begin{bmatrix} +\text{discrimination} \\ +\text{production} \end{bmatrix} \quad (3)$$

The above schematic representation of the stages simply means that at stage 1 neither discrimination nor production is acquired. At stage 2 discrimination singularly develops. The acquisition of production represents the final stage, or stage 3, of development.

A fourth possibility is:

$$\begin{bmatrix} -\text{discrimination} \\ +\text{production} \end{bmatrix}$$

However, this category is theoretically untenable if the position is taken that auditory skills precede the acquisition of articulation. Eilers and Oller (1976) tested a group of young children less than 2½ years of age on a variety of phonetic contrasts. Discrimination of contrastive pairs, as well as production of both members of a pair, was assessed. An analysis of the individual responses falling within the category

$$\begin{bmatrix} -\text{discrimination} \\ +\text{production} \end{bmatrix}$$

were rare, no doubt reflecting inherent subject variability. The bulk of the children's responses, as expected, fell in categories 1, 2, or 3.

The group marked

$$\begin{bmatrix} +\text{discrimination} \\ -\text{production} \end{bmatrix}$$

is of theoretical interest because two contradictory hypotheses can be entertained. First, this group can be considered to be at that point in development at which discrimination has been acquired. Production has not developed prior to this time, presumably because discrimination (or some level of discrimination) must precede production. Possibly, the capacity to produce articulatory responses has developed, but memory constraints impair the recall of correct productions (Winitz & Bellerose, 1972). The absence of production in this group reflects developmental constraints, but it does not deny a relationship between discrimination and production. Second, this group can be used to deny a relationship between discrimination and production because both functions are not present at the same point in time. This interpretation was apparently taken by Locke and Kutz (1975), but it appears to be incorrect.

The remaining two groups

$$\begin{bmatrix} -\text{discrimination} \\ -\text{production} \end{bmatrix} \quad \text{and} \quad \begin{bmatrix} +\text{discrimination} \\ +\text{production} \end{bmatrix}$$

indicate early and late developmental stages. At the early stage neither speech discrimination nor articulation has developed. At the final stage both of these skills have developed. At these two stages, discrimination and production are correlated.

Auditory Knowledge and Phonological Disorders

Beginning with Haas' (1963) paper on the application of phonological analysis to the disorder of articulation, and heightened by Compton's (1970) subsequent report, interest in using phonological theory to describe articulation errors continues to grow. Stated simply, a phonological analysis is used to gain insight about the organization of a child's articulation errors. Phonological rules are tools used to describe the underlying sound system of a child's articulation errors.

Consider a child who voices all voiceless sounds. The organizing phonological principle or rule is:

$$[-\text{voice}] \rightarrow [+\text{voice}]$$

This rule states that all voiceless sounds are voiced (e.g., /t/→/d/, /f/→/v/). According to this rule, this child's phonological disorder or underlying pattern is that all voiceless sounds are misarticulated as voiced.

The factors contributing to a phonological disorder remain to be determined in each case. Possible causes are auditory discrimination, auditory knowledge, and articulatory control. These three possible causes will be examined in detail by making reference to two children studied by Compton (1970).

Tom, the first child Compton (1970) observed, had several sounds in error, as follows:

Standard Sound	Sounds Produced
/ʃ/	[ʃ] [s] [*sk*][1] [k]
/s/	[s] [*sk*] [k]
/z/	[s] [*sk*] [k]
/tʃ/	[s] [*sk*] [k]
/f/	[*sk*] [k]
/k/	[k]
/dʒ/	[d] [g]
/d/	[d] [g]
/g/	[g]

Ingram (1976) revised Compton's rules and a portion of these are presented below:

Fronting: { /tʃ/ , /ʃ/ } →[s] (1)

Stopping:
/f/→/k/ (obligatory) (2a)
/s/→/k/ (optional) (2b)
/dʒ/→/d/ (obligatory) (2c)

Assimilation: [/t/ , /d/] → [[k] , [g]] / — Vowel [/k/ , /g/] (3)

Each of the three rules will now be considered. Rule 1 specifies that the substitution for /tʃ/ and /ʃ/ are fronted, or made further forward in the mouth, by the substitution of /s/ for these two sounds. There are three parts to rule 2. In each case, the substitution is by a stop, hence the use of the term *stopping*. In rule 2a the /f/ becomes a /t/. It is an obligatory rule in that /f/ must always be changed into a [t]. Rule 2b is an optional rule in that /s/ can be substituted by other sounds in addition to [t]. Rule 3 requires that [d] always be substituted for /dʒ/.

Rule 3 expresses regressive assimilation for the sounds /t/ and /d/. Ingram surmised that in some of Compton's test words, such as 'dog,'' the initial consonant assimilated the place of articulation of the final consonant. Ingram's rule of assimilation means that /t/ becomes [k] and /d/ becomes [g] when they appear in the context, respectively, vowel + k and vowel + g. Thus, in words such as dog the /g/ influences the pronunciation of /d/ causing it to be articulated as [g] (e.g., [gɔg]).

The discussion of Tom's rules has been lengthy. The purpose was to show the interpretation Compton and Ingram drew of the errors underlying Tom's

[1][*sk*] is defined in Compton (1970) as a cluster in which the [s] segment is relatively long in length with a wider distribution of energy across the frequency spectrum. The [*sk*] apparently developed during clinical training in an attempt to teach [s] is the context of the word "sock."

phonological system. To these investigators, Tom's errors reflect an organizational pattern, not merely a list of unrelated sound substitutions.

Does Tom's phonological system reflect an inability to make phonetic discriminations? Compton's answer is decidedly no. He first notes that one of the articulatory productions for /ʃ/ is [ʃ], a correct production, but that /s/, /z/ and /tʃ/ are not realized as [ʃ]. Following Compton's logic, the pair of sounds /dʒ/ and /g/, and the pair /f/ and /k/, would also be discriminable. The fact that [d], in addition to [g], is substituted for /dʒ/, suggests that /dʒ/ and /g/, and /f/ and /k/ are respectively discriminable.

Perhaps the only sounds not discriminable from each other are within the triplet /s/, /z/, and /tʃ/, and those within the pair /dʒ/ and /d/. However, the source of Tom's deviant phonological system, Compton believes, is not that of an inability to discriminate the phonological contrasts of English.

There are other interpretations regarding Tom's discrimination. For example, the large number of free variants for each of the standard sounds, four in the case of /ʃ/, may indicate that the discrimination of sound segments has not been completely realized. Rather, the discrimination may be based on one or two phonetic features. This is partial learning. That is, several but not all features of a stimulus complex are discriminated (Ingram, 1974). Thus, in the case of /ʃ/ and /s/, Tom may be using one or two features, rather than the full set of possible features to discriminate between these two sounds. Additionally, the discriminations may be context restricted: /ʃ/ in some contexts is distinguished from /s/, but in other contexts it is not. Sufficient data are not provided by Compton to resolve this issue of partial discrimination.

Another alternative is that the discriminations are in the process of being refined and that production reflects learning at an earlier period. Tom, for example, may have acquired /s/ for /ʃ/ as an approximation which then became stabilized through use. Later, when /ʃ/ and /s/ could be discriminated, /ʃ/ and /s/ productions were correctly realized in new words, but early production errors and perhaps, discrimination errors, were retained. Examples in phonetic learning reflecting this process have been reported by child language investigators (Moskowitz, 1973; Slobin, 1971).

Much of what was said above is speculation. It is almost impossible to know Tom's perceptual skills from Compton's report because detailed discrimination testing was not made.

An interesting observation by Goldstein and Locke (1971) suggests that phonological disorders, in some instances, may not be a function of poor auditory discrimination. Their (Goldstein & Locke, 1971) investigation revealed that when children misarticulate sounds, they exhibit greater difficulty in making category assignments of substitution errors. In each case, the subjects were to respond yes, meaning correct, or no, meaning incorrect, when the target sound or its substitution was spoken in isolation, a singular, noncontrastive item, accompanied by a correct pictorial referent. The children missed very few items for the target sound whether or not they articulated the sound correctly. However, they said "yes" to many of the substitutions. For example, for the picture *thumb*, items like /θ ʌ m/ were overwhelmingly regarded as "thumb" whether correctly articulated or not, whereas misarticulations like /f ʌ m/ were regarded as correct some of the time. The children were apparently more "tolerant" of incorrect forms. They seemed to know what a correct item was, but additionally, they often regarded an incorrect item as belonging to the same phonological category as the correct item.

The Goldstein and Locke (1971) finding suggests two possible interrelationships between auditory knowledge and phonological disorders. The first relationship, as mentioned above, is that auditory discrimination can be in an inconsistent state. Target productions may be easily and correctly identified, whereas there is indecision regarding off-target productions.

A second possible interrelationship between auditory knowledge and phonological disorders is more abstract (or central) than perception. Auditory knowledge refers to the underlying mental organization of the sound system. A phonological analysis is a procedure, or set of techniques, used to describe underlying mental arrangements. When the underlying system departs from the standard of the community language, it is called a phonological disorder.

Auditory discrimination tests cannot be used to measure auditory knowledge, because the ability to distinguish between sounds does not necessarily mean that the standard, or adult, phonological system has been acquired. It is generally known that the ability to make discriminations in a dialect of English different from that of the speaker does not imply similarity of auditory knowledge. Similarly, young children with articulation disorders may capably make the discriminations requested in a speech discrimination test, but they may not utilize these contrasts in their speech.

Now, the question of testing auditory knowledge is to be addressed. It is assumed, of course, that speech sound discrimination would in all cases be assessed first. Following this assessment, auditory knowledge would be tested. One approach would be to ask the question, "Which is the correct way to say it?" However, the reliability of this technique is questionable when dealing with young children.

Another approach might be to devise testing procedures in which the comprehension of meaningful utterances is examined. Assume a child substitutes stops for fricatives, [+fricative] → [−fricative]. Discrimination testing shows further that discrimination of stops and fricatives in isolated, nonsense syllables can be made. If discriminations cannot be made in words, sentences, or in understanding a short story composed of minimal pair stop–fricative contrasts, it may be inferred that the stop–fricative contrast is not functional.

Finally, phonological descriptions that specify underlying rules may reflect neither auditory discrimination nor auditory knowledge, but articulatory control. Compton's subject, Tom, displayed a serious phonological disorder because of the multiple instances of articulatory substitutions. His ability to articulate the involved sounds was not completely impaired. As may be recalled, Tom misarticulated seven sounds. Of these seven sounds, /z/, /tʃ/, /f/, and /dʒ/ never appeared in the record as a correct production or error substitution; however, three of the seven sounds, /s/, /ʃ/, and /d/, appeared in the record as correct articulatory productions. There is the possibility, then, that articulatory ability, as well as auditory knowledge, was reflected in Tom's phonological system. However, this conclusion should be made with caution. Of the four sounds that were not articulated correctly by Tom, the voiced or voiceless cognate of two was articulated correctly. The counterpart of [z], [s], and of [f], [v], was produced correctly. Additionally, Tom was able to produce voiced and voiceless fricatives. Therefore, there is no reason to believe that Tom lacked the articulatory control to produce [z] or [f]. The defective affricates seem to argue for a different conclusion. Tom could produce [ʃ], hence there is no reason to believe that its voiced counterpart, [ʒ], could not be produced. However, the ability to articulate the blending of a stop + fricative may not have been within Tom's skill level. This question cannot be answered without further information. Most importantly, Tom's imitation skills were not assessed, and, therefore, we do not know precisely what his production skills were. In conclusion, Tom's non-standard phonological system may have reflected a developmental lag in auditory discrimination, articulation control, and/or organizational structure.

Jim, a second child for whom Compton (1970) developed phonological rules, is also of interest. The results of Jim's treatment program could be interpreted as a measure of validity for using phonological rules to describe articulation errors. Jim had a peculiar disorder of the nasal sounds, /m, n, ŋ/, in final position as follows:

/m/ → { omitted; /m/ lengthened followed by /p/ in unstressed syllables; /m/ lengthened followed by /b/ in stressed syllables }

/n/ → { omitted; /n/ lengthened followed by /t/ in unstressed syllables; /n/ lengthened followed by /d/ in stressed syllables }

/ŋ/ → { omitted; /ŋ/ lengthened followed by /k/ in unstressed syllables; /ŋ/ lengthened followed by /g/ in stressed syllables }

Treatment was administered to Jim for five weeks on the /m/ sound only. At the conclusion of treatment /n/ and /ŋ/ were articulated correctly, a result which indicates that treatment of the /m/ sound had affected a change in the general rule governing final nasals. This finding suggests strongly that the phonological rules describing Jim's use of final nasals were correct.

The focus of treatment for Jim was articulatory control. Emphasis was placed on production practice. The following questions can now be asked: "What if Jim were given auditory training exclusively? Would he have learned to alter the phonological patterns governing the deviant pronunciation of nasals?" It is difficult to say. Were this result to be the case, we could be more certain that auditory knowledge, rather than articulatory control, was the underlying cause of Jim's phonological disorder.

A phonological disorder is an underlying sound system that is deviant from the community language. Rules are used to describe this disorder. Auditory knowledge is the term used to refer to the underlying mental processes of a phonological disorder. Difficulty in making auditory discrimination may be an additional factor contributing to disorders of phonology. In some cases auditory knowledge may not be deviant, but articulatory control may be lacking. In such cases it is probably inappropriate to interpret a phonological disorder as a deviation in the use of phonological rules.

Auditory Knowledge and Articulatory Control

The teaching of articulatory targets has been the major aim of traditional articulation teaching routines. Three procedures have generally been used to teach articulatory productions: (1) *phonetic placement*—the major articulator is moved by the clinician into target position; (2) *verbal instruction*—phonetic descriptions are given by the clinicians as to how and where to move the major articulators: and (3) *imitation*—the child repeats the clinician's utterances.

Although the approach of these three instructional routines differs, the major goal of each is the teaching of an articulatory target. The targets which are aimed for are usually those described by place, manner, and voicing features. Thus, the target for [s] is tongue to alveolar ridge, and frication with no voicing.

A common element of the three methods is auditory stimulation because most clinicians hold to the belief that more than one avenue of instruction should be provided in order to accommodate individual differences. Thus, auditory stimulation is usually given when articulation is taught by either the verbal instruction or phonetic placement method.

Most recently, Kent (1976) has provided a summary statement of the complexities involved in the execution of the skilled motor movements of speech.

Several models of speech motor control were discussed. It is instructive to note that the control of speech can be viewed as hierarchical, that is, there are several levels of planning dominated by a rhythmic structure and culminating in a motor command unit (Kent, 1976). The workings of these various levels illustrates the great complexity involved in the production of speech articulations.

Two important questions must now be raised in light of the realization that the complexity of the speech production system is staggering: (1) What production techniques can be developed to teach articulatory responses? and (2) What are the procedures by which these techniques can be taught? The production techniques might emphasize only target values, or target values and transitional movements. However, whatever the technique, it is only a method for displaying or illustrating the target sound. It cannot be regarded as a complete set of instructional rules that describes or explains a particular speech motor act. The underlying processes that govern articulatory movements and the articulatory movements themselves (not the static articulatory targets) are too complex to describe in complete detail. Furthermore, our knowledge of these rules is largely limited.

In the traditional stimulus–imitation procedure, the clinician says a sound, syllable, or word, asks the student to imitate, and then evaluate the imitation. Target values might initially be stated and the feedback might involve instructions (e.g., "move the tongue a little more forward," or "keep the tongue down.") It is unclear at this time as to whether information of this kind is useful. Nevertheless, the primary tracking system is auditory. Without the auditory component, a student would not be able to compare the clinician's stimulus and his own response, and learn to track the clinician's stimulus with his own articulators.

The stimulus–imitation method appears simple on the surface because it appears not to be an elaborate and explicit teaching system. As a teaching methodology, however, it is consistent with the premise that articulatory responses are largely implicitly learned. The imitation–response method gives the learner the necessary experiences to develop the motor programs required for the execution of speech articulations.

Auditory Knowledge and Conversational Speech

The final and most difficult stage of articulatory treatment is learning to produce sounds in sentences, easily and freely, in conversational speech (Winitz, 1975). In this section, the point will be made that conversational speech requires a different set of auditory and articulatory skills than deliberate speech or speech in isolation.

The syntactic and semantic components of language are important influences on conversational speech. There is evidence to suggest that semantic and syntactic influences govern certain phonological representations. Belasco (in press) has used the example of "amina," [áminə], which has the meaing "I am going to," to show that certain phonetic forms are syntactically constrained. It can be used as a substitution in the sentence: "I am going to go home now," but not in "I am going to New York," where in the latter sentence "to" is a preposition. Similarly, Chomsky (1976) has discussed the fact that the elision "wanna" (want to) can appear in "Who will want to go," but not in "Who will Bill want to go." The rule for the correct use of "wanna" is complex and relates to the underlying subject to which it can be attached. Thus, if one is to understand English, one must be able to decode the phonetic variations that occur in conversational speech. One must know, for example, that [kʌp ə] means "cup of," [æn] means "and," and [si əm] means "see him."

Cooper, Sorensen, and Paccia (1977) have recently concluded that non-phonetic aspects of a sentence will affect the duration of word segments. They examined duration across clause boundaries using one of four conjunctions, "and," "but," "since," and "if," for the sentence, "Dick will take the jeep ______ Clark will take the truck." They obtained the interesting finding that the major grammatical form classes (Dick—Clark, $take_1$—$take_2$, and jeep—truck) showed durational correlations across the clause boundary. The authors concluded that the regulation of speech timing is not restricted to phonetic factors, but that grammatical coding also governs speech timing.

The above observations indicate that the production of speech cannot be independent of syntactic and semantic influences. Similarly, the perception of conversational speech is dependent upon grammatical knowledge. Knowing the intent of a sentence is probably as important as hearing the indivual speech sounds that make up the words in a sentence.

Language is filled with redundancies. If a word is unclear in a sentence because some of its sounds have been masked, there are any one of a number of linguistic cues that will make the word readily intelligible. For example, in the following sentence one letter has been visually masked with a small square: The librarian dropped the two book■ on the floor. It is clear from context that *s* is the missing letter for a number of reasons. The word "two" cues plurality, and the letter *k*, in *book*, indicates that the following letter is *s*. Warren (1970) found that when a phoneme is excised from a speech sample and replaced by noise, listeners are often not aware that it is missing. In speech, the meaning of sentences is usually so clear that the omission of phonemes often goes undetected.

It might seem that listeners never mishear sentences. They do, although there is no record as to their frequency. Garnes and Bond (1975) collected a large number of observations on misperceptions in conversational speech and some of their examples are given below:

$\theta \rightarrow$ f
Death in Venice → Deaf in Venice
n → nd
Sunny weather → Sunday weather
syllable change
*Fri*ar *Tuck* Pizza → Ken*tucky* *Fri*ed Pizza
deletion of word boundary
six to seven → sixty-seven

The above errors purport to be rule governed in that they reflect linguistic regularity. Listeners misperceive sounds or segments, but they do so in terms of the regularities imposed by the community language. When a sound or cluster is misperceived, it is replaced by another sound or cluster that is allowable in English. For example, "I'm covered with chalk dust" was reported by Garnes and Bond (1975) to be heard as "I'm covered with chocolate," an amusing, but sensible replacement.

Grammatical influences are strong when phonetic signals are unclear or ambiguous (Cole, Jakimik & Cooper, 1978; Garnes & Bond, 1977; Winitz, LaRiviere, & Herriman, 1973). In the Cole *et al.* (1978) investigation, listeners did well in detecting purposely misprounced sounds in story contexts in which the words were changed to nonwords (e.g., boy → poy, and five → vive). Not all errors were detected, however, suggesting that in some instances errors, like Warren's phoneme deletions, go undetected because meaning predominates.

In the Cole *et al.* (1978) investigation, words were altered to make nonwords. This alteration is a special kind of ambiguity because the resultant error is not a different word, only a "poor" pronunciation of the same word. Under these circumstances considerable ambiguity in the phonetic signal can be tolerated, especially when the errors are few and/or nonadjacent. However, the intended meaning of a message can be destroyed when adjacent or neighboring words are altered, even when the errors are nonwords.

Experiments in which excerpts (words and phrases) are taken from conversational speech have also been examined for intelligibility (Lieberman, 1963; Pickett & Pollack, 1963; Winitz & LaRiviere, 1979). Under these conditions the excerpts are often misunderstood. In the Winitz and LaRiviere (1979) investigation, a single word and the adjacent neighboring words, excerpted from conversational speech, were tested for intelligibility.

Sentences such as: "The boy found the tack under the chair," produced near

100% intelligibility of all words. However, when a word from the middle of a sentence such as *tack* in the sentence above, was presented to listeners, the average intelligibility score was 75%. When adjacent words were presented to form a three-word phrase, (e.g., "the tack under") intelligibility climbed only 7% to 82%. It can be seen, then, that without knowledge of meaning and perhaps tempo (Verbrugge, Strange, Shankweiler, & Edman, 1976), intelligibility drops to a point that is unusuable for understanding speech.

The large amount of redundancy present in conversational speech can be an aid to the articulatory-defective child who exhibits perceptual or production deficits. This point is dramatically punctuated in an important study (Miyawaki, Strange, Verbrugge, Liberman & Jenkins, 1975) of Japanese speakers who spoke fluent English, but demonstrated poor discrimination between /r/ and /l/. The stimuli were several synthetically generated patterns ranging from /r/ through /l/ and the listeners' task was to assign each stimulus to either the /r/ or /l/ phoneme. This assignment proved to be almost impossible for the Japanese to perform, despite the fact that they were fluent speakers of English. Additionally, their pronunciation of /r/ and /l/ was apparently not used correctly in English (Fujimura, 1977; Liberman, 1977).

The question may be asked: How do these Japanese individuals understand English without the ability to make /r/–/l/ distinctions? They apparently resort to context, linguistic and nonlinguistic, to derive the meaning of phonetically ambiguous sentences. Only rarely would an /r/–/l/ ambiguity occur, which could not be disambiguated by context.

Of interest to speech clinicians is the fact that a certain amount of phonological ambiguity can be tolerated by listeners, especially when only a small number of phonological contrasts are involved. Perhaps it is for this reason that errors in conversational speech are difficult to correct. Children with articulation errors have probably learned to rely on nonphonetic cues to resolve phonetic or phonological ambiguities.

There is another important difference between conversational speech and nonconversational speech. In deliberate speech, time constraints are minimized. The duration of sounds is sometimes increased, coarticulatory influences are less, and elisions are not as frequent. Utterances said in isolation are similar in form to those said in deliberate speech.

Coarticulatory influences have been found to stretch across syllables and sentences (Daniloff & Moll, 1968; Lewis, Daniloff, & Hammarberg, 1975). Coarticulatory movements are complex physiological gestures, which maximize the number of phonemes that can be uttered per unit time interval.

It appears, then, that at least two factors contribute to the complexity of acquiring articulatory responses in conversational speech. Children with articulation errors may simply fail to attend to their defective phonetic productions because the meaning component capably enables them to understand speech and to be understood. Also, the training they have received with deliberate and isolated speech utterances may not have provided them with the right kinds of auditory and production experiences to correct errors in conversational speech.

Until recently the question as to whether children make more errors in isolated contexts than in sentences was viewed within the framework of the more general concept of phonetic context (Winitz, 1969). There is now reason to believe, however, that articulation within sentences involves much more than simply phonetic context. A study by Faircloth and Faircloth (1970) examined the articulation of an 11-year-old child in which the conclusion was made that conversational speech produced consistently more errors than speech tested in isolation. However, the testing conditions were not equivalent for the two types of speech samples. The isolated words were elicited by auditory stimulation and imbedded in a carrier phrase that minimized coarticulatory influences.

Nevertheless, two additional studies have been conducted that seem to support the Faircloth and Faircloth position. Syntactic contexts were varied by Schmauch, Panagos, and Klich (1978). Examples for /s/ are: The *s*aw (noun phrase); The boy could have used the *s*aw (active sentence); The wood was cut by the *s*aw (passive sentence). Significantly more errors were obtained in the two sentence contexts than in the noun phrase. This finding seems to confirm an earlier suggestion by Winitz (1975) that sentential cues can function as eliciting stimuli, increasing the probability that a word will be mispronounced. In the above sentences the words "boy," "used," "cut," and "wood" act to intensify the word "saw" as an experiential concept, increasing the likelihood that an earlier (defective) pronounciation will be used. An alternative hypothesis might be that little attention was paid to the auditory component of /s/ because the meaning of the target word was clear. These two factors need to be sorted out in future investigations.

Dubois and Bernthal (1978) provided convincing evidence that articulation errors are greater in conversational speech than in isolated word contexts. This study controlled for two important factors: (a) sample size: a large number of articulatory-defective children were sampled; and (b) method of instruction: the isolated words and sentences were assessed in a similar way. Dubois and Bernthal used two procedures to assess conversational speech. One procedure involved *telling* a story about a picture and the other *retelling* a story (with a picture present) told by the experimenter. Isolated word contexts produced the least number of errors and telling a story produced the greatest number of errors. This finding suggests that speech errors are directly related to the complexity of the speech act. Conversational speech appears to represent a special kind of speech activity that is more complex than deliberate speech.

As suggested above, there are several factors that can contribute to the difference between deliberate speech and conversational speech. Among these are rate, phonetic context, and redundancy. Furthermore, it is not at all convincing that correcting articulatory patterns in isolated speech alters the underlying phonological system. It is apparent, then, that much research needs to be done on conversational speech and articulation errors.

Operant Conditioning and Programming

It is often difficult to describe a school of thought in a single statement or set of statements. The operant approach is no exception. Even the term operant is sometimes a source of confusion. An operant is usually defined as a response that is the focus of conditioning in an experiment or training session. Articulatory responses, such as lip closing or tongue to alveolar ridge, can be regarded as operants.

Those who regard themselves as operant conditioners adhere to a set of practices and theoretical positions. Often the term programmed instruction is associated with this school of thought, although not exclusively (Winitz, 1969, 1975). Programmed instruction, according to Costello (1977, p. 3), is a set of procedures in which there is a "systematically designed remediation plan which specifies a priori the teaching and learning behaviors required of both the teacher and the learner." The behaviors of the "teaching act" are dictated by principles that have been found to be particularly effective in teaching skills to animals and humans.

The use of programmed instruction is noncontroversial. What is controversial are the tasks or skills that individual investigators may use to teach articulation and language. Program devlopers take three considerations into mind: (a) the general properties of linguistic processes as we understand them today, (b) interpretations placed on information obtained from studies on native language acquisition, and (c) results obtained from articulation-training studies.

A property of most instructional systems is a statement of goals. This important function is not an exclusive attribute of operant conditioning. Historically, there has been disagreement on choice of goals, but not on whether goals should or should not be established in advance of instruction. One could hardly teach without an assessment of goals.

Operant procedures have been applied to the teaching of discrete articulatory responses by a process called *successive approximation or shaping*. Through a series of graduated steps or response approximations, a terminal response is taught from a rudimentary response that is initially within the skill level of the learner. For example, Winitz (1969) has suggested that /k/ or /g/ can be learned from a snorelike sound. The graduated steps include shaping a snore, which is usually made on inspiration, into a expirated snorelike sound. From there this fricativelike sound is shaped into a stop. The snore was suggested as the first stop in the program because snoring and the production of /k/ or /g/ have similar velarlike qualities. Another example is provided by McLean (1976) who recommends teaching the [ɜ] (by beginning with the vowel /i/, going to /I/ and then to [ɜ]). He notes, however, that others recommend teaching this sound by using the intermediate sounds /g/ or /θ/.

Other programmed routines have been suggested by McLean (1976) and Costello (1977), some of which (Leonard & Webb, 1971; McLean, 1970; Ryan, 1971) have been tested and found to be effective, in that subjects are able to learn the goal responses of a variety of articulation behaviors. However, as Costello is quick to note, specific steps in these programs have not been tested. All that is known is that the programs have been effective, but their effectiveness has not been assessed relative to other programs. More importantly, however, we do not know whether learning is a function of the program or a function of having experience with sounds including the attendant feedback that provides information as to the quality of performance and the attainment of goals.

Even the criterion of correctness has not been well established. Costello and Onstine (1976) provide a detailed set of instructions for teaching two children a set of articulatory responses of which the criterion of correctness, as well as the schedule of instruction, is arbitarily determined. Repetition is important, but exactly how many trials should be given and how many correct responses should be determined prior to moving to a following step is currently unknown. One solution is to provide probes (or advanced testing) to determine whether subjects can jump steps or move to another step prior to perfect mastery of the previous step. Substantial research is needed in this direction.

Often subjects fail to produce a response after being given considerable repetition. A program that includes alternative steps is called a branching program. However, current branching programs are largely intuitively based (Costello & Onstine, 1976; McLean, 1976). There are no branching programs in the treatment of articulation that have a firm foundation in research.

Operant conditioners stress production. Discrimination practice or auditory experience is often neglected. Perhaps auditory training is not emphasized because articulation errors are viewed only as a defect in the production of motor movements. Only when the totality of factors are taken into consideration can effective programs of treatment be developed. Central to all programs of articulatory treatment is auditory experience. Its role in the development of programs and training routines needs serious consideration.

References

Aungst, L. F., & Frick, J. V. Auditory discrimination ability and consistency of articulation of /r/. *Journal of Speech and Hearing Disorders*, 1964, *29*, 76–85.

Bankson, N. W., & Byrne, M. C. The effect of a timed correct sound production task on carryover. *Journal of Speech and Hearing Research*, 1972, *15*, 160–168.

Belasco, S. Aital cal aprene las lengas estrangièras. [Comprehension: The key to second language acquisition.] In H. Winitz (Ed.), *The comprehension approach to foreign language instruction*. Rowley, Mass.: Newbury House, in press.

Chomsky, N. *Aspects of the theory of syntax*. Cambridge, Mass.: M.I.T. Press, 1965.

Chomsky, N. On the nature of language. In S. R. Harnad, H. D. Steklis, & J. Lancaster (Eds.), *Origins and evolution of language and speech*, (Vol. 280). Annals of the New York Academy of Science, 1976.

Cole, R. A., Jakimik, J., & Cooper, W. E. Perceptibility of phonetic features in fluent speech. *Journal of the Acoustical Society of America*, 1978, *64*, 44–56.

Compton, A. J. Generative studies of children's phonological disorders. *Journal of Speech and Hearing Disorders*, 1970, *35*, 313–339.

Cooper, W. E., Sorensen, J. M., & Paccia, J. M. Correlations of duration for nonadjacent segments in speech: Aspects of grammatical coding. *Journal of the Acoustical Society of America*, 1977, *61*, 1046–1050.

Costello, J. Programmed instruction. *Journal of Speech and Hearing Disorders*, 1977, *42*, 3–28.

Costello, J., & Onstine, J. M. The modification of multiple articulation errors based on distinctive feature theory. *Journal of Speech and Hearing Disorders*, 1976, *41*, 199–215.

Daniloff, R. G., & Moll, K. Coarticulation of lip rounding. *Journal of Speech and Hearing Research*, 1968, *11*, 707–721.

Dubois, E. M., & Bernthal, J. E. A comparison of three methods of obtaining articulatory responses. *Journal of Speech and Hearing Disorders*, 1978, *43*, 295–305.

Edwards, M. L. Perception and production in child phonology: The testing of four hypotheses. *Journal of Child Language*, 1974, *1*, 205–220.

Eilers, R. E., & Oller, D. K. The role of speech discrimination in developmental sound substitutions. *Journal of Child Language*, 1976, *3*, 319–329.

Eimas, P. D. Linguistic processing of speech of young infants. *Language perspectives—acquisition, retardation, and intervention*. In R. Schiefelbusch & L. L. Lloyd (Eds.), Baltimore: University Park Press, 1974.

Faircloth, M. A., & Faircloth, S. R. An analysis of the articulatory behavior of a speech-defective child in connected speech and in isolated-word responses. *Journal of Speech and Hearing Disorders*, 1970, *35*, 51–61.

Fujimura, O. Personal communication. 1977.

Garnes, S., & Bond, Z. S. Slips of the ear: Errors in perception of casual speech. *Proceedings of the 11th Regular Meeting of the Chicago Linguistic Society*, 1975, 214–225.

Garnes, S., & Bond, Z. S. The relationship between semantic expectation and acoustic information. In

W. U. Dressler & O. E. Pfeiffer (Eds.), *Phonologica 1976*, Innsbruck, Austria, Thomas Herock, 1977.

Goldstein, J. I., & Locke, J. L. Children's identification and discrimination of phonemes. *British Journal of Disorders of Communication*, 1971, *6*, 107–112.

Guess, D. A functional analysis of receptive language and productive speech: Acquisition of the plural morpheme. *Journal of Applied Behavioral Analysis*, 1969, *2*, 55–64.

Haas, W. Phonological analysis of a case of dyslalia. *Journal of Speech and Hearing Disorders*, 1963, *28*, 239–246.

Halle, M., & Stevens, K. N. Speech recognition: a model and a program for research. In J. A. Fodor & J. J. Katz (Eds.), *The Structure of Language*, Englewood Cliffs, N.J.: Prentice-Hall, 1962.

Ingram, D. The relationship between comprehension and production. In R. L. Schiefelbusch & L. L. Lloyd (Eds.), *Language perspectives—acquisition, retardation, and intervention*. Baltimore: University Park Press, 1974.

Ingram, D. *Phonological disability in children*. New York: Elsevier, 1976.

Johnson, W. *People in quandries*. New York: Harper, 1946.

Johnston, J. M., & Johnston, G. T. Modification of consonant speech-sound articulation in young children. *Journal of Applied Behavioral Analysis*, 1972, *5*, 233–246.

Kent, R. D. Models of speech production. In N. J. Lass (Ed.), *Contemporary issues in experimental phonetics*. New York: Academic Press, 1976.

Ladefoged, P. *A phonetic study of West African languages*. London: Cambridge University Press, 1964.

Lenneberg, E. H. Understanding language without ability to speak: A case report. *Journal of Abnormal and Social Psychology*, 1962, *65*, 419–425.

Leonard, L. B. The nature of deviant articulation. *Journal of Speech and Hearing Disorders*, 1973, *38*, 156–161.

Leonard, L. B., & Webb, C. An automated therapy program for articulatory correction. *Journal of Speech and Hearing Research*, 1971, *14*, 338–344.

Lewis, J., Daniloff, R., & Hammarberg, R. Apical coarticulation at juncture boundaries. *Journal of Phonetics*, 1975, *3*, 1–8.

Liberman, A. M. Personal communication, 1977.

Liberman, A. M., Cooper, F. S., Harris, K. S., & MacNeilage, P. F. A motor theory of speech perception. *Proceedings of the speech communication seminar*. Stockholm, 1962. Stockholm: Royal Institute of Technology, 1963.

Liberman, A. M., Cooper, F. S., Shankweiler, D. P., & Studdert-Kennedy. Perception of the speech code. *Psychological Review*, 1967, *74*, 431–461.

Lieberman, P. Some effects of semantic and grammatical context on the production and perception of speech. *Language and Speech*, 1963, *6*, 172–187.

Locke, J. L., & Kutz, K. J. Memory for speech and speech for memory. *Journal of Speech and Hearing Research*, 1975, *18*, 176–191.

MacNeilage, P. F., Rootes, T. P., & Chase, R. A. Speech production and perception in a patient with severe impairment of somesthetic perception and motor control. *Journal of Speech and Hearing Research*, 1967, *10*, 449–467.

McDonald, E. T. *Articulation testing and treatment: A sensory motor approach*. Pittsburgh: Stanwix House, 1964.

McLean, J. E. Extending stimulus control of phoneme articulation by operant techniques. In F. L. Girardeau & J. E. Spradlin (Eds.), *A functional analysis approach to speech and languge: ASHA Monograph*, 1970, *14*, 24–47.

McLean, J. E. Articulation. In L. L. Lloyd (Ed.), *Communication assessment and intervention strategies*. Baltimore: University Park Press, 1976.

McReynolds, L. V. Articulation generalization during articulation training. *Language and Speech*, 1972, *15*, 149–155.

McReynolds, L. V., & Bennett, S. Distinctive feature generalization in articulation training. *Journal of Speech and Hearing Disorders,* 1972, *37,* 462–470.

McReynolds, L. V., Kohn, J., & Williams, G. Articulatory-defective children's discrimination of their production errors. *Journal of Speech and Hearing Disorders,* 1975, *40,* 327–338.

Menyuk, P., & Anderson, S. Children's identification and reproduction of /w/, /r/, and /l/. *Journal of Speech and Hearing Research,* 1969, *5,* 39–52.

Milisen, R. A rational for articulation disorders. *Journal of Speech and Hearing Disorders. Monograph supplement,* 1954, *4,* 6–17.

Miyawaki, K., Strange, W., Verbrugge, R., Liberman, A. M., Jenkins, J. J., & Fujimura, O. An effect of linguistic experience: The discrimination of /r/ and /l/ by native speakers of Japanese and English. *Perception and Psychophysics,* 1975, *18,* 331–340.

Monnin, L. M., & Huntington, D. A. Relationship of articulatory defects to speech-sound identification. *Journal of Speech and Hearing Research,* 1974, *17,* 352–366.

Moskowitz, A. I. The acquisition of phonology and syntax: a preliminary study. In K. J. J. Hintikka, J. M. E. Moravcsik & P. Suppes (Eds.), *Approaches to natural language.* Dordrecht, Holland: D. Reidel Publishing Co., 1973.

Pickett, J. M., & Pollack, I. Intelligibility of excerpts from fluent speech: Effects of rate of utterance and duration of excerpt. *Language and Speech,* 1963, *6,* 151–164.

Poole, I. Genetic development of articulation of consonant sounds in speech. *Elementary English Review,* 1934, *11,* 159–161.

Prather, E. M., Hedrick, D. L., & Kern, C. A. Articulation development in children aged two to four years. *Journal of Speech and Hearing Disorders,* 1975, *40,* 179–191.

Prins, T. D. Analysis of correlations among various articulatory deviations. *Journal of Speech and Hearing Research,* 1962, *5,* 152–160.

Ruder, K. F., Smith, M., & Hermann, P. Effects of verbal imitation and comprehension on verbal production of lexical items. In L. McReynolds (Ed.), *Developing systematic procedures for training children's language: American Speech and Hearing Association Monographs,* 1974, *18,* 15–29.

Ruder, K. F., Hermann, P., & Schiefelbusch, M. Effects of verbal imitation and comprehension training on verbal production. *Journal of Psycholinguistic Research,* 1977, *6,* 59–72.

Ryan, B. P. A study in the effectiveness of the S-pack in the elimination of frontal lisping behavior in third grade children. *Journal of Speech and Hearing Disorders,* 1971, *36,* 390–396.

Schmauch, V. A., Panagos, J. M., & Klich, R. J. Syntax influences the accuracy of consonant production in language-disordered children. *Journal of Communications Disorders,* 1978, *11,* 315–323.

Shelton, R. L., Johnson, A. F., & Arndt, W. Monitoring and reinforcement by parents as a means of automating articulatory responses. *Perceptual and Motor Skills,* 1972, *35,* 759–767.

Slobin, D. I. Data for the Symposium. In D. I. Slobin (Ed.), *The ontogenesis of grammar.* New York: Academic Press, 1971.

Snow, C. E., & Ferguson, C. A. (Ed.), *Talking to the children: Language input and acquisition.* Cambridge: Cambridge University Press, 1977.

Snow, K. A comparative study of sound substitutions used by "normal" first grade children. *Speech Monographs,* 1964, *31,* 135–141.

Spriestersbach, D. C., & Curtis, J. F. Misarticulation and discrimination of speech sounds. *Quarterly Journal of Speech,* 1951, *37,* 483–491.

Templin, M. C. Certain language skills in children, their development and interrelationships. *Institute of Child Welfare,* Monograph Series, No. 26. Minneapolis: University of Minnesota Press, 1957.

Van Riper, C. *Speech correction, principles and methods.* Englewood Cliffs, N.J.: Prentice-Hall, 1978. (Originally published, 1939.)

Verbrugge, R. R., Strange, W., Shankweiler, D. P., & Edman, T. R. What information enables a

listener to map a talker's vowel space? *Journal of the Acoustical Society of America*, 1976, *60*, 198–212.
Warren,.R. M. Perceptual restoration of missing speech sounds. *Science*, 1970, *167*, 392–393.
Weir, R. H. *Language in the crib*. The Hague: Mouton, 1962.
Williams, G. S., & McReynolds, L. V. The relationship between discrimination and articulation training in children with misarticulations. *Journal of Speech Hearing Research*, 1975, *40*, 401–412.
Winitz, H. *Articulatory acquistion and behavior*. Englewood Cliffs, N.J.: Prentice-Hall, 1969.
Winitz, H. *From syllable to conversation*. Baltimore: University Park Press, 1975.
Winitz, H. (Ed.). *The comprehension approach to foreign language instruction*. Rowley, Mass.: Newbury House, in press.
Winitz, H., & Bellerose, B. Relation between sound discrimination and sound learning. *Journal of Communication Disorders*, 1967, *1*, 215–235.
Winitz, H., & Bellerose, B. Effect of similarity of sound substitutions on retention. *Journal of Speech and Hearing Research*, 1972, *15*, 677–689.
Winitz, H., & LaRiviere, C. Factors contributing to the recovery of monosyllabic words excerpted from natural speech. *Journal of Phonetics*, 1979, *7*, 225–233.
Winitz, H., & Preisler, L. Discrimination pretraining and sound learning. *Perceptual and Motor Skills*, 1965, *20*, 905–916.
Winitz, H., LaRiviere, C., & Herriman, E. Perception of word boundaries under conditions of lexical bias. *Phonetica*, 1973, *27*, 193–212.
Wright, V., Shelton, R. L., & Arndt, W. B. A task for the evaluation of articulation change. III. Imitative task scores compared with scores for spontaneous tasks. *Journal of Speech and Hearing Disorders*, 1969, *12*, 875–885.

9

R. W. Rieber

PHONATORY AND RESONATORY PROBLEMS

FUNCTIONAL VOICE DISORDERS

Introduction

The human voice is one of the most obvious indicators of an individual's reaction to himself and the world around him. Even in today's space age, despite all the scienitific instrumentation for measuring the condition of the astronauts in our space program, the medical staff uses the quality of the astronaut's voice as a primary source of information about his physical and emotional state.

In spite of the prevalence of functional voice disorders and the ease with which one can recognize them, little research has been done in this area. This chapter will not attempt to recover the body of perceptual research dealing with voice as it is related to personality. The reader is referred to Kramer (1964) and Diehl (1959) for a comprehensive review of this area.

The purpose of this chapter is to review the literature dealing with functional voice disturbances and to describe the current trends of research in this field.

Functional voice disorders are problems that are related to a deviation from normal in terms of usage of the vocal mechanism as distinguished from disturbances of structure or innervation.

It should not be inferred that a sharp demarcation can be drawn between these two diagnostic categories. It is rather intended that they should be considered as

R. W. Rieber • Department of Psychology, John Jay College, CUNY, and Columbia University, College of Physicians and Surgeons, New York, New York 10019

extremes of a continuum of vocal dysfunctions. In many instances, continual misuse of a normal vocal mechanism may result in the appearance of an organic voice disorder. In other instances it is possible that organic disturbances, such as acute laryngitis, may give rise to a functional disturbance of phonation, which will persist after the disappearance of the organic condition. Brodnitz (1962) makes reference to this pheonomenon in terms of organic aphonia, which may result in functional disturbances of voice production after the original cause of the phonia has been removed.

Any discussion of deviations from the normal implies a set of criteria as to what is within the range of normal. It has been pointed out, however, that standards for normal vocal functioning are, at best, vague and ill-defined, and that possibly a majority of Americans do not meet the standards of normal voice production (Brodnitz, 1962).

Problems in Nosology and Evaluation

Classification

Johnson, Darley, and Spriestersbach (1963) gave a comprehensive survey of the various procedures used in the evaluation of phonation. The material in this chapter is presented in the theoretical framework that functional voice disturbances may be related predominantly to either hyperfunctioning or hypofunctioning of the phonatory and resonatory mechanisms. This framework was orginially suggested by Froeschels (1940, 1943) and has been used by Brodnitz (1962) and others in categorizing these disturbances. Those disturbances that will be considered as related to hyperfunctioning include disturbances related to excessive tension, movement, or strain in the entire vocal mechanism, or in specific portions thereof. This division of classification has also been referred to by other researchers as hyperkinetic and hypokinetic disturbances (Perkins, 1957). In this discussion of the hyper- versus hypofunctioning, Froeschels refers to six specific localized areas of hyperfunctioning as: "1) the coup de glotte; 2) the strangulated voice; 3) the 'knödel' or constriction of the lower oro-pharynx; 4) the hyperinteraction of the soft palate—giving rise to a functional hyporhinolalia; 5) stiffening of the anterior part of the tongue; and 6) stiffening of the lips."

Brodnitz (1962) in his description of the hyper- versus hypofunctioning categories, refers to the observable acoustic symptoms, rather than to a specific anatomical location.

Definitions of various kinds of voice qualities such as breathiness, harshness, stridency, brassiness, are somewhat difficult to clarify through written description. Fairbanks (1960) has defined four specific categories of abnormal voice quality as "hyper": nasal, harsh, breathy or hoarse. Murphy (1964) states in reference to voice quality "all of these terms are commonly used in referring to vocal characteristics. Not the least of our problems is that of attempting to describe in writing the complex acoustical patterns called voice, and to work toward a usable concensus of opinion covering the meanings of the individual terms. . . . The terminoloogy of abnormal voice is largely a language of metaphor." For the purpose of this chapter, specific descriptions will be given under the various categories of disturbances described under hyperfunctioning and hypofunctioning.

Pitch, Loudness, and Quality Measurement

A number of procedures have been suggested for determining the natural pitch level. Zaliouk (1963) describes a tactile approach where the speaker feels vibrations in the nose and maxillary bones. Fairbanks (1960) has devised a procedure that measures the habitual pitch level that corresponds closely to an individual's natural pitch level. Natural pitch level is a reference point that can be used to compare an individual's habitual pitch level. The habitual pitch level is that range of pitches that clusters around the optimum pitch. Steer and Hanley (1957) have developed instrumentation that allows for a precise measurement of an individual's pitch level. Darley (1964) recommends that the discrepancy between natural and habitual pitch levels be routinely evaluated. Vocal instability manifested by tremor during phonation is often diagnostic of neurological disturbances. However, Brown and Simonson (1963) point out that patients manifest tremor during propositional speech. Darley (1964) points out that examination of phonation is essentially an abbreviated form of experimental therapy. Darley also indicates that the clinician should bring about adjustments in terms of pitch, loudness, and quality, noting the effect upon each other.

Berg (1956) demonstrated in his study the fact that there is greater efficiency of voice production at the higher pitch levels that can be produced easily. This study was later confirmed and elaborated upon by Perkins and Yanagihara (1968). According to these studies, vocal adjustments that produce the greatest acoustic energy with the least amount of efforts are the most optimal at any particular pitch. Perkins (1971) points out that a three-way interaction between loudness, pitch, and constriction is probably necessary to account for physical injury to the vocal chords.

Phonatory Problems due to Hyperfunction

Disturbances related to hyperfunctioning of the voice mechanism include disturbances of phonation due to excessive subglottic pressure, tension of the muscles of the larynx, and excess tension in the muscles that shape or alter the resonating cavities.

The Harsh Voice

Many individuals, for a variety of causes, tend to use an excessive amount of vocal intensity accompanied by a strident quality of vocal tone. Such individuals are likely to remind the listener of the voice characteristics of an army first sergeant "reasoning with" a group of raw recruits. Physiologically, this type of voice quality represents a combination of increased subglottic pressure and tension in the muscles of the larynx and resonating cavities. In most instances the pitch level is low and phonation may be characterized by a glottal fry (Moore & Von Leden, 1958; Moser, 1942). This phenomenon consists of a low frequency crackling sound. Cleeland (1948) had reference to this as "voice quality x." This variety of phonatory quality is characteristic of so-called normal speakers when under the influence of strong emotions, particularly that of anger. It has been theorized by Arnold (1962) and Rees (1958) that the excessively loud, strident voice can be correlated with the kind of personality that is typically aggressive, hostile, domineering, and punitive. This distinctive vocal quality has variously been described in such terms as harsh or gutteral and has been related to the physiological dysfunctioning involved by the use of descriptive terms such as "hyperkinetic dysphonia."

Hollien, Moore, Wendahl, and Michel (1966) proposed that this condition may not be always considered a pathological form or mode of laryngeal function especially in well-produced, low, male-pitched voices. This stresses the importance of a careful differential diagnosis in this area.

Spastic Dysphonia

Another diagnostic entity involving hyperfunctioning of the phonatory mechanism is the condition referred to as spastic dysphonia. In this condition, attempts at voluntary phonation result in spasms of the laryngeal sphincter, accompanied by increases in tension of the other muscle groups involved in the phonatory act. Arnold (1959) describes the symptoms of spastic dysphonia as follows: "With distorted face and various parakinesias of the muscles of the head . . . the sufferer strains all muscles of respiration during his vain attempts at communication. . . . Pneumo-phonic coordination is altered." Observers of, and researchers into the phenomena of spastic dysphonia are nearly unanimous in relating the symptoms of spastic dysphonia to those observed in stuttering, and in pointing out the psychoneurotic aspects of the disability (Arnold, 1962). It is frequently observed that patients unable to use the phonatory mechanism properly for the purposes of phonation may be able to perform other functions such as coughing, laughing, crying, or perhaps even singing, without difficulty.

Although the onset and early stage of spastic dysphonia are considered to be psychogenic or psychoneurotic in origin, and therefore of a functional nature, the kind of abuse and strain imposed on the phonatory mechanism by attempted phonation may result in changes in the tissues of the larynx of an obviously anatomical nature. Those changes may include redness and hypertrophy of the vocal folds and, in rare cases, nodules and contact ulcers of the vocal folds (Aronson, Peterson, & Litin, 1964; Freud, 1962). E. Freud has described the problem of convergence and opposition of the ventricular bands during phonation. The results of this abnormal positioning of the ventricular bands include voices that "sound thick or unclear and whining due to altered resonance conditions which the sound waves encounter in the upper part of the larynx."

Hoarseness

A hoarse voice quality has been described as combining the acoustic quality of harshnsss and breathiness. It is usually manifest in individuals who speak at low pitch levels. Williamson (1946) states, in his study of 72 cases of hoarse voice, that hoarseness is generally a product of faulty learning. In hoarseness, the vocal folds generally do not approximate closely, the thyroid cartilage may be tilted upward, there may be a hard glottal attack, pitch is generally lowered and the voice frequently breaks and requires "throat clearing." The hoarseness may be associated with phonation using the ventricular bands, rather than the vocal folds (Freud, 1962). There may be a resulting inability to raise the intensity level above that required for ordinary conversation. Although hoarseness frequently appears as a result of hyperfunctioning of the laryngeal sphincter, it often gives rise to disturbances of voice characterized as hypofunction, or lack of adequate energizing of the vocal mechanisms as hypofunctions, or lack of adequate energizing of the vocal mechanisms (Diehl, White, & Burk, 1959).

Pubertal Dysfunction of the Voice

One of the stages of human development most notable for change accompanied by psychological upheaval is that of adolescence. Among the many characteristic and obvious changes occurring during this period is the change of voice, most noticeable among males. Although the change in voice is attributable to physiological changes within the larynx, these changes, or lack of them, acquire considerable psychological significance; particularly since they occur at an age when the individual is intensely conscious of body and its development, as well as his interpersonal relationships with others. Of particular significance to the male adolescent is the lowering of vocal pitch that results in a characteristically male voice quality. Therefore, the vocal changes take on a strongly sexual connotation, and delay or difficulty associated with the change may have a marked impact on the psyche of the individual involved.

Since the enlargement of the larynx requires a relearning of neuromuscular coordination patterns, many instances occur in which proper use of the phonatory mechanism does not accompany the growth. As a result such phonemena as voice breaks, pitch variations, improper resonance for the new vocal pitch, and abnormal quality may be observed.

Among these phenomena the voice break is most common. It usually consists of sudden involuntary shifts of vocal pitch from a low to a high level. The range of the shift usually encompasses about one octave. Some boys, in an attempt to avoid these breaks will adjust their vocal pitches, either up or down. If the pitch change is extensive, abuse of the vocal folds because of inefficient use of the entire phonatory mechanisms may result.

A variety of factors may result in the retention of an abnormally high vocal pitch, or falsetto. These factors include: abnormal endocrine function, faulty vocal habit, attempts to avoid pitch breaks of unaccustomed low pitch, and personality deviations. The most serious instances of falsetto voice are those that occur as a result of psychogenic factors such as emotional immaturity, inability to assume a dominant masculine role, and homosexual difficulties.

Often associated with pubertal changes are defects in voice quality such as breathiness or huskiness. Because of the speaker's unfamiliarity with his or her now phonatory mechanism, the usual adolescent boy or girl may attempt to continue to speak at the old accustomed pitch that may no longer be the optimum pitch level for the new size of the larynx and vocal folds, or he may, in attempting to compensate for the changes actually overcompensate and make changes in pitch that are excessive and therefore inefficient. Inexperience with the newly enlarged vocal mechanism may cause the individual to approximate the vocal folds inexactly or in a strained manner, thus producing a voice of defective quality.

Pitch Disorders

For every human vocal mechanism there is a fundamental voice pitch to which that mechanism is "tuned." This so-called optimum pitch is a function of such variables as the length, width, and thickness of the vocal folds, the dimensions of the chest cavity, the size of the larynx, pharynx and head cavities, size and thickness of the bone structure of the upper chest and head, as well as the other factors. Phonation at or near this optimum pitch requires the least effort in terms of subglottic pressure and energizing the adductor and tensor muscles of the larynx to produce vocal sounds that are most pleasing to the listener.

Failure to maintain the pitch range of the voice about this optimum pitch will necessitate increases in subglottic pressure, increases in tension of the adductor and tensor muscles of the larynx, or combinations of both, thus creating undue stress on the entire phonatory system. The result is generally a strained, tense quality of the voice and inadequate vocal intensity, and if persisted in for a sufficiently long period of time, or to a sufficiently intense degree, will result in physical changes in the larynx. In many individuals, this improper pitching of the voice has been determined to be related to poor pitch discrimination, anxiety, emotional tension, hyperthyroidism, and other organic and psychological disturbances.

Phonatory Problems due to Hypofunction

The hypofunctional phonatory problem is often the final stage of development that started as hyperfunction. Brodnitz (1962) points out that where hypofunction appears directly without a primary stage of hyperfunction, careful neurological history and examination are essential. If no organicity can be demonstrated, psychogenic factors may be causing the disorder. Hypofunction manifests itself clinically in terms of a general weakness of the vocal mechanism.

Breathiness

Breathiness is one of the major acoustic characteristics of hypofunction. Individuals with breathiness manifest improper closure of the glottis or bowing of the cords that results in a wastage of air. Breathiness can result from either improper use of the vocal mechanism or from a pathological condition in the laryngeal area. A shallow type of respiration may be present, which may be associated with clavicular breathing. Inadequate subglottal pressure may result in an inadequate level of loudness. Breathiness is often associated with paretic hoarseness as well as chronic harshness.

Phonasthenia

The term phonasthenia refers to a voice that is exceedingly weak. Psychogenic factors are usually responsible for this type of voice problems. Murphy (1964) feels that excessively weak voices often are a manifestation of personal depression and are usually accompanied by sensations of laryngeal discomfort. He also points out that this disorder is rarely found in children.

West and Ansberry (1968) discussed specific structural abnormalities that are associated with the reduction of the volume of the voice. The structural abnormalities are such things as a narrowed epiglottis, small facial arch, and enlarged palatine tonsils. The failure to find any of these above-mentioned conditions would strongly point to the posssibility of a functional cause of this symptom of a phonatory disorder.

Hysterical Aphonia (Conversion Aphonia)

Hysterical aphonia is a voice problem found primarily in females between puberty and menopause. The word hysterical is derived from the term *uterine* (womb). It had been believed since the earliest times that there were symptoms of a general nature peculiar to women. From this followed that these symptoms must originate in the only organ peculiar to women, the womb. A disorder of the uterus was believed to indirectly lead to hysterical symptoms. One of the most dramatic of these symptoms is suffocation or difficulty in breathing. Thus hysteria was referred to as *suffocation of the mother,* because it was most commonly manifest by a gagging in the laryngeal area. When the predominant hysteric symptom is a sudden loss of voice, usually accompanied by anxiety or disappointment, we call the problem hysterical aphonia. Usually there is a complete loss of voice. In some cases, however, the patient is capable of phonating in a whisperlike voice. The psychodynamics of hysterical aphonia are quite important in understanding this complex problem. Although the psychodynamic aspects of this problem seem to vary from case to case, psychic trauma and the consequential denial of this trauma, via loss of voice, seems to be common theme of the disorder. Bangs and Freidinger

(1950) and Van Riper and Irwin (1959) discuss some of the early literature in this area. Bangs and Friedinger (1949) also mention some of the early studies and suggest that a differential diagnosis between true and hysterical aphonia can be made if the vocal folds do come together along the midline during coughing. Douglas (1955) reported a case of loss of voice after the administration of anesthesia. He interpreted this as a use of the symptom to protect the patient from expressing his true feelings.

Barton (1960) describes the frequently observed whispering syndrome in cases of hysterical aphonia. He points out that whispering is sometimes erroneously labeled hoarseness and that his subjects were not only unconcerned about their problem but almost seemed relieved by it. He recommends psychoanalysis as the best method of treatment and cautions laryngologists not to treat the patient in the hope that he will come around.

Guze and Brown (1962) reviewed the case histories of 12 patients with functional voice disorders. Their results are summarized as follows:

1. Functional voice problems are more common in adult females.
2. Aphonia is more likely to be associated with psychiatric illness than dysphonia.
3. Aphonia and dysphonia may be seen with an absence of any psychiatric illness.
4. Hysteria is associated with aphonia rather than dysphonia.
5. Life stress and emotional reactions influence the voice in two-thirds of their patients.

Several recent papers in the literature stress new approaches in the treatment of hysterial aphonia. Rose (1962) reported on the use of hypnosis and suggestion.

Walton and Black (1960) report on the application of learning theory in the treatment of a case of hysterical aphonia. The authors found support for the idea that the aphonia developed originally as a conditioned avoidance response that, because it satisfied a need, became reinforced. A method of treatment was evolved based on learning theory constructs of reactive inhibition: the treatment was successful removing the symptom completely. A follow-up one year later showed no relapses or alternative symptoms.

Wolski and Wiley (1965) describe a program of treatment utilizing the services of psychiatry and speech therapy. The authors utilized an indirect approach to therapy, rather than emphasis upon the mechanics of speech. The treatment was successful and the authors recommend their program for other kinds of functional aphonia.

Brodnitz (1965) gives a thorough discussion of how psychological factors effect phonation through the analysis of laryngeal hypofunction and hyperfunction. Hyperfunction and hypofunction are extremes of a single continuum. At the extreme of the hypofunctional end is hysterical aphonia. Brodnitz, in another article, described this type of individual as an individual who has given up the will to speak. At the hyperfunctional extreme, we find the condition of spastic dysphonia. Some of the more recent findings regarding this problem are summarized in Lawrence (1979).

Phonatory Problems due to Abnormal Resonance (Supraglottal Cavities)

The use of the supraglottic cavities resulting in vocal quality is the element of speech that enables the listener to distinguish one speaker from another. The way in which one resonates the vocal tones emanating from the larynx among men. Resonance, and the disturbances of resonance, are therefore a vital aspect of a study of functional disturbances of the voice. West, Ansberry, and Carr (1960) state that, with the possible exception of the lisp, excessive nasality (one of the common disturbances of resonance) is the most prevalent of all speech defects.

Disturbances of resonance can be divided, for purposes of study into two major categories: Excessive nasal resonance (hyperrhinolalia) and inadequate nasal resonance (hyporhinolalia), or hyperrhinophonia and hyporhinophonia.

Hyperrhinolalia Aperta

In so-called normal American speech, there is a certain amount of nasal resonance of all voiced sounds in addition to the three nasals represented by /m/, /n/, and /ŋ/. In essence, therefore, the distinction between normal and excessive nasal resonance is not one of kind, but rather one of degree. Calnan (1953) and Bloomer (1953) report the absence of complete palatopharyngeal closure in speakers who did not demonstrate excessive nasal resonance. Kaltenborn (1948) relates the distinction between nasal and nonnasal resonance to the relationship between the size of the oronasal part and the opening into the mouth between the tongue and velum and concluded that nasality is the result of having too large an opening into the nasopharynx compared with the size of the opening into the oral cavity. Adler (1960) relates the presence of this type of hypernasality to a resonatory coupling between the oral and nasal cavities and indicates that excessive nasality occurs when a critical ration between the size of the opening of the nasopharynx and the size of the opening of the nares is reached.

Hyperrhinolalia Clausa

When the anterior portion of the nose is constricted, an increase in the degree of nasality is observed. This phenomenon is caused, according to West *et al.* (1960) by the action of the nasal cavities as a *cul de sac* or closed cavity resonator. This *cul de sac* resonance resulting in excessive nasality is frequently referred to as hyperrhinolalia clausa.

Hyporhinolalia

In instances where the nasopharyngeal port does not open sufficiently to allow any nasal resonance, the result is a damping of the high frequency overtones, and the vocal quality tends to sound "flat" or "hollow." This lack of adequate nasal resonance is called hyporhinolalia. This condition is frequently associated with disturbances of articulation characterized by high posterior tongue placement and low tongue-tip placement, and resulting distortion of linguaalveolar consonants.

Assimilation Nasality

Among many speakers who normally maintain production of phonemes free of excessive nasal resonance, there may occur nasal emission during the production of vowel sounds, immediately preceded by or followed by nasal consonants /m/, /n/, or /ŋ/. This type of nasality is frequently referred to as associated or assimilation nasality. Fairbanks (1960) describes this kind of nasality and states that while vowels followed by nasal consonants and those preceding nasal consonants are affected, those preceding the nasal consonant are usually more severely affected. This phenomenon is undoubtedly related to the anticipatory lowering of the soft palate and relaxation of the velopharyngeal closure.

Therapy

With somewhat Gilbertian logic, Gilbert himself to the contrary, much voice therapy has not been an attempt "to let the punishment fit the crime," but rather, in many cases, making the "criminal fit the punishment." Many therapists dealing with functional voice cases tend to maintain a rather consistent therapeutic approach regardless of the etiology or symptomatology displayed by the patient.

We thus frequently observe that some therapists approach all voice problems with "breathing," others with relaxation exercises, still others with voice placement and resonance training, while some, believing in the shotgun approach, try a little bit of everything.

In planning a therapeutic program for a person who has a functional voice disorder it is vital that the therapist bear in mind the interrelationship between voice and personality. In cases where the patient is considered to be emotionally disturbed, psychotherapy may be indicated with some bearing on the voice disturbance, either alone, or in conjunction with voice therapy. In any event, the quality of the relationship between the therapist and patient is of prime importance. Murphy (1964) suggests making the therapy setting one in which the client is encouraged to "predict new events before they are explained to him. . . . The individual needs to learn that he can find his own best solutions."

Treatment of Hyperfunction

The functional disorders of the voice created by hyperfunction of the vocal mechanism are frequently related to excessive tension of the entire body as well as to psychological tension. This relationship suggests that one approach to the problem of hyperfunctioning is to teach the process of relaxation. This procedure has been recommended in the past by such writers as Edmund Jacobson and Manser and Finlan (1951). Murphy states "Today routing exercises are used very little by speech pathologists, except in cases of muscular incoordination, weakness or inactivity." Brodnitz (1962) comments on the same approach: "Any system of exercises that endeavors to 'relax' the individual can be of only limited value." Many practitioners indicate, however, that relaxation, indirectly induced, may be of help in teaching the mechanics of normal phonation.

The chewing method, developed by Froeschels in the early 1930s, and advocated by such writers as Weiss and Beebe (1952), Froeschels and Jellinek (1941) and others, is described by Brodnitz (1962) as "using a twin-function of speaking (i.e., chewing)" and interprets much of the resistance to the use of the technique as due either to the misconception that it is a panacea, or to the lack of agreement for the theoretical rationale for its use.

Psychotherapy in its many forms has been used in the treatment of many voice disorders related to hypofunctioning. Mention only will be made here of some of the applications of psychotherapy for functional voice disorders. Heaver (1959) describes a case of spastic dysphonia and states "there can be no doubt that . . . psychiatric treatment should be the therapy of choice for these patients." Rousey (1961) describes the use of hypnosis as a form of psychotherapy used in the treatment of cases of aphonia. In discussing nodules and polyps of the vocal folds, Arnold (1962) states: "(Nodules) . . . are a secondary organic sign of a primary disorder of psycho-somatic or socio-economic origin. . . . Attention to emotional factors and the need for systematic re-education of the entire personality pattern are important."

Treatment of Hypofunction

Many of the approaches toward the reeducation of hyperfunction disturbances of the voice may be equally appropriate to the treatment of hypofunctions as well. The approach of psychotherapy may, for example, be used with equal effectiveness with many disorders of hypofunction.

For specific disorders of hypofunctional origin other approaches have been utilized. The technique of pushing exercise (Froeschels, Kastein, & Weiss, 1955) for hypofunctioning of the vocal folds, soft palate, and insufficient use of the diaphragm and muscles of exhalation is one such approach. Exercises for tonal energizing (Manser & Finlan, 1951; Perkins, 1957) have also been recommended for the treatment of this variety of disturbance.

In the clinical use of therapeutic techniques for the amelioration of functional voice disturbances, the therapist would do well to be familiar with the wide variety of approaches to be used and make use of those which are suitable to the needs of the patient as well as those which fit comfortably within the therapist's own theoretical frame of reference.

Conclusion

Within the scope of functional disorders of the voice there exist many areas in which research gives promise of fruitful results. As has been previously stated, there are many aspects of functional voice disturbances that have not been thoroughly examined. The conclusion of this chapter will be devoted to a consideration of several of these possible areas for study.

1. The study of interrelationships among etiological factors and vocal symptomatology. Some of the questions that might be examined include: What relationship, if any, exists between poor auditory discrimination or poor pitch perception and hoarseness, harshness, nasality, or pitch disorder? Is there a causal relationship between such personality factors as anxiety, hypertension, hypotension, aggressiveness, or punitiveness, and specific vocal symptoms?

What is the relationship between the adolescent's self-concept and sexual identification and pubertal disturbances of the voice? What, if any, are the sociological influences bearing on functional disturbances of phonation and resonance?

2. The study of interrelationships between phonatory techniques and habits and vocal symptomatology. Some of the questions that might be investigated in this area include: What, if any, is the relationship between habitual pitch and harshness? Between pitch and hoarseness? Between pitch and nasality? What, if any, is the relationship between vocal intensity and hoarseness? Between vocal intensity and harshness? Between vocal intensity and nasality?

3. The study of trends in the direction of change from functional or organic problems. Is there a relationship between continued harsh or hoarse phonation and eventual organic damage to the vocal folds? Do other hyperkinetic functional disturbances ultimately lead to organic pathologies?

4. The scientific measurement of physical and biological factors involved in phonation. Are there significant differences between normal speakers and those who demonstrate functional voice problems in terms of muscle action potentials or subglottic pressures, or movements of the intrinsic and extrinsic laryngeal muscles?

5. The study of developing vocal patterns in infants and young children. Studies in this area may be concerned with questions such as: are functional disturbances of the voice related to specific patterns of vocal development? Can accurate predictions be made of the incidence of vocal disorders? If so, by what age can the prediction be accurately made?

6. The investigation, through controlled experimentation, of the relative efficacy of the variety of therapeutic techniques and approaches, including the use of psychotherapy, relaxation techniques, chewing therapy, pushing exercises, and the many other approaches commonly used.

7. The study of the observable or measurable aspects of vocal quality. Is there a set of measurable variables by which we may objectively differentiate among such qualitative aspects of voice such as hoarseness, huskiness, breathiness, stridency, brassiness, tenseness, or throatiness? What are these variables, and how do they affect the clinician's judgment of the vocal quality?

The foregoing represent but a sampling of the large number of problems that represent as-yet-unanswered, or incompletely answered, questions faced by the clinican working with the functional disorders of the voice.

References

Adler, S. Some techniques for treating the hypernasal voice. *Journal of Speech and Hearing Disorders*, 1960, *25*, 300–302.

Arnold, G. E. Spastic dysphonia. I. Changing interpretations of a persistent affliction. *Logos*, 1959, *2*, 3–14.

Arnold, G. E. Vocal modules and polyps: Laryngeal tissue reaction at habitual hyperkinetic dysphonia. *Journal of Speech and Hearing Disorders*, 1962, *27*, 205–217.

Aronson, A. E., Peterson, H., & Litin, E. M. Symptomatology in functional dysphonia and aphonia. *Journal of Speech and Hearing Disorders*, 1964, *29*, 367–380.

Bangs, J. L., & Freidinger, A. Diagnosis and treatment of a case of hysterical aphonia in a fourteen year old girl. *Journal of Speech and Hearing Disorders*, 1949, *14*, 313–317.

Bangs, J. L., & Freidinger, A. A case of hysterical dysphonia in an adult. *Journal of Speech and Hearing Disorders*, 1950, *15*, 316–323.

Barton, R. T. The whispering syndrome of hysterical dysphonia. *Annals of Otology, Rhinology, & Laryngology*, 1960, *69*, 156–164.

Berg, J. van den. Direct and indirect determination of the mean subglottic pressure. *Folia Phoniatrica*, 1956, *8*, 1–24.

Bloomer, H. H. Observations on palatopharyngeal movements in speech and deglutition. *Journal of Speech and Hearing Disorders*, 1953, *18*, 230–246.

Brown, J. R., & Simonson, J. Organic voice tremor. *Neurology*, 1963, *13*.

Brodnitz, F. Functional voice disorders. In N. Levine (Ed.), *Voice and speech disorders (medical aspects)*. Springfield, Ill.: Charles C Thomas, 1962.

Brodnitz, F. R. *Vocal rehabilitation*. Rochester, Minn.: American Academy of Opthalmology and Otolaryngology, 1965.

Calnan, J. Movements of the soft palate. *British Journal of Plastic Surgery*, 1953, *5*.

Cleeland, C. E. *Definitions of a voice quality called X*. Unpublished doctoral dissertation, University of Denver, 1948.

Darley, F. L. *Diagnosis and appraisal of communication disorders*. Englewood Cliffs, N.J.: Prentice-Hall, 1964.

Diehl, C. F. Voice and personality: An evaluation. In D. Barbara (Ed.), *Psychological and psychiatric aspects of speech and hearing*. Springfield, Ill.: Charles C Thomas, 1959.

Diehl, C. F., & McDonald, E. T. Effect of voice quality on communication. *Journal of Speech and Hearing Disorders*, 1956, *21*, 233–285.

Diehl, C. F., White, R., & Burg, K. W. Voice quality and anxiety. *Journal of Speech and Hearing Research*, 1959, *2*, 282–285.

Douglas, B. L. Postanesthetic hysterical alphonia. *Oral Surgery, Oral Medicine, Oral Pathology*, 1955, *8*, 1270–1271.

Fairbanks, G. *Voice and articulation drillbook*. (2nd Ed.). New York: Harper & Row, 1960.

Freud, E. Functions and dysfunctions of the ventricular folds. *Journal of Speech and Hearing Disorders*, 1962, *27*, 334–340.

Froeschels, E. Laws in the appearance and the development of voice hyperfunctions. *Journal of Speech Disorders*, 1940, *5*, 1–4.

Froeschels, E. Hygiene of the voice. *Archives of Otolaryngology*, 1943, *38*, 122–130.

Froeschels, E., & Jellinek, A. *Practice of voice and speech therapy*. Magnolia, Mass.: Expression, 1941.

Froeschels, E., Kastein, S., & Weiss, D. A. A method of therapy for paralytic conditions of the mechanism of phonation, respiration, and glutination. *Journal of Speech and Hearing Disorders*, 1955, *20*, 365–370.

Guze, S. B., & Brown, O. L. Psychiatric disease and functional dysphonia and aphonia. *Archives of Otolaryngology*, 1962, *76*, 84–87.

Heaver, L. Spastic dysphonia. II. Psychiatric considerations. *Logos*, 1959, *2*, 15–24.

Hollien, H., Moore, P., Wendahl, R., & Michel, J. On the nature of vocal fry. *Journal of Speech and Hearing Research*, 1966, *9*, 245–247.

Johnson, W., Darley, F. L., & Spriestersbach, D. C. *Diagnostic methods in speech pathology*. New York: Harper & Row, 1963.

Kaltenborn, A. L. *An X-ray study of velopharyngeal closure in nasal and non-nasal speakers*. Unpublished M. A. thesis, Northwestern University, 1948.

Kramer, E. Personality stereotypes in voice: A reconsideration of the date. *Journal of Social Psychology*, 1964, *62*, 247–251.

Lawrence, van L. *Spastic dysphonia: State of the art*. New York: The Voice Foundation, 1979.

Manser, R. B., & Finlan, L. *The speaking voice*. New York: Longmans, Green, 1951.

Moore, G. P., & von Leden, H. Dynamic variations of the vibratory pattern in the normal larynx. *Folia phoniatrica*, 1958, *10*, 205–238.

Moser, H. M. Symposium on unique cases of speech disorders: Presentation of a Case. *Journal of Speech Disorders*, 1942, *7*, 173–174.

Murphy, A. *Functional voice disorders*. Englewood Cliffs, N.J.: Prentice-Hall, 1964.

Perkins, W. H. The challenge of functional disorders of voice. In L. E. Travis (Ed.), *Handbook of speech pathology*. New York: Appleton-Century-Crofts, 1957.

Perkins, W. H. Vocal function: assessment and therapy. In L. E. Travis (Ed.), *Handbook of speech pathology and audiology*. New York: Appleton-Century-Crofts, 1971, p. 513.

Perkins, W. H., & Yanagihara, N. Parameters or vocal production: I. Some mechanisms for the regulation of pitch. *Journal of Speech and Hearing Research*, 1968, *11*, 246–267.

Rees, M. Harshness and glottal attack. *Journal of Speech and Hearing Research*, 1958, *1*, 344–349.

Rose, J. T. The use of relevant life experience as the basis for suggestive therapy. *International Journal of Clinical Experimental Hypnosis*, 1962, *10*, 227–229.

Rousey, C. L. Hypnosis in speech pathology and audiology. *Journal of Speech and Hearing Disorders*, 1961, *26*, 3258–3267. (Cites 2 references to use of hypnosis as treatment of aphonia.)

Steer, M. D., & Hanley, T. D. Instruments of diagnosis, therapy, and research. In L. E. Travis (Ed.), *Handbook of speech pathology*. New York: Appleton-Century-Crofts, 1957.

Van Riper, C., & Irwin, J. V. *Voice and articulation*. Englewood Cliffs, N.J.: Prentice-Hall, 1959.

Walton, D., & Black, D. A. The application of modern learning theory to the treatment of chronic hysterical aphonia. In H. S. Eysenck (Ed.), *Behaviour therapy and the neurosis*. London: Pergamon, 1960.

Weiss, D. A., & Beebe, H. H. *The chewing approach in speech and voice therapy*. Basel and New York: Karger, 1952.

Wendahl, R. W., Moore, P., & Hollien, H. Comments on vocal fry. *Folia phoniatrica*, 1963, *15*, 251–255.

West, R. W., & Ansberry, A. M. *Rehabilitation of speech*. New York: Harper & Row, 1968.

West, R., Ansberry, M., & Carr, A. *The rehabilitation of speech*. (3rd Ed.) New York: Harper & Row, 1960.

Williamson, A. B. Diagnosis and treatment of 72 cases of hoarse voice. *Quarterly Journal of Speech*, 1946, *31*, 189.

Wolski, W., & Wiley, J. Functional aphonia in a fourteen-year-old boy. *Journal of Speech and Hearing Disorders*, 1965, *50*, 71–75.

Zaliouk, A. The tactile approach in voice placement. *Folia phoniatrica*, 1963, *15*.

E. Harris Nober

PSEUDOHYPACUSIC DISTURBANCE IN ADULTS

Introduction

In audiometry, descriptive terms like hearing loss, hearing disorder, hearing disturbance, hearing dysfunction, and dysacusis are used in specific contexts. More recently, the term "auditory" has gained favor over "hearing" since the former relates to a physiological system and not just a perceptual process *per se*. In the past, the main body of literature dealing with pseudohypacusic disturbance employed "hearing loss" in spite of the fact that this term may lack validity. Often, there is no physiological loss in pseudohypacusic disturbance. This author will reserve hearing loss to describe actual shifts in auditory sensitivity and use it in direct quotations from the literature. Auditory disturbance will be used in lieu of hearing loss since the former is more valid by definition alone. Hearing loss implies a reduction in auditory sensitivity or threshold shift and this may or may not be the case.

There are constraints and implications in all the terms, i.e., functional, psychogenic, nonorganic, pseudohypacusis, etc. For example, Goldstein (1966), in his discussion for support of Carhart's (1961) term "pseudohypacusis," used "hearing loss," "hearing impairment," and "auditory disorder," interchangeably. The term pseudohypacusis seems to be gaining support (Katz, 1978) if only out of desperation for a single term to reduce ambiguity (Martin, personal communication, Feb. 9, 1980), and for this reason it will be used in this manuscript.

A pseudohypacusic disturbance can range in magnitude from a distressed

E. Harris Nober • Department of Communication Disorders, University of Massachusetts, Amherst, Massachusetts 01003.

mother's complaint that her child ostensibly hears what he wants to hear to a severe and stifling psychic deafness of a terrified soldier. The former is a selective "tuning out" phenomenon that has been utilized by most of us at one time or another. However, at the other end of the continuum, the dynamics are more complex. Prior to World War II, there was almost no literature relevant to pseudohypacusic disturbance, but since the early 1950s pertinent research has appeared. Pseudohypacusis can occur at all ages, for both sexes and in different cultural, socioeconomic, and racial groups. The symptoms include false or exaggerated hearing loss for personal advantage or unconscious motivation (F. Martin, 1978). Most of the research appears to deal with one or more of three aspects of the problem. i.e., *classification, identification,* and *rehabilitation.*

A striking pattern in classification has been to dichotomize, i.e., organic vs. nonorganic or organic vs. functional, psychogenic vs. physiogenic. Some professionals have questioned the validity of dichotomized differentiation, i.e., organic vs. nonorganic, because they feel that all organic hearing impairments are also confounded with some psychological overlay. Fowler (1947) writes, "There is probably no such thing as a purely organic disability; every somatic disturbance tends to produce a psychic change." Farley, Waldrop, Derbyshire, Carter, Austin, McCormick and Mills (1960) also accepted this type of holistic view and defined hearing as "the entire sequences of events, initiated when a sound stimulus reaches the tympanum and concluded when the organism has completed a meaningful response to that stimulus." Many authorities who support this view still use some dualistic classification system but acknowledge the multiple complexities in any disorder. Pseudohypacusis does have the advantage of not dichotomizing the disturbance.

It is not primarily by preference that dichotomized classifications prevail. Part of the reason is the inability to develop definitive diagnostic techniques that isolate pseudohypacusis as an entity. This difficulty is reflected in the redundant assortment of terms that have evolved through the years and are still in current use. Many terms connote different things to different people (Williamson, 1974). Some of these terms reflect the professional areas from which they emanate. A brief review of these terms will provide an interesting historical perspective as well as the connotations commonly associated with them.

Functional Auditory Disturbance

Functional[1] auditory disturbance is designated for both malingering and psychogenic problems. It does not semantically preclude the possibility of some

[1] "Functional" is also used in medical parlance to describe abnormal physiologic or organic activity without tissue pathology and should not be confused with the psychogenic connotation implied here.

organic involvement as does the term "nonorganic." Furthermore, psychologists and medical personnel often use the term functional. Clearly, functional imples that *at the present time* organic pathology cannot be demonstrated. Davis and Silverman (1978) prefer functional dysacusis as a separate class of central auditory impairment.

Nonorganic Auditory Disturbances

Nonorganic auditory disturbance parallels functional auditory disturbances in connotation but ostensibly rules out the possibility of an undetected organic component. Hopkinson (1978) asserted that the term gained popularity in post-World War II Veterans Administration programs. Farley *et al.* (1960) were concerned and warned that the term nonorganic assumed that the psyche was independent of the organism and actually antagonistic, i.e., if one was operating, the other was not. While the term nonorganic, as currently used, is quite inclusive and even encompasses the malingering syndrome, it is more closely related to the term psychogenic.

Psychogenic Auditory Disturbance

Psychogenic auditory disturbance denotes psychological origin. It is one of the more frequently used of the terms. Often, it is employed in contrast to the term "malingering," the latter a form of conscious deception. Psychogenic auditory disturbance indicates a severe emotional stress is being manifested as an auditory impairment. Davis and Silverman (1978) use the term to connote "an extreme form of functional or nonorganic dysacusis." Acoustic signals are transmitted to the brain but they are not consciously perceived and consequently the person is unaware that he can actually hear.

Included in psychogenic auditory disturbances is an extensive variety of subtypes or changes in personality. Fowler (1947) asserted that there were as many varieties as individuals involved. Below are some of the major subdivisions that categorize the different types:

Conversion Hysteria. A *conversion reaction* (also regarded as an *attention disorder)* is mostly associated with emotional conflicts related to war stress and is actually a rare entity in civilian life. When it occurs in civilian life, as an isolated syndrome, it is usually associated with a head injury or an auditory trauma. The hearing disturbance, which is often marked, represents an expression of profound emotional conflicts that associate with auditory imagery. Subconsciously, these people have repressed their hearing and are unaware that they can hear (Bailey & Martin, 1961; Ballantyne, 1960; Klotz, Koch, & Hackett, 1960). Some even refer to the hysteria syndrome as an *unconsious hearing loss*. These subjects ostensibly encounter the same sensory deprivation inconveniences associated with organic hearing loss.

Psychopathic Deafness. Fournier (1962) employed this term to describe mental confusion, severe mental retardation, and various psychoses that include hysteria as a general psychotic syndrome. It has not received widespread use in this country.

Psychic Deafness. Farley, McCormick, Oppfeli, and Mills (1962) used this term in much the same context as psychogenic deafness. They explained the interactions between the psyche and the auditory modality under three psychological conditions: (1) *availability*—that there is some model in the person's background where the experience and behavior of deafness fit into this pattern; (2) *necessity*, which refers to the need to resolve the problem in this fashion; (3) *maintenance*, which refers to the gratification elicited from this behavior.

Situational Deafness. Fowler (1947) described a variant of psychogenic overlay in which the situation is the cause of the trouble, i.e., war industry. He wrote that a change of the situation usually diminished the tensions and the symptom of deafness became lessened in severity. The duration of the extreme disability may be relatively short but the person continues to utilize these established patterns of deafness after the remission of the symptoms.

Depression Deafness. Coined by Ramsdell (Davis & Silverman, 1978) this term assumes a schizophrenic prototype where the individual isolated himself from reality by excluding auditory sensations. This represents a hypochrondriacal preoccupation with the subjective effects of deafness.

Ramsdell (1978) also described a typical and normal depression that is associated with organic hearing loss. There are feelings of sadness, deadness, and even insecurity. It is especially prevalent in adults who suddenly lose their hearing. The dynamics of this universal depression are explained as follows: The background sounds of our environment are subconsciously perceived and form a type of *primitive hearing*. This primitive hearing provides the affective background for our readiness to react. The person feels he or she is part of an active, living world. When the primitive contact is no longer present, the ongoing auditory interface with the environment is severed and a feeling of depression results. While there are no experimental studies to support this hypothetical construct, it has become a widely quoted explanation of the depression associated with sudden deafness. Information is also wanting on whether this reaction is associated with pseudohypacusic disturbances.

Pseudoneural hypacusis (Pseudohypacusis). This term was used by Brockman and Hoversten (1960) and later by Rintelmann and Harford (1963) and its currently modified spelling (pseudohypacusis) was originally used to describe functional auditory disturbance in children. Those children would manifest an audiologic picture that was inconsistent with their speech pattern and their ability to understand the speech of others. There is also little reliability on repeated audiometric tests, a major symptom of pseudohypacusic disturbance. Carhart (1961) used pseudohypacusis to connote simulated hearing loss. Others (Hopkinson, 1978; F. Martin, 1978) have switched to this term (also spelled pseudohypacusis) because it does not imply any basic causal factor.

Malingering

Malingering is the conscious and premeditated attempt to deceive others into believing there is an actual hearing loss when hearing acuity is normal, or to exaggerate the symptoms of an organic loss. The most prevailing underlying motivation is to secure financial compensation for disabilities associated with industrial or governmental activities. Johnson, Work, and McCoy (1956) estimated that the incidence ranged from 11% to 45% in Veterans Administration populations, and Gaynor (1974) predicts increases. Other sources give varying estimates (N. A. Martin, 1946; Morrissett, 1946; Truex, 1946) while Gibbons (1962) cited that one area of this country had a 90% incidence because of a marked economic depression. More recently Barelli and Ruder (1970) found 24% of 116 compensation cases had pseudohypacusis. It is now seen more often in conjunction with industrial claims and automobile accidents. Quite often—perhaps too often—it is used with children when the audiologic results are inconsistent or unreliable. This author agrees with F. Martin (1975) who asserted the term is "probably used far too often."

Identification

In spite of the extensive development of elaborate techniques to uncover a pseudohypacusic disturbance, there is still no single test that universally isolates the condition with total assurance. Many question whether there is a pseudohypacusic disturbance entity *per se* or whether it exists only with other problems. Dixon and Newby (1959) advised the systematic collaboration of an otologist, an audiologist, a psychologist, a social worker, and a psychiatrist. A definitive diagnosis necessitates the integrated findings of all these specialists. Most of the identification techniques fit into the domain of at least one or more of the above experts. For convenience this author will review identification techniques under three subheadings: (1) *psychological*, (2) *otological*, (3) *audiological*.

Psychological Techniques

Behavioral descriptions, such as tension, anxiety, inconsistency, evasiveness, and a multitude of similar terms that denote stress or neurosis, have been employed to describe people with pseudohypacusic disturbance. These descriptions often allude to the precipitating cause or even the resultant defense reactions. Symptomatically, pseudohypacusic disturbance and malingering are difficult to distinguish (Davis & Silverman, 1978). Fowler (1947) dwelled on this point and asserted that malingering represented a form of psychogenic disturbance because the malingerer was being driven to escape from overwhelming circumstances. Psychiatric explanations with school-age children focus on the conflict between the home and school environments (Barr, 1960), and on the need for recognition and attention (Ross, 1964). Certainly, the medical case history provides extremely pertinent clues including client reactions, medical and health history, source of referral, and other sociological data. Not infrequently, answers to case history questions are more emotionally exaggerated.

Behavioral–Social Patterns. Since these people characteristically have a predilection for stereotyped symptoms, it is often possible to identify some subjects from their behavioral patterns. Johnson *et al*. (1956) emphasized the merit of using behavioral cues. Thorne (1960) even developed an extensive checklist called *psycho-tell*. N. A. Martin (1940), Heller, Anderman, and Singer (1955), Saltzman (1949), and Weiss and Windrem (1960) also advocated the pertinence of subjective impressions from behavioral clues such as: nervousness, trembling, perspiration, compulsive reactions, exaggerated responses, unsolicited questions regarding compensation, exaggerated lip-reading capacity, curiously good speech for the degree of threshold shift, hesitancy, delayed responses, and inconsistency.

The last symptom, *inconsistency*, is so pervasive it is often considered a separate entity. Perhaps this point is best accentuated with Harbert's (1943) paradoxical statement that the most consistent aspect of pseudohypacusic disturbance is the inconsistency. The test–retest techniques, the cross-checks between otological and audiological findings, and the symptomatic behavior are based on the criterion of inconsistency and incompatability. Some use the descriptor "incompatibility" in lieu of inconsistency. Generally, in the literature, inconsistency is related to test–retest reliability or replication of results. Incompatability denotes a lack of agreement among tests such as the pure tone average at 500 Hz, 1000 Hz, 2000 Hz and the speech reception threshold using spondee words. Inconsistency and incompatability are especially prevalent in children when hearing problems identified during formal testing may not interfere with normal communication or be reflected in the speech patterns. Pseudohypacusic disturbance in children is based on unconscious repressions rather than deliberate preconceived attempts to deceive (Leshie, 1960) and is perhaps very different in motivation when compared to adults (Dixon & Newby, 1959).

Psychometric Studies. Studies here attempted to relate certain psychological correlates to pseudohypacusic disturbance in adults. Chaiklin and Ventry (1963) surveyed the literature in this area and concluded that people with pseudohypacusic disturbances are average or even below average in intelligence, are poorly adjusted, and are often hypochondriacs. But with children the reverse is true as they tend to be average or above average intelligence (Barr, 1960). Some of the other studies showed these people also exhibited neurotic profiles (Kodman, Sedlacek, & Powers, 1959) and an excess of somatic complaints (Ballantyne, 1960). Doerfler (1951) listed some of these complaints: pain, cutaneous hypersensitivity, menianethesis, blandness, loss of smell, loss of taste, vertigo, and amblyopia.

Otological Techniques

Inconsistencies in the case history, the medical examination, and specific behavioral patterns alert the otologist to a tentative diagnosis of pseudohypacusic disturbance. Some of his clues are: total deafness with normal labyrinthine functioning, good responses under sodium pentathol, discrepancies between tuning fork tests and other results, sudden onset of a progressive bilateral loss. Indeed, the sudden onset of a hearing disturbance is an extremely significant clue and very common in children. More recently, electronystagmography has been widely adopted by otologists to compare peripheral vertibular functioning to cochlear integrity.

Audiological Techniques

The first part of the audiologic evaluation consists of the standard pure tone air, bone, and speech tests. These tests are then repeated under varying conditions to determine the consistency of the replicated responses. After the audiologist administers the standard audiometric battery, which may include site of lesion tests as well, special procedures may be given to assess the pseudohypacusic component. Below is a description of these tests and how they are related to or are modified for the assessment of pseudohypacusic disturbance.

Pure Tone Audiometry

Repeated Air Conduction Threshold Test. Air conduction thresholds do not usually vary more than ±5 dB on repeated tests. A replicated air conduction threshold that exceeds the ±5 dB variance is usually one of the first suspect clues for detecting a pseudohypacusic disturbance (Heller *et al.*, 1955; Johnson *et al.*, 1956; Sataloff, 1957).

Ascending vs. Descending Methods. Harris (1958) compared the threshold difference between the ascending and descending methods and asserted that this difference also should not exceed 5 dB; he felt a difference of 10 dB or more could suggest a nonorganic loss. The ascending–descending measure above identified 81% of the pseudohypacusic cases in the Menzel (1960) study and 66% in the Ventry and Chaiklin (1965b) study. Kacker (1971), using a simulated hearing-loss design, found the ascending–descending measure more reliable. Other researchers found little or no difference in test–retest consistency (Berger, 1965; Shepherd, 1965).

Configuration of the Audiogram. Several attempts have been made to describe an air-conduction configuration that typifies a pseudohypacusic disturbance. A saucer-shaped audiogram was first described by Doerfler (1951) and altered by Carhart (1958). Allegedly, the saucer follows the 60 phon equal-loudness contour. Doerfler (1951) did not find the saucer audiogram was a good criterion. Kacker (1971), too, found the saucer audiogram unreliable as it identified only 15% of his simulated losses. Another audiogram pattern described in the literature is a "flat" loss (Fournier, 1958; Semenov, 1947). Barr (1960) described a bilateral flat perceptive loss at 60–80 dB for the typical pattern of school-age children. Still another common audiogram pattern consists of fragmented indications of severe deafness. Indeed, there is no pathognomonic audiogram pattern.

Bone-Conduction Thresholds. Functional auditory disturbances have bone-conduction thresholds that have been described as typically poorer than the air-conduction thresholds (Doerfler, 1951; Johnson *et al.*, 1956). Chaiklin and Ventry (1963), however, were unable to substantiate this air–bone relationship. Quite often an air–bone gap is noted, which may exceed 60 dB; here the bone lines were found to represent the organic component if there was one (Hopkinson, 1973). Nober (1967, 1970) also suggested the bone-conduction thresholds should not exceed the low frequency "cutile" (vibrotactile) thresholds.

Speech Reception Thresholds. Another criterion is provided by speech reception thresholds repeated for test–retest reliability. Normally, variations in speech reception threshold should not exceed ±6 dB (Carhart, 1946a,b; Ruhm & Carhart,

1958; Tillman & Jerger, 1959). Discrepancies in excess of this value become suggestive of a pseudohypacusic component. Menzel (1960) and Chaiklin and Ventry (1963) reported that approximately 40% can be identified with this criterion. Subjects with pseudohypacusic disturbance tend to give legitimate speech reception thresholds. This is especially true of children. The inexplicable tendency to give valid speech reception thresholds and invalid air-conduction thresholds provides an effective clue for identifying pseudohypacusic disturbance. This leaves poor agreement between the speech reception thresholds and the pure tone average for 500 Hz, 1000 Hz, and 2000 Hz (Carhart, 1946a; Jerger, Carhart, Tillman, & Peterson, 1959) or even the better two of these three frequencies (Fletcher, 1950). European audiologists refer to this pure tone average-speech reception threshold relationship as the *Carhart test*. The prevalence for failure on the Carhart test is well documented in the literature (Brockman & Hoversten, 1960; Carhart, 1952; Dixon & Newby, 1959; Glorig, 1954; Juers, 1956; Portman & Portman, 1961; Rintelmann & Harford, 1963; Ventry & Chaiklin, 1965a). Some researchers report a disproportionate number of "nonresponse" designations and half word speech reception threshold responses.

Discrimination Tests. These speech tests are given suprathreshold and are generally ineffective as indices for detecting pseudohypacusic disturbance. Hence, they are in limited use although Johnson *et al*. (1956) and Carhart (1960) did report significantly lower discrimination scores than would be predicted from the air-conduction thresholds. There are also a large number of nonresponse errors and poor consistency among words. While scores may be similar, on retest different words are missed (Hopkinson, 1978).

Calearo (1957) introduced a discrimination test only for detecting unilateral pseudohypacusic disturbances. It was based on the research of Miller and Licklider (1950) and Cherry and Taylor (1954) who found that intelligibility varies directly with the interruption rate, i.e., low rates give low scores and high rates give high scores. The interruptions (a constant 50/50 on–off ratio) produced rapid improvement in intelligibility scores when increased from one per second to 10 per second; scores varied from 50% to 100%, respectively. Calearo (1957) alternated sentences between the two ears at a level 30 dB above the threshold of the better ear; the bad ear represented the off phase. He claimed impressive success with this technique in identifying pseudohypacusic disturbance. Reliability of discrimination scores can be checked by checking the percentages against established values associated with the "articulation curves" (Hopkinson, 1978). At 0 dB SL, the discrimination scores are about 25%, at 5 dB they are 50%, and at 40 dB 100%. Snyder (1977) has noted a pseudohypacusic response syndrome.

Story Tests. This test essentially delivers a story in three parts, i.e., one part is given 20 dB above the better ear, a second about 10 dB below the poorer ear, and a third part bilaterally (Hopkinson, 1978). Response in the poorer ear indicates at least that much hearing. It is basically a test for unilateral loss. In essence, the test is a formalized version of a questioning interview an audiologist would conduct through a two-channel audiometer, simulating that used in the Veterans Administration battery.

Unconditional Reflexes. These are specific, unlearned, physiological reactions to a stimulus. Several auditory reflexes have been utilized for detecting pseudohypacusic disturbance:

Lombard Voice Reflex. This reflex is based on the phenomenon of auditory feedback. There is a subconscious, almost reflexive, modulation of the person's speech power output relative to the environmental noise level. As the environmental noise level varies, the individual accordingly adjusts his speech level to maintain a relatively constant signal-to-noise ratio; thus, a 10 dB increase in environmental noise level would precipitate an approximate 10 dB increase in the speech level output in most normally hearing persons.

Several variations of the Lombard test can be used depending on whether the pseudohypacusic disturbance is binaural. Usually, in cases of binaural disturbance the masking noise is applied binaurally (Taylor, 1946); with unilateral disturbance most authorities advocate masking the better ear (Harbert, 1943; Pitman, 1943; Saltzman, 1949) although Heller *et al.* (1955) suggested masking both ears. As subjects show wide variations in Lombard reflex responses, the tendency is to interpret the results conservatively.

Auropalpebral Reflex. The auropalpebral reflex or the cochleopalpebral reflex is more commonly associated with hearing tests for infants. The reflex is considered to be extrinsic since its occurrence does not alter the sensitivity of the ear. Approximately 70–90 dB or more is required to trigger it (Galambos, Rosenberg, & Glorig, 1953). Any response below 70–90 dB may suggest recruitment. This level also applies to the intraural muscle reflex. The auropalpebral reflex has limited value for detecting a pseudohypacusic disturbance. Fournier (1958) discussed some of the reasons for its limitations: the reflex is not universally found in hearing persons; the stimulus must be significantly above thresholds so the absolute threshold is not readily ascertained; finally, it may not be mediated through the cortex. However, the test has been successfully used to determine the auditory response of newborn infants (Wedenberg, 1956, 1963).

Intraural Middle Ear Muscle (Stapedius) Reflex. This reflex results from a bilateral contraction of the tensor tympani and the stapedius muscles. This is an intrinsic reflex as it increases the stiffness of the ossicular chain coupling and subsequently causes a mechanical reduction of sensitivity for the low frequencies. The acoustic reflec is one measure in the impedance battery of middle ear testing.

In addition to reducing low-tone sensitivity, the reflex increases the impedance of the tympanic membrane (Metz, 1946), which can be measured by the use of an electroacoustic bridge (Thomsen, 1955). A 220-Hz tone is presented to the supposedly deaf ear at 70–90 dB above audiometric zero (sensation level). If an impedance change occurs in the good ear it indicates sufficiently acuity in the poorer ear to have triggered the bilateral contractions. Jepson (1953, 1963) indicated success but pointed out that it can be used only with people who claim to have severe or complete hearing loss. Indeed, the alleged loss would have to exceed the 70–90 dB required to elicit the reflex. Lamb and Peterson (1967) described this procedure as imprecise and at best a qualitative procedure. See Chapter 3 for details of impedance audiometry.

Delayed Auditory Feedback. Delayed auditory feedback has served a multitude of useful purposes. Briefly, the speech signal is returned to the speaker at a delay of approximately 200 msec (Black, 1955). Reactions vary markedly; i.e., disruption of the speaker's fluency, variation in rate, hesitancy in rhythm, increase in intensity, partial or total breakdown in syntax, test withdrawal, exaggerated affection such as surprise, laughing, and fear.

The delayed auditory feedback was readily adaptable as a test of pseudohypacusic disturbance (Gibbons & Winchester, 1957; McGranahan, Causey, & Studebaker, 1960; Spilka, 1954; Tiffany & Hanley, 1962), but it never achieved widespread acclaim. Eventually, delayed feedback techniques that used pure tones and a key tapping procedure were developed (Causey & McGranahan, 1959; Chase, Sutton, Fowler, & Ruhm, 1961a; Chase, Sutton, & Rapi, 1961b; Cooper & Stokinger, 1976; Karlovich & Graham, 1966; Rapin, Costa, Mandel & Fromowitz, 1963; Ruhm & Cooper, 1962, 1963, 1964). In the Ruhm and Cooper technique, the subject is required to tap a temporal pattern with pure tones under the following conditions: (1) the tone is fed back simultaneously, and then (2) the tone is returned with a delay and at an intensity level supposedly below threshold. The significant clues are the tapping rate, the pressure applied, and the number of errors between synchronous and delay conditions. They reported good results.

Lateralization. This denotes the ability to trace a sound spatially from one ear to the other through 180° plane. One lateralization test described by Kodman (1961) is based on transcranial transmission. Pure tones and speech stimuli should be perceived when the intensity in the deaf ear exceeds the air-conduction intercranial transmission attenuation, i.e., 40–50 dB Kodman (1961) describes obtaining pure tone thresholds, a speech reception threshold, and a discrimination score from the contralateral ear, which represent shadow responses.

Another test is the swinging tone or swinging voice test. It is more specifically a location test since it involves the sudden and periodic shifting of a tone or key word to the ear with the alleged pseudohypacusic impairment. The intensity level is varied from above to below the threshold of the supposed bad ear (Carhart, 1960; Goetzinger & Prowd, 1958; L. Watson & Tolan, 1949) until it can be demonstrated that the stimulus was received in the deafened ear. This technique has met with limited success.

Masking. Masking refers to a change in the listener's perception of one stimulus due to the intervening effect of a sound stimulus—the masking stimulus. The masking stimulus can be presented during, before, or after the onset of the signal. Several audiometric techniques based on masking are employed for detecting pseudohypacusic disturbance:

Stenger Test. In diotic stimulation (the identical stimulus is presented simultaneously to both ears), the signal is heard only in the ear that perceives it louder. If the tone intensity is perceived as equal in the two ears it is not localized to either ear but rather to the center of the head. The Stenger test is based on this principle and is ideal for unilateral disturbances. It was originally used with tuning forks but is not given with an audiometer. A modified version of the test uses speech material and is naturally called the Speech Stenger Test (L. Watson & Tolan, 1967). It uses the same principle but substitutes words where tones were used. The Stenger test is one of the more successful tests for detecting unilateral pseudohypacusic disturbance (Goetzinger & Prowd, 1958; Menzel, 1960; Newby, 1958; Peck & Ross, 1970; Taylor, 1949). Clinicians often use both the pure tone and speech techniques since the pure tone test is useless in cases with diplacusis. Azzi (1951) attempted to improve the efficiency of the test by pulsing two tones per second. The periodic interruptions were initiated to deter the subject from perceiving the intensitive difference-limen in the good ear. The Stenger test has also been successfully performed on a Bekesy audiometer (J. Watson & Voots, 1964). A discussion of this technique is included in the section dealing with Bekesy tracings.

Doerfler–Stewart Test. This test, suitable for binaural loss, is based on the relationship between speech intelligibility and the signal-to-noise ratio. Normal hearing individuals can attain excellent intelligibility scores with signal-to-noise ratios as low as −10 dB, i.e., the speech signal is 10 dB less intense than the noise signal. This is generally true for persons with organic hearing loss also, although in many instances of sensorineural impairment the complications imposed by the dysacusis prevent the applicability of this principle.

Doerfler and Stewart (1946) superimposed noise on spondee words and systematically varied the signal-to-noise ratio. Several years later the test norms and procedure were reported in a monography (Doerfler & Epstein, 1956). Although Ventry and Chaiklin (1965a) felt the norms were too general and the procedure inadequate, the test is still in use. Even though this test is one of the more effective tests it is still only successful in approximately 58% of the cases (Menzel, 1960). Researchers, using normal subjects simulating hearing loss, reported the Doerfler–Stewart test was not a valid measure with this group.

Hood's Masking Test. The shifting, with effective masking, of the auditory threshold in decibels linear to the decibel magnitude of the masking stimulus is the basis of Hood's masking test. J. Hood (1959) applied this to his subjects with pseudohypacusic disturbance and reported good success. This test also has not achieved widespread acclaim.

Tone-in-Noise Test. This is a modified Doerfler–Stewart test monaural approach (Pang-Ching, 1970). It uses only one criterion, the pure tone threshold differences in quiet and in noise and the relationship of the changes.

Bekesy Audiometry. This procedure involves a markedly different psychophysical technique known as the *method of adjustment*. With this technique the subject controls and manipulates the intensity rather than the examinee. Several varieties of hearing procedures have evolved. One procedure compares threshold tracings made in two ways: (1) with the tone presentation periodically interrupted or pulsed, and (2) with a continuous tone presentation. The relationship between the interrupted and continuous tracings were classified into four (and later five) distinct patterns (Jerger, 1961). Normal subjects and subjects with organic hearing impairment should fit into at least one of the four patterns. The interrupted tracings are equal to slightly better than continuous tracings in all four patterns. A fifth Bekesy pattern (Type V) was described (Jerger & Herer, 1961; Resnick & Burke, 1962; Rintelmann & Harford, 1963; Stein, 1963), where the interrupted tone was consistently poorer than the continuous one. Type V was ostensibly symtomatic of a pseudohypacusic impairment. Stein (1963) found that 57% of his subjects with pseudohypacusic disorders exhibited a Type V pattern; 30% of his subjects gave patterns that could not be classified; and the rest gave Type II or Type IV tracings. He concluded that collectively the Type V and unclassified group identified 87% of these pseudohypacusic subjects. Peterson (1963) and Rintelmann and Harford (1963) reported some success with this technique on school-age children. They felt it offered some qualitative support to the results of the other audiometric tests, but was not useful in ascertaining the degree of true hearing. Other researchers challenged the concept of subject-controlled audiometry (Hattler, 1968; Hopkinson, 1965; Melnick, 1967; Price, Sheperd, & Goldstein, 1965; Stark, 1966), resulting in a redefinition of the Type V (Rintelmann & Harford, 1967). Hattler (1970) accentuated the Type V pattern by changing the 50% on–off program to a 1:4 on–off program, i.e., 200 msec on to 800 msec off.

J. Watson and Voots (1964) described a modification of the Bekesy audiometer that enabled them to perform the Stenger test with Bekesy tracings. They reported high clinical dependability with this procedure. Still a further adaptation of Bekesy audiometry is based on the Harris (1958) ascending–decending test. In this procedure called Bekesy Ascending–Descending Gap Evaluation, Hood, Campbell, and Hutton (1964) employ a fixed frequency for three relay threshold tracings: (1) continuous ascending, (2) pulsed ascending, and (3) pulsed descending. They found normal or organic hearing losses gave nearly identifical thresholds but the pseudohypacusic ears gave greater differences among techniques.

Sensorineural Acuity Level. The sensorineural acuity level test was developed by Jerger and Tillman (1960) for ascertaining the sensorineural component of a hearing loss. A fixed amount of masking is presented through a bone oscillator placed on the forehead. The threshold shift imposed on normal ears serves as a source of reference for comparison with the threshold shift of the defective ears. The relationship between the normal and abnormal threshold shifts determines the sensorineural loss.

Rintelmann and Harford (1963) reported an *Air–sensorineural acuity level gap*. This is the sudden shift in the air-conduction threshold due to the introduction of white noise. Actually, it is a Doerfler–Stewart effect for pure tones and is rarely used.

Electrophysiologic Audiometry. The common factor in this group of test procedures is the nature of the response mode, i.e., a physiological activity recorded by some electronic means. For example, one mode measures a change in skin resistance to an induced current. In this instance, the change called the electrodermal response is mediated through the sympathetic branch of the autonomic nervous system. In another mode of physiological response, brain wave activity, which is mediated through the central nervous system, is recorded as electroencephalographic patterns. These are fed into an averaging computer in time-locked sequence to the tonal stimuli until a statistically averaged evoked response is evolved. There are other electrophysical measures, to be sure, such as electrocochleography and electronystagmography, but the first two have a historical foundation for assessment of pseudohypacusic disturbance and therefore will be outlined in greater detail. Their imposing virtue is that the electrophysiologic responses are supposedly objective since they do not require conscious subject participation.

Electrodermal Audiometry. Formerly called the galvanic skin reflex and before that the psychogalvanic skin reflex (Nober, 1958), electrodermal audiometry is used as part of the standard Veterans Administration test battery. The test was ideal for assessing potential or alleged malingering since it is an "objective" technique that does not require conscious client participation and cooperation. Briefly, the technique employs a classical Pavlovian conditioning format, which pairs a tone stimulus with a mild shock. In electrodermal audiometry, the response is the change in skin resistance (mediated through a Wheatstone bridge arrangement) to the conditioned tone. The result is a measurable change in flow of current between two electrodes connected to the fingers of one hand. Current flow changes are traced by a stylus onto a graph as an analogue recording. Without conditioning, a tone presented above threshold as an unconditioned stimulus may evoke a small electrodermal change, but this extinguishes rapidly after a few repeated trials, making it impossible to elicit an audiogram. Therefore, conditioning is used. By pairing an unconditional shock with the stimulus tone (reinforcement trials), the "resistance to extinction" or "habit strength" is enhanced and, in addition, the electrodermal reflex is greater in magnitudes near threshold (Nober, 1958). Thus, hearing assessment is possible since these reinforcement trials (usually 40%) enable testing several frequencies at near-threshold levels. This procedure is more effective with adults than young children.

In assessing pseudohypacusic disturbances, Chaiklin and Ventry (1963) were successful in identifying 80% of their cases with electrodermal audiometry but others have reported less success. Goldstein (1956) reported at least 50% failures and concluded that electrodermal audiometry was a poor instrument for identifying pseudohypacusic impairment. Grave (1966), using a simulated malingering research design, also reported poor results.

Shortly after electrodermal audiometry became part of the standard Veterans Administration battery for detecting malingerers, the technique was adapted to speech audiometry, i.e., the "electrodermal speech reception threshold (EDSRT)" (Hopkinson, Katz, & Schill, 1960; Katz & Connelly, 1964; Ruhm & Carhart, 1958; Ruhm & Menzel, 1959). In this adaptation, the shock is paired with a special spondee word, although several words are given from a list. Hopkinson *et al*. (1960) then modified the test to an "instrumental avoidance conditioning" technique by advising the client in advance the shock was avoidable if the words were repeated. The electrodermal audiometry technique is being used much less in current-day audiometry because it is difficult to gain client permission to induce shock. Many institutional human-subject committees will not allow it. Electrodermal audiometry was never very successful with young children, brain-damaged persons, or emotionally disturbed persons.

Evoked Response Audiometry. The evoked response audiometry technique employs an averaging computer to program electroencephalograph recording of ongoing brain wave activity. Prior to the advent of using the averaging computer it was called "electroencephalographic audiometry." Brain wave impulses are transmitted by electroencephalograph electrodes carefully placed at different head locations. However, this continuous outflow of varying cortical impulses produces relatively small changes in amplitude and is difficult to identify as a charge in electrical potential that was associated with an auditory stimulation, particularly if the stimulus level was near threshold.

The detection of an auditory electroencephalograph response to a tonal stimulus was enhanced—indeed revolutionized—by the application of electronic computer. The computer memory made it possible to add only the auditory brain responses that were time locked to a rapid series of tonal impulses fed to the ear through an earphone. As the computer's memory summates and stores the small but related cortical–auditory activity bits, the cumulative amplitudes grow into a characteristic and recognizable response pattern. Concurrently, the rest of the cortical impulse activity continues "at random" (relative to the time-locked tonal presentations) with its positive and negative components. Since the computer summates both the positive and negative background brain wave activity, which is random in respect to the time-locked tonal presentations, the sum total of this random plus–minus background averages zero. Hence, only the cumulative evoked response activity evolves to form the characteristic waveform configuration (Davis, 1969; Goldstein, 1969). This waveform has an early response (Derbyshire, Fraser, McDermott, & Bridge, 1956; Geisler, 1964), a late response (McCandless & Best, 1964), and a contingent negative variation (Davis, 1969). The significance of these components is still under intense exploration.

Evoked response audiometry requires highly specialized skill and equipment and is usually restricted to medical settings. It has been employed for assessing pseudohypacusic disturbance (Cody & Bickford, 1965; Cody, Griffing, & Taylor, 1968; McCandless & Lentz, 1968). While the technique is questionable for this population, McCandless and Lentz (1968) reported they obtained some threshold information in their pseudohypacusic group. Clearly, evoked response audiometry is not in widespread use for assessing pseudohypacusis.

Summary

In summary, it appears that no one test, examination, or technique can independently isolate a pseudohypacusic disturbance with definite success. An elaborate battery of audiologic tests and otologic confirmation are mandatory before a positive diagnosis can be made. It is interesting that, long before the extensive array of tests evolved, Glorig (1954) asserted that tests were inadequate and an admission of intent to defraud was the only positive answer. In some ways this is still true. In recent years, much of the research was conducted on normal subjects who were simulating pseudohypacusis. While there may be some psychophysical merit and convenience to this approach, conclusions may be fraught with misleading assumptions (Hopkinson, 1978) is spite of the spartan attempt to have the data interpreted conservatively.

Treatment

Malingering

The treatment would naturally be structured around the diagnosis, especially for pseudohypacusic problems. Chaiklin and Ventry (1963) asserted that progress in this area is dependent on our attitude, that is, it is important to abandon the idea that malingering must be identified and instead to adopt a nonjudgmental acceptance of the patient and his problem. F. Martin (1975) feels the term is probably used too often anyway and unless the malingerer admits to fraud, the amount of organic and nonorganic hearing loss is difficult to determine in absolute terms.

Several approaches have been explored to assist the malingerer. One simple technique used by Johnson *et al*. (1956) and Gibbons (1962) was to reveal the test inconsistencies to the malingerer in a casual and noncommittal fashion. They suggested to the malingerer that perhaps inattentiveness or misunderstanding intervened. Then they repeated the tests with success in two-thirds of the cases. Apparently, the fear of potential exposure prodded these people to reconsider their equivocal plan.

A technique similar to the above was described by Hardy (1948) and by Harbert (1943). The clinician and patient openly discussed the psychological forces that precipitated their behavioral patterns. They hoped the revelation would engender sufficient insight to resolve the problem.

Pseudohypacusic Disturbance

Several of the traditional therapeutic techniques have been explored with these people. N. A. Martin (1946) also had comparable success.

Little is known about the effectiveness of hypnosis, shock therapy, and psychoanalysis with these patients. There is a great dearth of literature concerning therapy. On occasion, individual case reports appear but none of these were scientifically controlled studies. There is virtually no precedented therapeutic approach for the rehabilitation of the person with a pseudohypacusic disturbance.

Davis and Silverman (1978) say that "positive suggestion that the hearing loss is only temporary . . . [is] the most effective treatment" for hysterical deafness. They also recommend that these patients need to emphasize listening skills. These patients do not benefit from speechreading alone.

References

Azzi, A. Le prove per suelare la simulazione di sordita. *Review of Audiology*, 1951, *Prat 1*, 5–6.

Bailey, H. A., Jr., & Martin, F. N. Non-organic hearing loss, Case Report. *Laryngoscope*, 1961, *71*, 209–210.

Ballantyne, J. *Deafness*. Boston: Little Brown, 1960.

Barelli, P. A., & Ruder, L. Medico-legal evaluation of hearing problems. *Eye, Ear, Nose, Throat Monographs*, 1970, *49*, 398–405.

Barr, B. Non-organic hearing problems in school children: Functional deafness. *Acta Otolaryngologica*, 1960, *52*, 337–346.

Berger, K. Nonorganic hearing loss in children. *Laryngoscope*, 1965, *75*, 447–457.

Black, J. W. The persistance of the effects of delayed side tone. *Journal of Speech and Hearing Disorders*, 1955, *20*, 65–68.

Brockman, S. J., & Hoversten, G. H. Pseudoneural hypacusis in children. *Laryngoscope*, 1960, *70*, 825–839.

Calearo, C. Detection of malingering by periodically switched speech. *Laryngoscope*, 1957, *67*, 130–136.

Carhart, R. Individual differences in hearing for speech. *Annals of Otology, Rhinology, & Laryngology*, 1946, *55*, 233–266. (a)

Carhart, R. Monitored live-voice as a test for auditory acuity. *Journal of the Acoustical Society of America*, 1946, *17*, 339–349. (b)

Carhart, R. Speech audiometry in clinical evaluation. *Acta Otolaryngologica*, 1952, *41*, 18–48.

Carhart, R. Audiometry in diagnosis. *Laryngoscope*, 1958, *68*, 253–277.

Carhart, R. The determination of hearing loss. *Veterans Administration Department of Medicine and Surgery Information Bulletin*, *18*. 10–115. Washington, D.C.: U.S. Government Printing Office, 1960.

Carhart, R. Tests for malingering. *Transactions of the American Academy of Opthamology and Otolaryngology*, 1961, *65*, 437.

Causey, G. D., & McGranahan, L. M. Determination of organic hearing threshold by means of delayed sidetone. *Journal of the American Speech and Hearing Association*, 1959, *1*, 106 (abstr.).

Chaiklin, J. B. The relation among three selected auditory speech thresholds. *Journal of Speech and Hearing Research*, 1959, *2*, 237–243.

Chaiklin, J. B., & Ventry, I. M. Functional hearing loss. In J. Jerger (Ed.), *Modern developments in audiology* (Chapter 3). New York: Academic Press, 1963.

Chase, R. A., Sutton, S., Fowler, E. P., Jr., & Ruhm, J. B. Low sensation level delayed clicks and keytapping. *Journal of Speech and Hearing Research*, 1961, *4*, 73–78. (a)

Chase, R. A., Sutton, S., & Rapi, J. Sensory feedback influences on motor performance. *Journal of Auditory Research*, 1961, *3*, 212–223. (b)

Cherry, C. E., & Taylor, W. K. Some further experiments upon the recognition of speech with one and two ears. *Journal of Acoustical Society of America*, 1954, *26*, 554–559.

Cody, D., & Bickford, R. Cortical audiometry: An objective method of evaluating auditory acuity in man. *Mayo Clinic Proceedings*, 1965, *48*, 273–287.

Cody, D., Griffing, T., & Taylor, W. Assessment of the newer tests of auditory function. *Annals of Otology, Rhinology, & Laryngology*, 1968, *77*, 686–705.

Cooper, W. A., Jr., & Stokinger, T. E. Pure tone delayed auditory feedback: Effective of prin. exp. *Journal of the American Audiology Society*, 1976, *1*, 64–68. (a)

Davis, H. The phenomenon of evoked responses in general. *Evoked Response Audiology*, 1969, *6*, 69.

Davis, J., & Silverman, S. *Hearing and deafness*. New York: Holt, Rinehart, & Winston, 1978.

Derbyshire, N., Fraser, H., McDermott, M., & Bridge, A. Audiometric Measurements by Electroencephalography. *Electroencephalography and Clinical Neurophysiology*, 1956, *8*, 467–478.

Dixon, R. F., & Newby, H. A. Children with nonorganic hearing problems. *AMA Archives of Otolaryngology*, 1959, *70*, 619–623.

Doerfler, L. G. Psychogenic deafness and its detection. *Annals of Otology, Rhinology, & Laryngology*, 1951, *60*, 1045–1948.

Doerfler, L. G., & Epstein, A. *The Doerfler–Stewart (D–S) test for functional hearing loss*. (Monograph). Washington, D.C.: Veterans Administration, 1956.

Doerfler, L. G., & Stewart, K. Malingering and psychogenic deafness. *Journal of Speech and Hearing Disorders*, 1946, 181–186.

Farley, J., Waldrop, W. W., Derbyshire, A. H., Carter, R. L., Austin, D. F., McCormick, C., & Mills, P. J. Psychic factors in hearing loss. *Annals of Otology, Rhinology, & Laryngology*, 1960, *69*, 731–776.

Farley, J., McCormick, C., Oppfeli, C., & Mills, P. The functions of psychic factors in hearing loss. *Audiology*, 1962, *1*, 125–133.

Fletcher, H. Method of calculating hearing loss for speech from an audiogram. *Journal of Acoustical Society of America*, 1950, *22*, 1–5.

Fournier, J. E. The detection of auditory malingering. *Transl. Beltone Inst. Hear. Res.*, 1958, *8*, 1–23.

Fournier, J. E. The problems of psychogenic deafness. *International Audiology*, 1962, 123–124.

Fowler, E. P., Jr. *Medicine of the ear*. New York: Williams and Wilkins, 1947.

Galambos, R., Rosenberg, P. E., & Glorig, A. The eyeblink response as a test for hearing. *Journal of Speech and Hearing Disorders*, 1953, *18*, 373–378.

Gaynor, E. B. Hearing and non-organic hearing loss. *Archives of Otolaryngology*, 1974, *100*, 199–200.

Geisler, C. D. Discussion of C. D. Geisler average responses to clicks in man recorded by scalp electrodes (Technical Report 380, Research Laboratory of Electronics, M.I.T., November 4, 1960). *Annals of the N.Y. Academy of Science*, 1964, *112*, 218–219.

Gibbons, E. W. Aspects of traumatic and military psychogenic deafness and simulation. *International Audiology*, 1962, *1*, 151–154.

Gibbons, E. W., & Winchester, R. A. A delayed side tone test for detecting uniaural functional deafness. *AMA Archives of Otolaryngology*, 1957, *66*, 70–78.

Glorig, A. Malingering. *Annals of Otology, Rhinology, & Laryngology*, 1954, *63*, 802–815.

Goetzinger, C. P., & Prowd, G. O. Deafness: Examination techniques for evaluating malingering and psychogenic disabilities. *Journal of the Kansas Medical Society*, 1958, *39*, 95–101.

Goldstein, R. Effectiveness of conditioned electrodermal responses (EDR) in measuring pure-tone thresholds in case of non-organic hearing loss. *Laryngoscope*, 1956, *66*, 119–130.

Goldstein, R. *Pseudohypacusis*. *Journal of Speech and Hearing Disorders*, 1966, *31*, 341–352.

Goldstein, R. *Use of averaged electroencephalic response (AER) in evaluating central auditory function*. Paper presented to Annual Convention of the American Speech and Hearing Association, Chicago, November 12, 1969.

Grave, T. G. Performance of jaive and role-playing pseudo-malingerers on an unconditioned EDR audiometric test. *Journal of Auditory Research*, 1966, *6*, 337–350.

Harbert, F. Functional and simulated deafness. *U.S. Navy Medical Bulletin*, 1943, *41*, 458–471.

Hardy, W. G. Special techniques for diagnosis and treatment of psychogenic deafness. *Annals of Otology, Rhinology, & Laryngology*, 1948, *57*, 65–95.

Harris, D. A. A rapid and simple technique for the detection of nonorganic hearing loss. *AMA Archives of Otolaryngology*, 1958, *68*, 758–760.

Hattler, K. W. The Type V Bekesy pattern: The effects of loudness memory. *Journal of Speech and Hearing Research*, 1968, *11*, 567–575.

Hattler, K. W. Lengthened-offtime: A self-recording screening device for nonorganicity. *Journal of Speech and Hearing Disorders*, 1970, *35*, 113–122.

Heller, M. F., Anderman, B., & Singer, F. *Functional otology*. New York: Springer, 1955.

Hood, J. Modern marking techniques and their application to the diagnosis of functional deafness. *Journal of Laryngology and Otology*, 1959, *73*, 536–543.

Hood, W. H., Campbell, K. A., & Hutton, C. L. An evaluation of the Bekesy ascending descending gap. *Journal of Speech and Hearing Research*, 1964, *7*, 123–132.

Hopkinson, N. T. Type V Bekesy audiograms: Specification and clinical utility. *Journal of Speech and Hearing Disorders*, 1965, *30*, 243–251.

Hopkinson, N. T. Speech tests for nonorganic hearing loss. In J. Katz (Ed.), *Handbook of clinical audiology*. Baltimore: Williams and Wilkins, 1972, pp. 374–394.

Hopkinson, N. T. Speech tests for pseudohypacusis. In J. Katz (Ed.), *Handbook of clinical audiology*. Baltimore: Williams and Wilkins, 1978, pp. 291–303.

Hopkinson, N. T., Katz, J., & Schill, H. A. Instrumental avoidance galvanic skin response audiometry. *Journal of Speech and Hearing Disorders*, 1960, *25*, 349–357.

Jepson, O. Intratympanic muscle reflexes in psychogenic deafness (impedance measurements). *Acta Otolaryngologica Supplement*, 1953, *109*, 61–69.

Jepson, O. Middle-ear muscle reflexes in man. In J. Jerger (Ed.), *Modern developments in audiology*. New York: Academic Press, 1963, pp. 194–237.

Jerger, J., & Herer, G. Unexpected dividend in Bekesy auiometry. *Journal of Speech and Hearing Disorders*, 1961, *26*, 390–391.

Jerger, J. F., & Tillman, T. W. A new method of clinical determination of sensorineural activity level (SAL). *Archives of Otolaryngology*, 1960, *71*, 948–955.

Jerger, J. F., Carhart, R., Tillman, T. W., & Peterson, J. L. Some relations between normal hearing for pure tones and for speech. *Journal of Speech and Hearing Research*, 1959, *2*, 126–140.

Johnson, K. O., Work, W. P., & McCoy, G. Functional deafness. *Annals of Otology, Rhinology, & Laryngology*, 1956, *65*, 154–170.

Juers, A. L. Pure tone threshold and hearing for speech: Diagnostic significance of inconsistencies. *Laryngoscope*, 1956, *66*, 402–409.

Kacker, S. K. Bekesy audiometry in simulated hearing loss. *Journal of Speech and Hearing Disorders*, 1971, *36*, 506–510.

Karlovich, R., & Graham, J. T. Effects of pure tone synchronous and delayed auditory feedback on keytapping performance to a programmed visual stimulus. *Journal of Speech and Hearing Research*, 1966, *9*, 596–603.

Katz, J. (Ed.), *Handbook of clinical audiology*. Baltimore: Williams and Wilkins, 1978.

Katz, J., & Connelly, R. Instrumental avoidance vs. classical conditioning in GSR speech audiometry. *Journal of Auditory Research*, 1964, *4*, 171–179.

Klotz, R. E., Koch, A. W., & Hackett, T. P. Psychogenic hearing loss in children: A preliminary report. *Annals of Otology, Rhinology, & Laryngology*, 1960, *69*, 199–205.

Kodman, F., Jr. Lateralization method for evaluating non-organic deafness. *Annals of Otology, Rhinology, & Laryngology*, 1961, *70*, 224–233.

Kodman, F., Jr., Sedlacek, G., & Powers, T. A preliminary study of personality characteristics of auditory malingerers. *Acta Otolaryngologica*, 1959, *50*, 455.

Lamb, L. E., & Peterson, J. L. Middle ear reflex measurements in pseudophypacusis. *Journal of Speech and Hearing Disorders*, 1967, *32*, 46–51.

Leshie, G. H. Childhood nonorganic hearing loss. *Journal of Speech and Hearing Disorders*, 1960, *25*, 290–292.

Martin, F. Nonorganic hearing loss. In *Introduction to audiology*. Englewood Cliffs, N.J.: Prentice-Hall, 1975, pp. 335–362.

Martin, F. *Introduction to audiology*. Englewood Cliffs, N.J.: Prentice-Hall, 1978.

Martin, N. A. Psychogenic deafness. *Annals of Otology, Rhinology, & Laryngology*, 1946, *55*, 81–89.

McCandless, G. A., & Best, L. Summed evoked responses using pure tone stimuli. *Journal of Speech and Hearing Research*, 1964, *9*, 266–272.

McCandless, G., & Lentz, W. Evoked response audiometry in nonorganic hearing loss. *Archives of Otolaryngology*, 1968, *87*, 123–128.

McGranahan, L., Causey, D., & Studebaker, G. A. Delayed sidetone audiometry. *Journal of the American Speech and Hearing Association*, 1960, *2*, 357.

Melnick, W. Comfort level and loudness matching for continuous and interrupted signals. *Journal of Speech and Hearing Research*, 1967, *10*, 99–109.

Menzel, O. J. Clinical efficiency in compensation audiometry. *Journal of Speech and Hearing Disorders*, 1960, *25*, 49–54.

Metz, O. The acoustic impedance measured on normal and pathological ears. *Acta Otolaryngologica Supplement*, 1946, *63*, 1–253.

Miller, G. A., & Licklider, J. C. R. The intelligibility of interrupted speech. *Journal of the Acoustical Society of America*, 1950, *22*, 167–173.

Morrissett, L. E. The aural rehabilitation program of the U.S. Army for the deaf and hard of hearing. *Annals of Otology, Rhinology, & Laryngology*, 1946, *55*, 821–838.

Newby, H. A. *Audiology: principles and practice*. New York: Appleton-Century-Crofts, 1958.

Nober, E. H. GSR magnitudes for different intensities of shock, conditioned tone and extraction tone. *Journal of Speech and Hearing Research*, 1958, *1*, 316–324.

Nober, E. H. Vibrotactile sensitivity of deaf children. *Laryngoscope*, 1967, *77*, 2128–2146.

Nober, E. H. Cutile air and bone conduction thresholds of the deaf. *Exceptional Children*, 1970, 571–579.

Pang-Ching, G. The Tone-in-noise test: A preliminary report. *Journal of Auditory Research*, 1970, *19*, 311–327.

Peck, J. E., & Ross, M. A comparison of the ascending and descending modes for administering the pure-tone Stenger test. *Journal of Auditory Research*, 1970, *10*, 218–220.

Peterson, J. L. Non-organic hearing loss in children and Bekesy audiometry. *Journal of Speech and Hearing Disorders*, 1963, *28*, 153–155.

Portmann, M., & Portmann, C. *Clinical audiometry*. New York: Charles C Thomas, 1961.

Pitman, L. K. A device for detecting simulated unilateral deafness. *Journal of American Medical Associates*, 1943, *121*, 752–753.

Price, L., Sheperd, D. C., & Goldstein, R. Abnormal Bekesy tracings in normal ears. *Journal of Speech and Hearing Disorders*, 1965, *30*, 139–144.

Ramsdell, D. A. The psychology of the hard of hearing and the deafened adult. In H. Davis & R. S. Silverman (Eds.), *Hearing and deafness*. New York: Holt, Rinehart & Winston, 1978, pp. 499–510.

Rapin, I., Costa, L., Mandel, I., & Fromowitz, A. Effect of varying delays in auditory feedback on key-tapping of children. *Perceptual Motor Skills*, 1963, *16*, 489–500.

Resnick, D. M., & Burke, K. S. Bekesy audiometry in nonorganic auditory problems. *Archives of Otolaryngology*, 1962, *76*, 38–41.

Rintelmann, W., & Harford, E. The detection and assessment of pseudohypacusis among school-age children. *Journal of Speech and Hearing Disorders*, 1963, *28*, 144–152.

Rintelmann, W. F., & Harford, E. Type V Bekesy pattern: Interpretation and clinical utility. *Journal of Speech and Hearing Research*, 1967, *10*, 733–744.

Rosenblith, W. Some quantifiable aspects of the electrical activity of the nervous system. *Review Modern Physics*, 1957, *31*, 532–545.

Ross, M. The variable intensity pulse count method (VIPCM) for the detection and measurement of the pure-tone thresholds of children with functional hearing losses. *Journal of Speech and Hearing Disorders*, 1964, *29*, 477–482.

Ruhm, H. B., & Carhart, R. Objective speech audiometry. A new method based on electrodermal response. *Journal of Speech and Hearing Research*, 1958, *1*, 169–178.

Ruhm, H. B., & Cooper, W. A., Jr. Low sensation level effects of pure-tone delayed auditory feedback. *Journal of Speech and Hearing Research*, 1962, *5*, 185–193.

Ruhm, H. B., & Cooper, W. A., Jr. Some factors that influence pure-tone delayed auditory feedback. *Journal of Speech and Hearing Research*, 1963, *6*, 223–237.

Ruhm, H. B., & Cooper, W. A., Jr. Delayed feedback audiometry. *Journal of Speech and Hearing Disorders*, 1964, *29*, 448–455.

Ruhm, H. B., & Menzel, O. J. Objective speech audiometry in cases of nonorganic hearing loss. *Archives of Otolaryngology*, 1959, *69*, 212–219.

Saltzman, M. *Clinical audiology*. New York: Grune and Stratton, 1949.

Sataloff, J. *Industrial deafness*. New York: McGraw-Hill, 1957.

Semenov, H. Deafness of psychic origin and its response to narcosynthesis. *Transactions of the American Academy of Ophthalmology and Otolaryngology*, 1947, *51*, 326–348.

Shepherd, D. C. Nonorganic hearing loss and the consistency of behavioral auditory responses. *Journal of Speech and Hearing Research*, 1965, *8*, 149–163.

Snyder, J. M. Characteristic patterns of etiologic significance from routine audiometric tests and case history. *Maico Audiology Library Series*, 1977.

Spilka, B. Some vocal efforts of different reading passages and time delays in speech feedback. *Journal of Speech and Hearing Disorders*, 1954, *19*, 37–47.

Stark, J. Jerger types in Fixell-frequency Bekesy audiometry with normal and hypacusic children. *Journal of Auditory Research*, 1966, *6*, 135–140.

Stein, L. Some observations on Type V Bekesy tracings. *Journal of Speech and Hearing Research*, 1963, *6*, 339–348.

Taylor, G. An experimental study of tests for the detection of auditory malingering. *Journal of Speech and Hearing Disorders*, 1949, *14*, 119–130.

Thomsen, K. A. Employment of impedance measurements in otologic and otoneurologic diagnostics. *Acta Otolaryngologica*, 1955, *45*, 159–167.

Thorne, B. Psycho-tell. An aid in the estimate of functional auditory disorders. *Acta Otolaryngologica*, 1960, *72*, 622–630.

Tiffany, W. R., & Hanley, C. N. Delayed speech feedback as a test for auditory malingering. *Science*, 1962, *115*, 59–63.

Tillman, T. W., & Jerger, J. F. Some factors affecting the spondee threshold in normal hearing subjects. *Journal of Speech and Hearing Research*, 1959, *2*, 141–146.

Truex, E. H., Jr. Psychogenic deafness. *Connecticut State Medical Journal*, 1946, *10*, 907–915.

Ventry, I. M., & Chaiklin, J. B. Multidiscipline study of functional hearing loss. *Journal of Auditory Research*, 1965, *5*, 179–272. (a)

Ventry, I. M., & Chaiklin, J. B. Evaluation of pure-tone audiogram configurations used in identifying adults with functional hearing loss. *Journal of Auditory Research*, 1965, *5*, 212–218. (b)

Watson, J. E., & Voots, R. H. A report on the use of the Bekesy audiometer in the performance of the Stenger test. *Journal of Speech and Hearing Disorders*, 1964, *29*, 36–46.

Watson, L. A., & Tolan, T. Hearing tests and hearing instruments. Baltimore: Williams and Wilkins, 1949.

Watson, L., & Tolan, T. Hearing loss and hearing instruments. New York: Hafner, 1967.

Wedenberg, E. Audiotory tests on newborn infants. *Acta Otolaryngologica*, 1956, *46*, 447–461.

Wedenberg, E. Objective auditory tests on non-cooperative children. *Acta Otolaryngologica*, 1963, *175*, 1–32.

Weiss, H., & Windrem, E. Some methodological problems in identificates of functional deafness. *Archives of Otolaryngology*, 1960, *72*, 240–247.

Williamson, D. Functional hearing loss: A review. *Maico Audiology Library Series*, 1974, *12*, 33–34.

11

Bruce Quarrington

STUTTERING

The Syndrome

Dysfluencies occurring in the otherwise orderly flow of speech are common and ordinarily receive little attention from the listener who is striving to anticipate and understand the message of the speaker. Some hesitations of speech or other variations from fluency of themselves convey meaning, but these, like gestural behavior, are read by the auditor as part of the total message. The waving hands, the stumbling speech, and the high pitch and intensity of voice are familiar characteristics of excitement or other emotional state and therefore are not considered abnormal if they are consistent with the message and the circumstances of the speaker. When, however, the rhythm of speech is disrupted in such a way that understanding of the messages is hindered, attention focuses on these sources of confusion or noise and the judgment is made that some speech abnormality is present. Stuttering and stammering are terms frequently used by listeners to designate a wide variety of these incomprehensible variations in speech rhythm. For the professional worker who has some responsibility to do something about deviances in speech there is an initial need to classify the type of speech disorder present and, accordingly, more precise definitions of stuttering and other disturbances of speech rhythm are needed.

Fluency abnormalities may assume a wide variety of forms characteristic of a particular individual. Some of these forms or features may be infrequently encountered in other nonfluent individuals. For example, a speaker might display several abnormalities of speech the most striking of which is the frequent use of the phrase "Well now then . . . ," which may be repeated two or three times before a more meaningful utterance is delivered. Most listeners might interpret this behavior as

Bruce Quarrington • York University, Downsview, Ontario, Canada M3J 1P3.

simply an unfortunate style of speech whereas most speech pathologists would suspect it to be an abnormal feature related to the presence of a stuttering difficulty. The function of this behavior might be that of permitting the individual to postpone, until a certain readiness is experienced, the attempt to say the first word of the utterance he wishes to deliver. This or some similar form of behavior is found in a number of individuals with stuttering difficulties, but not in all. The understanding of this particular behavior and its modification may be important in the therapeutic management of such individuals. It is not a behavior that is characteristic of stutterers, but where it is observed, the likelihood is high that other types of stuttering behaviors will be found. There are many behaviors that, like the one presently under discussion, might be considered signs of the stuttering syndrome. Understandably, in any classificatory system, greatest interest inheres in the identification of the smallest set of distinctive features, not only for simplicity of the classificatory operation, but also because there is reason to believe that those features common to all cases within a diagnostic category are likely to be more closely linked to the etiological factors of significance in the disorder.

Wingate (1976), in searching for those behaviors that are characteristic of stuttering and therefore close to the immediate source of stuttering, has identified repetitions of syllables or sounds (elemental repetitions) and prolongations of sounds beyond their appropriate duration as the core features of stuttering. It is important to note that these repetitions may not always be audible to the listener but may be present, largely or entirely, as silent repetitive attempts to form the sought sound, or silent hesitation or struggles preceding the emission of a word, or silent breakages of voicing occurring within a spoken word. These behaviors are seen as constituting the "basic stutter" and some of these behaviors must be present in order that stuttering may be judged to be present. Other behaviors such as the frequent interjection of sounds, words, or phrases may be regarded as secondary mannerisms or accessory verbal features present in some stutterers.

While Wingate's description of the basic stutter and the associated definition of stuttering and accessory features appears to be in accord with much of the literature and though it has won rather general acceptance there are certain diagnostic difficulties that merit attention here. The reader is referred to the classic works of Van Riper (1971, 1973) for a more extensive treatment of the following issues and many others to be considered throughout this chapter.

A first difficulty to be noted is the core behaviors of stuttering at the level of description considered here do not make a qualitative distinction between stutterers and those experiencing other fluency disorders, or indeed, between stutterers and individuals who should be considered fluent speakers. Elemental repetitions and prolongations may be found in the speech of most speakers although at a significantly reduced level compared to most stutterers. Repetitions also occur at relatively high frequencies in the condition called "cluttering," which will be considered shortly. Are the distinctions then quantitative ones, and, if so, what precisely is the frequency of occurrence necessary for the judgment of stuttering to be made? This apparently reasonable question is difficult to answer. While all stutterers will show, at certain times or under certain conditions, the core behaviors of the disorder, it is quite possible that in much of their everyday speech these particular features will not be prominent. To take an extreme example, there are individuals who appear to have learned ways of avoiding the display of the core behaviors of stuttering in most social situations although they believe that without the deliberate use of secondary mannerisms such as avoiding feared sounds or words that their speech would reveal stuttering (Douglass & Quarrington, 1952; Freund, 1932). Other stutterers may be less successful in concealing or avoiding core stuttering behaviors, but still show relatively low levels of core behaviors with a clear preponderance of secondary mannerisms.

There may be several ways, not necessarily mutually exclusive, of dealing with such difficulties and preserving the notion of a syndrome of stuttering that is qualitatively distinct from the condition of normal nonfluency and other forms of fluency disorder. One line of argument would accept the designation of core stuttering behaviors but insist that repetitions and prolongations must be studied in greater detail to reveal their differences from similar phenomena in normal speakers and others.

The core stuttering behaviors include silent or inaudible motor behavior indicative of repetitions and prolongations. Unlike their vocalized counterparts, these behaviors appear to be so rare in normal speakers and those with other conditions of nonfluency that they permit essentially a qualitative differentiation. It has been noted that stutterers in their audible elemental repetitions and prolongations of consonants typically coarticulate the schwa or neutral vowel instead of the vowel sound appropriate for the attempted word. This characteristic appears to be rare in the repetitions or prolongations of normal speakers. Stromsta (1965) employed spectrograms of repetitions and prolongations of young children. Some of the children studied showed absent or deviant juncture formants, whereas others showed close approximations of normal transitions. Followed over a ten-year period, the former group showed a persistence of stuttering while the latter group became fluent at some time during this period. Such evidence suggests that finer characterization of the core behaviors may reveal differences of a qualitative nature between stutterers, normal speakers, and individuals affected with other disorders of fluency.

Van Riper (1971) should be consulted for an extensive discussion of the foregoing and other features that may differentiate normal from stuttering dysfluencies. A practically oriented discussion dealing with the differentiation of normally nonfluent children from incipient stutterers has been offered by Adams (1977).

A second line of argument would hold that stuttering, unlike some disorders, should not be regarded from a static diagnostic viewpoint, but must be considered in a developmental context. From this vantage point it would be held that stuttering, at its onset, may have a more or less characteristic form corresponding rather closely to the core stuttering behaviors set out by Wingate (1976). As stuttering persists in the individual, various means of coping with these difficulties may be adopted, which elaborate or modify the abnormalities of speech in some common directions, and also in some highly idiosyncratic ways. Some of these lines of development operate to make it difficult to observe the core stuttering behaviors. Some stutterers develop complex sorts of motor struggle during speech, which appear unrelated to repetitions or prolongations of sounds or syllables. In some cases the beginning of an utterance may be signaled by violent head shaking and foot stamping prior to any more direct attempt to vocalize. Other stutterers may show very different secondary mannerisms, such as the extensive use of complex interjections, circumlocution, or the substitution of expressions or words for feared words. The point to be made here is that in some stutterers these developments may either reduce the actual occurrence or the observability of core stuttering behaviors to an extremely low frequency level. It is apparent, then, that in some stutterers the diagnosis of stuttering may be made on the basis of the presence of these common or idiosyncratic secondary mannerisms of speech, many of which are pathognomic of stuttering.

Wingate (1976) refers to these individual developments of stuttering behavior as accessory features rather than as secondary mannerisms. He believes that the latter widely used term is too closely tied to the developmental theory of stuttering propounded by Bleumel (1932). Bleumel's theory attributed considerable theoretical and clinical importance to the distinction of a primary stuttering of childhood that consisted of easy effortless repetitions without awareness, from a fully developed secondary state of stuttering that was characterized by awareness, prolongations, silent struggle, and all other secondary mannerisms. This particular developmental view has served speech pathology well, but now appears to be an oversimplified account of developmental realities. It is evident that stuttering at its onset in some children may appear in the form of inaudible repetitions or prolongations and may be accompanied by other behavior suggesting awareness (Bloodstein, 1958, 1960a). It is also clear that some children may, in the first months of stuttering, show the sudden awareness and equally sudden disappearance of secondary mannerisms formerly thought to be manifestations of chronic or secondary stuttering. In some of these cases all evidence of stuttering may vanish despite the fact that a number of secondary mannerisms had briefly appeared. Also, in a few adults, it appears that stuttering has persisted for many years but does not appear to have changed significantly in form from that present in childhood. Observations such as these do not accord well with the developmental inevitability implied in Bleumel's theory. If, however, one clearly dissociates the term secondary mannerism from its historical source, the term has the value of indicating behaviors that have some coping component and are not accessory in the sense of being trivial, since these behaviors may for some stutterers constitute their main concern and may be perceived by some professional workers as requiring major therapeutic attention.

A second aspect of the developmental context within which the diagnosis of stuttering may be made concerns the age at which the onset of stuttering was observed. It is possible that age limits may be specified beyond which an increase in the frequency and variety of core stuttering behaviors should not be assumed to represent the onset of stuttering but some other condition or developmental status.

Stuttering is a disorder that typically appears to have its onset in the preschool years. Many investigators from many countries have offered evidence of this. In England, Morley (1957) reported that 89% of stuttering observed in children had its onset at or before the age of five. The comparable figure from the longitudinal study of Andrews and Harris (1964) is 69%.

Since stuttering typically has an onset around age three (Johnson, 1955), and since it is known that many children show a prominence of repetitive speech behaviors around age two (Metraux, 1950), the question arises as to the reality of a disorder of stuttering in early childhood. Is it not possible that the three-year-old identified as showing stuttering is simply showing the dysfluencies found in most children at a somewhat earlier age? Does such a child represent a simple variation in developmental rate, or indeed is the judgment of stuttering even more arbitrary and based more on the standards and judgments of significant adults than on the actual speech development and speech performance of the child? These are possibilities that have concerned a number of workers, and that prompted Johnson and associates (1959) to view the onset of stuttering as a complex psychosocial phenomenon in which parents with relatively high expectations for speech in their children reacted negatively to what were essentially normal variations in fluency. It would appear likely that in children particularly sensitive to parental concerns this would lead to anxiety and deterioration of speech performance that would in turn enhance parental concern in subsequent parent–child interactions. Johnson, in two major studies (1955; Johnson and associates, 1959), sought support for this diagnosogenic theory of stuttering by comparing sizable groups of children who had been labeled by one or both parents as stutterers for a rather brief time, with children of comparable sex, age, and socioeconomic status not so labeled by their parents. In his discussion of his own findings, Johnson *et al.* (1959) believed some supporting evidence had been found, but others analyzing the data he provided have come to different conclusions. McDearmon (1968), studying the parental responses to detailed questions regarding the child's speech behavior at onset, identified 69.5% of the children labeled as stutterers to have some of the core stuttering behaviors whereas only 10.4% of the control children were characterized by their parents in a comparable way.

Glasner and Rosenthal (1957), interviewing the mothers of children registered for grade one, carefully collected information regarding the extent to which repetitions, hesitations, and prolongations had been observed in the children before asking if the parents judged that the child had expressed a period of stuttering or was still stuttering. The results indicate that a child reported to have shown only one type of dysfluency would be identified as a stutterer 29% of the time while with two symptoms this rose to 65% and with all of them to 94%. These findings suggest that most parents, when identifying their children as stutterers, are for the most part responding to behaviors they recognize to be other than dysfluencies of the early language development of most children. Other aspects of Johnson's psychosocial view of stuttering development may have merit and will be considered shortly.

Two difficulties in the professional diagnosis of stuttering in young children should be noted. In children with evidence of delayed language development it is possible that stuttering behaviors, often of transitory nature, will appear during periods of rapid growth of expressive performance (Van Riper, 1971). Hall (1977) has reported some cases of older language-disordered children in therapy in whom stuttering has appeared with improved language and articulatory performance. In the young child with delayed language development are these elemental repetitions and prolongations normal developmental dysfluencies or stuttering? While it is true these episodes of dysfluency are often transitory, it is also evident that there is an unknown but sizable likelihood that secondary mannerisms will develop rapidly in some of these children. Therefore, it would appear to be a mistake to attribute routinely little significance to the dysfluencies of the language delayed and the language disordered.

A second difficulty pertains to the cyclical nature of stuttering in most young children. It is well-known that stuttering behaviors increase and decrease in frequency and severity in rather regular fashion, with a typical periodicity of two or three months. In some children, this oscillation of stuttering behaviors will vary between complete freedom from stuttering to very frequent stuttering, including the brief appearance of some secondary mannerisms. The source of this periodicity is unknown. In some instances, where stuttering completely dissipates at an early age, one might be inclined to relate such oscillations to periods of rapid language growth followed by periods of consolidation. In cases where stuttering persists, however, the periodicity may continue indeterminately. Cyclical variations in stuttering are not uncommon in adult stutterers (Quarrington, 1956) and tend to have period lengths comparable to those found in childhood. Obviously since periodicity is common, one must exercise care when a child fails to show much or any evidence of stuttering behaviors at one admission interview and particular weight must be given to the information regarding speech behavior elicited from the parents.

Differential Diagnoses

So far a brief examination has been made of the central features of stuttering and some of the problems associated with assessing the presence of the stuttering syndrome. A brief survey of the main problems of differential diagnosis will be considered using the information provided.

Perhaps the most difficult problem is presented with the disorder of cluttering, which also has a typical onset in early childhood. Cluttering is a disorder characterized by a rapid rate of utterance delivery in which there are frequent omissions of words or syllables, the intrusion of irrelevant sounds or syllables, the inversion of sounds and syllables, and by repetitions and prolongations of sounds and syllables. Comprehension of cluttered speech is difficult even for those closely associated with the affected individual. Although this condition has been known for many years it has received relatively little clinical attention and, until recently, little scientific investigation (Rieber, Smith, & Harris, 1977). Since some of the behaviors associated with cluttering correspond to the core stuttering behaviors, other similarities and differences have been sought. Kussmaul (1891) was the first to emphasize what appear to be critical differences from stuttering. Clutterers were said to be largely unaware of their speech difficulties, and if demands for increased attention to speech were made, that their fluency increased briefly. Other differences that have been demonstrated include a higher frequency of a marginally pathological EEG electroencephalograph for clutterers (Luchsinger & Landholt, 1951) and differential response to certain medications. Langova and Moravek (1964) reported that with the major tranquilizer chlorpromazine most stutterers showed an increase in stuttering and none showed increased fluency, whereas most clutterers showed increased fluency with this drug. Weiss (1964) has offered a comprehensive clinical account of cluttering in which he stressed the likelihood of a neurologically based deficit in language functions, since delays in the acquisition of speech, reading, and writing skills are commonly associated with cluttering. It would appear that stuttering and cluttering are conditions with different etiologies, prognostic considerations, and different implications for optimum management. Unfortunately, there are some complications.

If stuttering and cluttering were independent disorders, and because they are both rather rare conditions, one would expect the conjoint occurrence of the two conditions to be extremely rare. Several investigators however, believe a rather strong association exists (Freund, 1952). Langova and Moravek (1964) believed that cluttering occurred in 31% of the stutterers they investigated and over half their sample of clutterers warranted an associated diagnosis of stuttering. Weiss (1964) conceived the relationship between these disorders to be that cluttering is the more general condition, while stuttering represents one consequence of certain attempts to overcome cluttering behaviors. Van Riper (1971), as it will be noted later, appears to consider cluttering to constitute the basis for one of four types of stuttering. One source of confusion here, however, is that some stutterers who have been free of cluttering behaviors may develop the secondary mannerism of speaking rapidly following stuttering, particularly stuttering at the commencement of an utterance. In such cases the telescoping of words and some of the other features of cluttering may be apparent in their stutter-free speech.

At this time, evidence is not available to dissipate the confusion. Jaffe, Anderson, and Rieber (1973) have suggested one way of conceiving stuttering and cluttering as polar opposite conditions on a complexly defined continuum of speech rate, the extremes of which are both characterized by fluency failure. In this view individuals are seen as capable of oscillation between these two extremes, the nature of deautomatized speech depending on the nature of the imbalances of temporal and sequential organizational processes of speech production and the individuals' reactions to these sources of fluency failure. Perhaps the safest position with the present state of the art is to assume that stuttering and cluttering are discrete disorders with distinctive requirements for management, but which may on occasion occur in the same individual. In such a case, management decisions must be based on the saliency of the particular behaviors in the affected individual.

There are a small number of conditions, usually associated with an onset in adolescence or later life, that warrant some attention in differentiation from stuttering as it is discussed here. In most instances these conditions are not difficult to distinguish from stuttering, but if one wishes to blur these distinctions and accept all these conditions as instances of stuttering, then the properties and dynamics of stuttering may appear to be very complex. There is little doubt that some of the many theories of stuttering that have been advanced are based on cases that, by the criteria employed here, would not be considered cases of stuttering. There are, for example, rare cases of neurotic individuals whose symptomatology may include persistent stumbling on particular words or in expressing meanings that either reveal some acute personal conflict or some feared conflict with the auditor. The suddenness of the appearance of such symptoms, the clarity of their psychological significance, and the dramatic nature of their resolution in psychotherapy or with some shift in intrapsychic dynamics may suggest the general value of a psychodynamic model for understanding stuttering. It is important to recognize that there are such cases, but they have little to teach us about stuttering. Yet another neurotic condition sometimes blurring a clearer conception of stuttering is that of conversion hysteria, which is a condition in which the symptoms formed often simulate those of some physical disorder. In rare instances the symptoms simulate stuttering. Differential diagnosis is not difficult in such cases since the simulation is poor when tested against the realities of stuttering. The stuttering-like behavior is usually a parody of the sound, syllable, or word repetitions, rarely involving inaudible behavior or any of the secondary mannerisms such as word avoidance or substitution. In such cases the stuttering-like behaviors appear fully developed in the midst of a period of stress or anxiety and typically have an obvious payoff in modifying the life situation and the stress experienced by the individual. An example from my experience involved a young woman who conducted much of the business of her father (who was an independent plumber) by telephone. She was fearful of her father and managed her duties poorly. The appearance of an unusually rapid and stereotyped high-pitched repetition of the initial sounds of many words in all situations quickly brought about removal from her job, psychiatric treatment, and a decision to enter upon another line of work. Not long after this the "stuttering" diminished rapidly and normal fluency returned.

Such cases raise the question as to the role of imitation in the development of stuttering. Froeschels (1943) recognized a condition which he termed pseudotoni in which stuttering-like behaviors were encountered without the usual tension and explosive release associated with typical stuttering. In such cases he believed that the presence of a parodied model could be identified and the neurotic usage of the pseudostuttering could be discerned. While this sort of imitation may occur in some individuals, in the opinion of Van Riper, imitation as a source of true stuttering in children is at best an extremely rare phenomenon (Van Riper, 1971).

A more important and more elusive diagnostic problem resides in the area that has come to be generally known by the term "communication apprehension." While most individuals will, under certain conditions, experience something akin to stage fright, some anxious freezing that interferes temporarily with optimum verbal performance, there are some individuals who show in most social situations a reticence, an audience sensitivity, or an oral communication apprehension that importantly determines the quantity and quality of their verbal performance. Many individuals with stuttering, articulatory, and voice disorders, but not all, may be said to have communication apprehension. It is also clear, however, that the majority of individuals with this trait have no problems in the basic speech skills (Phillips, 1968). Extrapolating from a number of studies of communication-apprehensive individuals, McCroskey (1977) suggests the communication-apprehensive individual is typically "an introverted individual who lacks self-esteem and is resistant to change, has a low tolerance for ambiguity and is lacking in self-control and emotional maturity" (p. 84). A variety of studies of those with communication apprehension indicate that they develop complex skills in avoiding situations in which demands for speech are likely and that when required they will speak less and with less relevancy. It is the clinical impression of the present writer that when faced with a clear demand to speak, some of these individuals will show silent blockages without evidence of speech effort and also some of the audible repetitions and prolongations of the set of core stuttering behaviors.

The proportion of individuals experiencing communication apprehension is estimated to be in the area of 20% when case-finding testing is carried out in school and other samples (McCroskey, 1977). Under ordinary circumstances these individuals tend to become visible when confronted with new situations and new speaking requirements which they have not yet learned to evade. This may account for what is sometimes reported as an increase in the incidence of stuttering in the first year of the various phases of schooling. For example, Cooper (1972), studying over five thousand children in junior and senior high schools largely by means of questionnaire, found the highest prevalences occurred in the first year of junior and senior school (3.1% and 3.4%, respectively), whereas the average prevalence for all other years of junior and senior high school was 1.9%.

In many circumstances and in many ways the individual with marked communication anxiety resembles a stutterer. The avoidance of speaking situations and the techniques for reducing the expectations that others may have of them for speech are among the familiar secondary psychological mannerisms of stutterers.

The key question is can such an individual become a stutterer? Can the individual with communication apprehension yet initially adequate basic speech skills develop the core stuttering behaviors and show some of the characteristics of secondary verbal mannerisms of stutterers? Evidence is lacking but a negative answer appears most likely. Cases in which stuttering-like behaviors have an onset in later childhood or adolescence without a clear neurotic context tend to have a brief duration of the stuttering-like behaviors. Communication apprehensiveness may persist without intervention, but the directions of development appear to be those of social isolation, withdrawal, etc., rather than in the direction of continued or evolving speech abnormality. Incidental support for this interpretation is offered by a study which examined the extent to which self-reported recovery from stuttering among high school students was confirmed when the parents were interviewed (Lankford & Cooper, 1974). In two-thirds of these cases the parents had not perceived their child to have ever stuttered. If, in fact, communication apprehension experienced in nonfamilial situations was reported as "stuttering," the foregoing discrepancies might be explained. It is of interest to note that in the Andrews and Harris (1964) study only five children showed an onset at age ten or later. All such cases showed a dissipation of stuttering behaviors in less than eight months. Were these actually cases of communication apprehension?

The foregoing is not intended to minimize the serious developmental consequences of communication apprehensiveness. Some of these consequences are similar to those of stuttering and may certainly be as serious. It is clear, however, that the management of communication apprehension by means of rather modest programs of systematic desensitization is largely effective (Goss, Thompson, & Olds, 1978; Paul & Shannon, 1966), while the management of stuttering appears to be a far more complex and perplexing matter.

The following are a few conditions where stuttering or stuttering-like behavior has been observed, and while not presenting a problem of a differential diagnostic nature these associations may be of some theoretical significance.

Stuttering-like behaviors are known to occur with relatively high frequency in aphasic conditions following brain damage. Many neurologists have made observations of this sort (Brain, 1961; Critchley, 1970; Head, 1926), but understandably these behaviors have received little attention in the face of the major problems of survival, language recovery, and rehabilitation that aphasic patients present. Farmer (1975) studied only the repetitive behaviors of aphasics and reported more repetitions of sounds, syllables, words, and phrases in aphasic patients than in a group of nonaphasic brain-damaged control cases. It is of interest that within the types of aphasic disorders the repetitions of an elemental nature were most prominent in those with Wernicke's syndrome. Since this type of aphasic disorder is characterized by fluent and rapid speech, it would be of interest to know if cluttering behaviors also occurred along with elemental repetitions.

Finally, it might be mentioned that among retarded individuals stuttering appears to be present to a greater extent than would be expected by chance. Some have argued that, in retardation, stuttering has a special character, lacking the usual development of secondary verbal mannerisms and consisting largely of audible repetitions and prolongations. Mysak (1976) has termed such behavior as gnosogenic dysautomaticity, believing it to occur mostly when speech is demanded of the retarded individual on a subject they know little about.

A recent study compared retarded individuals labeled as stutterers by speech pathologists with retarded individuals not so labeled (Bonfanti & Culatta, 1977). The type and frequency of dysfluencies were quite comparable in these two groups, which suggests that the initial labeling was arbitrary and probably best understood as a psychosocial phenomenon. It was observed that the relative saliency of dysfluency types differs from that found in normal 5-year-olds. It was also found that, unlike stutterers in the normal range of intelligence, ratings of stuttering severity were unrelated to the frequency of dysfluencies and that awareness and secondary features such as postponement and word avoidance were absent. These findings tend to support the view that dysfluencies in retarded individuals are very general and in most cases distinct from stuttering.

In individuals affected by Down's syndrome stuttering-like behaviors are reported to be very common. This, however, is a disorder in which language, articulatory, and voice disorders are also known to be unusually prominent (Zisk & Bialer, 1967), so that a very special fluency phenomenon may be involved. Cabanas (1954), who appears to be unique in the intensity of his investigations of these behaviors in Down's syndrome, concluded that the condition involves a high rate of utterance delivery and a lack of awareness of concern by the subject, which suggests that these behaviors may more closely resemble cluttering.

As it is viewed here, stuttering is a disorder that is distinguishable from some fluency disorders by the character of the dysfluencies manifested and in part by the developmental period within which these dysfluencies first appear. Appearance of the core stuttering behaviors before the age of two and after the age of nine show a low likelihood of persisting more than a few months. The prognostiic implications of these age considerations suggest that, beyond these roughly defined limits, one is dealing with phenomena that should be distinguished from stuttering. Stuttering in this sense is a disorder bounded on the one hand by normal developmental dysfluencies and on the other hand by the communication apprehensions and dysfluencies experienced by a sizable group of individuals in later childhood and adolescence.

Epidemiological Features

The central diagnostic characteristics of stuttering have been identified; it is thus appropriate to consider what is known regarding its incidence and differential prevalence. In this realm a number of investigations have been carried out, but differences in the way that stuttering has been defined or identified, and the frequent use of investigative procedures that are not sound, have resulted in rather inconsistent and inconclusive finds. Some brief discussion of these matters may be of help in assimilating studies that report widely varying values or different relationships.

Some investigators have attempted to develop information about various developmental features of stuttering by asking affected adolescents or adults for a retrospective account of their experience with speech. How adequate are such procedures? Since it is known that most stuttering begins in the preschool years, common sense would suggest that self-reported data pertaining to onset is likely to be inaccurate and is likely to place onset much later than other data sources such as parents. Aron (1958) has demonstrated this to be the case. A group of European adolescent stutterers in their own recollection of onset reported a mean age of 8 years, 5 months, and in reporting the age of onset they believed their parents had told them, gave a mean value of 7 years, 4 months. Both of these values are in sharp contrast to the mean age of onset of 4 years, 10 months as reported by their parents when interviewed by a therapist. One would also expect such self-reported data to underestimate the incidence of stuttering compared with other data sources, since short experiences with stuttering in the preschool years are not likely to be remembered. This appears to be the case. For example, Sheehan and Martyn (1970), employing this approach with university student subjects, obtained results suggesting an incidence of 2.9%, which is considerably lower than estimates of incidence issuing from studies employing parental report or professional observations as data sources. Van Riper (1971) points out that only longitudinal studies can reflect the time incidence of stuttering or the rate with which individuals have been classified as stutterers at some point in their lives. Andrews and Harris (1964), in a longitudinal study, followed a thousand children from birth to age sixteen and had health visitors (who saw the children regularly) refer children with suspected speech and language abnormalities to speech therapists associated with the project. During the period of study, 4.3% of the children were identified as stutterers using the implicit criteria of the speech therapists. Other investigators have asked parents about the appearance of stuttering in their children. Glasner and Rosenthal (1957) interviewed parents registering their children for grade one (aged five but not yet seven) in a catchment area including both urban and rural areas. According to parental judgments 15.4% of their sample were identified as stuttering or as having experienced a period of stuttering. Since an estimate of incidence would be some unknown amount greater than this prevalence, it is clear that the three studies considered here lead to very different estimates of the incidence of stuttering. These differences in findings are probably the result of many factors, such as the inclusion of cases involving normal developmental dysfluencies in some parental reporting, the omission of some brief stuttering episodes by health visitors, the demand characteristics of the parent interviews at school registration resulting in acquiescent responding in some parents, and several other factors including time differences in the various samples. It is difficult to know how to discount for such factors except by identifying those studies with the fewest evident shortcomings and accepting the values reported therein as the best estimates. In the present instance, the most adequate investigation to date appears to be that of Andrews and Harris (1964), which suggests the incidence of stuttering to be around 4%. Serious deficiencies have been reported with this study (Ingham, 1976), so that even this figure should be regarded as tentative.

Are individuals in particular societies or in particular segments of society more likely to be affected than others? Where sizable non-European cultural groups have been studied by speech pathologists, the incidence and other features of stuttering appear to compare with what has been found in most European and American societies. Aron (1962), for example, studying Bantu-speaking children in South Africa, found a prevalence of 1.26%, which is a comparable prevalence value for schoolage children in American studies (Van Riper, 1971). One study (Morgenstern, 1956), surveying a large sample of Scottish schoolchildren, identified stutterers by means of teacher report and therapist examination and related prevalence to several social indices. This study found a disproportionately large number of stutterers to have fathers in semiskilled occupations. This finding was unconvincingly interpreted as supporting the notion that stuttering would be more common in upwardly mobile families. This same study was able to compare its subsample of eleven-year-old stutterers with national norms for several social characteristics and found that the stutterers differed significantly in having a greater separation in years from older or younger sibs. This finding suggests that stutterers may, in the early years of development, be functionally less related to other children and particularly affected by relationships with the mother.

On much more substantial grounds, it may be asserted that stuttering is more prevalent in males than in females. Many studies ranging widely in sample characteristics and investigative procedures have reported sex ratios in the range of 2:1 to 4:1. Dickson (1965) surveyed American schoolchildren in several elementary and high schools for current or past experience with stuttering and found a sex ratio of 2:1 averaging over all grades. In Scottish schoolchildren, Morgenstern (1956) reported a ratio of 4.4:1. From the often-cited Andrews and Harris study (1964) the ratio of 2.5:1 was reported, while among Bantu schoolchildren a sex ratio of 3.2:1 was found (Aron, 1962). The significance of this sex-biased distribution of stuttering will be considered later as evidence of the contribution of genetic factors to the onset and persistence of stuttering.

Also on firm grounds is the assertion that stuttering is more likely to occur and to persist in children when stuttering is or has been present in other family members. Sheehan and Costley (1977) have reviewed the studies of familial association and concluded that one-fourth to one-third of stutterers reveal a familial history of stuttering that is obviously significantly greater than found in fluent control subjects.

It is, of course, unclear to what extent this association is attributable to genetic factors or to influences that the presence of stuttering in one member of a family may have on the management of speech dysfluencies. There is some evidence that suggests that modeling and other social influences that depend on direct interaction with a stutterer do not completely account for the relationship under consideration. Andrews and Harris (1964) compared 80 stutterers and control subjects and found a history of stuttering in 37.5% of the former and 1.25% of the latter. Examination of subjects where stuttering was present only in family members with whom the subject had no direct contact resulted in 16.25% of the experimental group being so characterized. Here, then, is further evidence suggesting that the role of hereditary factors in stuttering may be important.

The Onset and Persistence of Stuttering

Since most stuttering begins in the preschool years, it is understandable that parents play a particularly important role in the identification of stuttering and its earliest management. Earlier, evidence was cited that most parents who say their child stutters or did stutter appear to be referring to types of dysfluency occurring far less frequently in children not so identified. There is additional evidence that parents, in identifying the presence of stuttering, usually are referring to a clear deviation in their child's speech development and characteristically react to the presence of this deviation in a reasonable fashion.

First of all, in a certain proportion of children the onset is sudden and the new condition is dramatically presented to parents. Van Riper (1971) estimates that in his experience about 10% of children show a sudden onset of stuttering. In other children the onset is gradual and the exact onset is therefore elusive, but in many of these children the new condition is made evident to the parents by the fact that the elemental stuttering behaviors are occurring not when the child is in an excited or other emotional state but when apparently calm and at ease. Then too the absolute frequency of elemental stuttering behaviors in a large proportion of children parentally diagnosed as stutterers may exceed the frequency of comparable dysfluencies in fluent children by many times. For example, Floyd and Perkins (1974), studying a small group of preschool stutterers, found the mean frequency of syllable dysfluency to be eight times that of the mean frequency of dysfluency of a larger group of nonstuttering preschoolers. Although this is a qualitative difference, its magnitude is such that these differences would impress most observers with some rudimentary knowledge of the character of normal childhood speech. There is also evidence that parents base their judgment of stuttering on the presence of the number of forms in which dysfluency is manifest. This could be thought of as a severity favor that enters into parental judgments. Glasner and Rosenthal (1957) asked parents of grade one children if hesitations, repetitions, or prolongations had been observed in their children, after he asked them initially to make a summary judgment as to whether their child stuttered, had stuttered in the past, or had always been fluent. Their results show that the likelihood of the judgment of stuttering increased from 1 to 29 to 65 to 94 chances in 100 as the number of the stuttering behaviors specified increased from 0 to 3. This suggests that a severity factor in terms of the variety of dysfluency behavior enters into parental identification of stuttering.

In some children either the parents or a clinician will relate the onset of stuttering to some disruption in the child's life or to some other event of significance. Disruptive events would include physical illness or psychological trauma. Johnson and associates (1959) failed to find a variety of such factors distinguishing between young stutterers and control subjects. Imitation of stutterers is sometimes offered as an explanation of stuttering, but Van Riper (1971) in his studies of many beginning stutterers found but one case in which imitation appeared to be a probable cause. Wyatt (1969) developed a view of the precipitation of stuttering that lays stress on the cognitive and communicative developmental level of the child as these relate to disruptions in the communicative patterns of the child, particularly as these involve the mother. This view has strong clinical appeal but has not as yet received the sophisticated testing that is required. There is little evidence that the child at onset is anxious or disturbed. Andrews and Harris (1964) failed to find differences in the "emotional conflicts" of young stutterers and controls. Van Riper (1971) on the other hand was impressed by the greater emotionality of children seen shortly after the onset of stuttering. Glasner (1949), studying behaviors that might be attributed to "emotional overloading," found high frequencies of feeding problems, neurosis, and some rarer behaviors. These investigations suggesting disturbances in the beginning stutterer did not, however, have the opportunity of control comparisons. It is also possible that what was observed was the result of stuttering. Finally, the "symptoms" that Glasner studied are not all clearly behaviors that are the result of neurotic symptom formation, but can also be regarded as habit disorders, perhaps revealing something of the parental management or long-standing parent–child relationship, but again not necessarily related to the precipitation of stuttering.

In any disorder that has a typically gradual onset the identification of precipitating factors will be difficult. In stuttering there is the additional matter, as shall be seen, that a large proportion of cases of stuttering show a dissipation of the disorder in the first two years following onset. This means that greater interest adheres in the identification of factors within the child, or external to the child, which are associated with the persistence of stuttering than in factors associated with its onset. These two sorts of factors need not be the same or similar.

According to Johnson's (Johnson and associates, 1959) diagnosogenic theory of stuttering development, stuttering persisted and evolved because of parental behavior that communicated negative evaluations of dysfluent speech behavior to a child who would resonate with these concerns. To what extent does the literature support this view of the parental role in the persistence of stuttering? The Glasner and Rosenthal (1957) study cited earlier is sometimes regarded as a source of supporting evidence. This study revealed a severity factor in terms of the number of these specific dysfluency behaviors observed by the parents. The higher this severity factor, it was found, the more likely that parents reported actively attempting to control or modify the child's dysfluencies, the more likely that parents perceived the child's problems to have a serious emotional basis, and the more likely that the stuttering had persisted and was present at the time of interview. This can be construed as supporting the notion that parental concern resulting in active corrective behavior operates to increase severity of stuttering and the likelihood of persistence of stuttering. An equally likely interpretation, however, is that parents correctly appreciate the severity of a stuttering condition and its likelihood of becoming chronic and accordingly usually resort to more active corrective measures to assist the child. Some clarification would be offered it if were known whether parental stances that minimized or ignored stuttering or the various active corrective procedures yielded more or fewer cases of spontaneous recovery when the severity factor is taken into account by partial correlations. While the data provided (Glasner & Rosenthal, 1957), do not permit this to be done with exactness, if one assumes that the trends relating severity to stuttering status hold for the total data, then it is possible to make a rough estimate of the relationship. When this is done there appears to be a higher-than-expected recovery for those children whose parents claimed to ignore or minimize stuttering ($\chi^2 = 6.82, p = <0.01$). This is in accord with expectations from the viewpoint of the diagnosogenic theory, but it should be noted that the relationship is still a weak one and that the actual data indicate many instances of children who are actively corrected recovering and many instances where stuttering persisted at least to entrance into grade one despite the parental claim that they ignored or minimized the problem.

Other studies of personality, child-rearing attitudes, and behavior of the parents of stuttering children have not yielded evidence supporting the diagnosogenic view of stuttering. One might for example expect parents of stutterers to show evidence of being socially ascendant or controlling individuals. In fact, Goodstein and Dahlstrom (1956), using a personality questionnaire (Minnesota Multiphasic Personality Inventory, MMPI), found them significantly more anxious and less dominant than parents of control children. La Follette (1956), using another questionnaire, also found parents of stutterers to be more socially submissive. Abbot (1957), studying the mothers of stuttering children, assessed child-rearing attitudes and found only two significant scale differences. One of these, "Seclusion of the Mother," resembles a social submissiveness dimension while the other, "Fear of Harming Baby," suggests diffuse fears regarding their adequacy of caring for children. Quarrington (1974) has reviewed the literature and suggests that there is evidence that the parents of beginning stutterers are unusually instructionally passive in their relations with their stuttering child. Some of this evidence comes from Johnson's (Johnson and associates, 1959) study designed to test the diagnosogenic theory of stuttering development. The parents of a stuttering child in this study were, for example, significantly less willing to correct the grammar or the pronunication of their child than were the parents of control children (Quarrington, 1974).

In some studies of the parents of children who have been stuttering for several years, evidence of hostile or punitive attitudes and guilt feelings have been revealed (Holliday, 1957; Moncur, 1952). Such findings might suggest that where parents have such attitudes, stuttering is likely to become chronic, but it may also be interpreted as evidence of a malignant transformation of the mother or the mother–child relationship after stuttering has persisted for several years and imposed stresses on the mother and the whole family.

The stresses associated with having and caring for a stuttering child can be considerable and are rather obvious. An interesting source of stress frequently encountered by parents comes from the ready counsel of physicians or other professionals who, after the onset of stuttering, assure the parents that if ignored, stuttering will disappear. Statistically such advice is one good ground, but as stated, it ignores the fact that most stuttering in childhood shows a cyclical patterning. Accordingly, as children move into a period of relatively infrequent stuttering, the parental expectations of an orderly decline and disappearance of stuttering appear to be confirmed. As the cycle continues, however, parental puzzlement is likely to be followed by mistrust of professional advice, suspicion with regard to behavior of one's spouse, and personal feelings of guilt. These feelings are likely to be intensified in subsequent cycles as parents cast about for new coping strategies and go through periods of apparent success followed by a period of desperation as stuttering increases in subsequent cycles.

If there is weak evidence as to the role of parental behavior in the persistence of stuttering, what other factors are known to be related to persistence? Two factors, the type of initial stuttering behaviors and the presence of a familial history of stuttering appear to be related. Glasner and Rosenthal (1957) found that, for those children for whom parents reported the early presence of prolongations, the recovery rate by grade one was 30%, which was significantly lower than the recovery rate (59.2%) for stuttering children displaying all other types of dysfluencies ($\chi^2 = 5.91, p = < 0.02$). It is interesting that some studies of self-reported stuttering and recovery have yielded rather congruent findings in that individuals reporting their early dysfluencies to be largely easy repetitions and hesitations showed significantly higher rates of remission than those reporting blockages between words or prolongations (Dickson, 1971; Sheehan & Martyn, 1970). The last investigators suggest that the experience of strong fears tends to be associated with the early appearance of blockings, in contrast to the more moderate fears associated with hesitations and repetitions.

The relationship of a familial history of stuttering and the likelihood of its persistence is on rather weaker grounds. Studies of self-reported stuttering in high school students (Cooper, 1972; Cooper, Parris, & Wells, 1974) have indicated a significant relationship between spontaneous recovery and an absence of stuttering in the family. Sheehan and Martyn (1970), in similar studies of university students, found a tendency in this direction that fell short of statistical significance.

To recapitulate briefly, major interest appears to lie not in the identification of conditions associated with the onset of stuttering, but in the determination of factors associated with the persistence of stuttering into adolescence and adult life. Present evidence indicates that parental attitudes and management of early stuttering are not critically important in accounting for the persistence of stuttering. Factors found to be related to persistence include the form of early stuttering and the presence of stuttering in the family. Both of these suggest, minimally, that a subclass of stuttering may exist in which hereditary factors are of major importance. This possibility will be examined in greater detail shortly.

The Development of Stuttering

Since stuttering disappears in a matter of a few weeks, months, or years in some children and persists in others, the question arises as to whether persistent stuttering is a different disorder from stuttering that dissipates spontaneously. One approach to this question is to determine to what extent stuttering develops or unfolds in a similar way from individual to individual where it persists. The best generalizations of this sort would appear to come from the work of Bloodstein (1960b, 1969), who systematically collected data at intake pertaining to stuttering children of various ages and of various stuttering durations. Looking at the behaviors associated with different ages at intake, he concluded that a general pattern of development with four phases or stages could be identified.

The first or onset phase is considered to occur typically between the ages of two and six. Repetition is the most common dysfluency form usually occurring at the beginning of sentences, phrases or clauses. Related to this is the likelihood that the words on which stuttering is most likely to occur are pronouns, conjunctions, and prepositions. In most cases, the frequency of stuttering fluctuates in a cyclical fashion and also shows minor phasic increases when the child is excited or upset or is experiencing a situation of increased communicative demand such as the need to speak quickly or accurately. If frustration in speaking is experienced at all, it appears to occur only at times of particular difficulty.

In the second phase, the child clearly has the notion of himself as having difficulties in speaking but relatively little concern or embarrassment is evident. The main features of this phase are that stuttering is less restricted to initial words of sentences or phrases, is less likely to consist largely of repetitions, and is more likely to occur on verbs, nouns, adjectives, and adverbs. The cyclical variations in stuttering are less likely to include periods of fluency.

In phase three, there is mounting evidence of concern about stuttering in that the individual now believes that certain sounds or words present particular difficulties and that stuttering may be avoided to some extent by substituting less-feared words or words not stressing the feared sound. It should be mentioned that the sounds or words feared by stutterers show little agreement. The explicit sound or word fears of a particular individual tend to change from time to time. In many individuals these fears bear little relationship to words actually stuttered or to the words avoided by word substitution or by other means. Observers now see that the amount of stuttering manifested is greater in some situations or social relations than in others and this may be appreciated to some extent by the child.

In the fourth and final phase, clear fears of situations have emerged and various ways of avoiding these situations or of reducing the communicative demands in such situations have developed. Devices to avoid stuttering are used most of the time since the anticipation of stuttering is vivid.

The foregoing outline of developmental trends roughly characterizes the personal evolution of stuttering for many individuals, although the ages at which the various phases occur may vary widely. For many individuals, however, this scheme appears quite at variance with their actual development. There are, for example, some children whose onset of stuttering after age six is the sudden appearance of silent blockages, which persist without significant variation in form or severity for several years. Many other examples of developmental histories at variances with the general scheme might be cited. Faced with such diversity in development, there are two general approaches for coping with it. One, which might be termed psychological, would regard stuttering behaviors as learned phenomena. Variations in the earliest forms of stuttering and differences in subsequent stuttering development would be attributed to different learning conditions experienced and also to some extent due to individual differences in those characteristics having some role in the learning of stuttering behaviors of all forms. In this view, developmental schemes such as the one offered by Bloodstein (1961) reflect some uniformities in the social experiences of individuals learning to stutter and to be a stutterer. While this might be of some interest, what is clearly most relevant when considering how to help the stutterer would be the individual history of stuttering development, which presumably would reveal something of the organization of stuttering behaviors together with information as to contingencies which establish and which maintain the various stuttering behaviors. Some clinicians and investigators hold a view of this general nature. Other workers are inclined to a view that the diagnostic or medical model, as a way of making scientific discoveries, has a further role to play in the realm of stuttering. In this view, when diversity of developmental courses (prognosis) is encountered in individuals believed to have the same disorder, one concludes that the disorder is not unitary but an aggregate of several superficially similar disorders. Each of these disorders, when identified by a distinguishing configuration of symptoms, signs, and features, will be found to have a particular etiology, prognosis, and optimum form of treatment for those individuals who can be classified or diagnosed as belonging to one of these new subgroups of clinical disorder.

A number of workers have suggested new classificatory schemes for stuttering (Luchsinger & Arnold, 1965; St. Onge, 1963; Zaliouk & Zaliouk, 1965) and while some of these appear promising, they have all been based on impressions of uncertain substance and validity. Much more substantial and more theoretically sophisticated is the classification offered by Van Riper (1971). This scheme was based on the analysis of some 300 cases managed clinically, but more importantly, was based on the longitudinal study of 44 cases seen shortly after the time of onset and subsequently followed for a varying number of years. Four subgroups or developmental tracks were recognized, which accounted for the great majority of individual developmental histories. In each subgroup, Van Riper recognizes the important role of learning conditions in development, but the conditions that are of importance and their manner of influence differ from subgroup to subgroup. This then is a complicated classsificatory scheme but still one that can best be viewed from an older tradition of medical research.

Approximately one half the cases studied by Van Riper (1971) appeared to belong to one subgroup (Track I), which not too surprisingly resembles the general developmental scheme induced by Bloodstein (1969). Individuals belonging to this group show a gradual onset between 2½ to 4 years with easy effortless syllabic repetitions and no evidence of awareness. Repetitions occur most frequently on the first word spoken after a pause or on the most meaningful word of an utterance. Further changes in the frequency and form of dysfluency occur typically in an episodic or cyclical fashion. Following a period of fluency or near fluency, repetitions are likely to become more frequent and the tempo and number of repetitions per volley may increase. The neutral vowel may appear more frequently in these repetitions. This new state of affairs may persist for several weeks and gradually subside to another period of fluency. With each subsequent increase in stuttering frequency, there is the possibility that further changes will occur in the forms of dysfluency. Van Riper is persuaded that a critical developmental point occurs with the appearance of prolongations, at first as a concluding act following a volley of repetitions, but gradually creeping forward in order to occur before repetitions and largely to supplant them. Tension and unusual effort to speak become evident and the stuttering tremor appears around this time. The tremor reported to appear is peculiar to severe stuttering and consists of 7 to 9 per second low-amplitude oscillations of the tongue, jaws, or lips. The tremors may spread throughout the speech musculature and may at times include the eyelids, arms, or legs. These tremors appear to occur as the result of the individual's struggle to release a prolongation or to terminate repetitions that have persisted too long. They are an involuntary emergent result of such efforts and accordingly are experienced as an alarming and frightening loss of control.

Despite the period of reduced stuttering frequency and severity, or perhaps because of this, fearful anticipation of stuttering develops and with these, word substitution, circumlocution, and variations in the timing of word utterances appear. In subsequent development the fears and avoidance behaviors become more idiosyncratic and at the same time more stereotyped. Particular situations come to be feared or regarded as relatively easy and while there is a tendency for situations with high communicative demands to be feared by most stutterers (Bloodstein, 1949), particular life experiences make for considerable variation from individual to individual. Speaking on the telephone and giving instructions to a stranger are examples of situations with high communicative demands and ones that are feared by most stutterers. For some stutterers, however, these will be among their easiest speaking situations. Individuals whose stuttering development falls into the present track come to acquire a map of social relationships colored for the likelihood of stuttering that they employ in regulating much of their social activity and that serves as an important source for the conception of self and self-worth. In general, choices of group membership, friends, recreational activity, vocation, and marital partner, and choices of particular roles played in these various areas of life, all show the influence of the individual's conception of self as a stutterer. For the most part these influences restrict and distort what would appear to be optimum personal development.

It is Van Riper's view that individuals showing the present developmental pattern are those most likely to show spontaneous remission and those most likely to show a good response to speech therapy.

The foregoing has been a sketch of the modal form of stuttering. The next most frequently encountered subgroup, Track II developmental pattern, resembles the stuttering emerging from cluttering as conceptualized by several writers but particularly by Weiss (1964).

The onset of stuttering in this subgroup may be said to be gradual and to have commenced with the appearance of connected speech. There is then no period of fluency. Accordingly, dysfluencies begin early, but since most benchmarks of connected speech, such as sentence usage, are delayed a year or more, not as early as normal language acquisition schedules might suggest. Gestures, jargon, or a few holophrastic phrases have served communicative needs longer than is usually the case. Van Riper conceives of these children as impoverished in their expressive abilities and requiring more time than usual in establishing consistent articulation or phonemic sequences and standard syntactical patterns. The patience of auditors is of critical importance in providing models and in allowing time for the preparation, execution, and revision of speech efforts. If this is not present, then in addition to the hesitations and broken speech which is usual, the repetition of sounds, syllables, and words increases, which further reduces the intelligibility of their speech.

Under most circumstances, children showing this form of development appear to remain largely unaware of and unconcerned about their dysfluencies. In most cases the pathology of speech development remains at this status with speech that is mildly stuttered but mainly cluttered. In some instances persistence is associated with certain additional abnormalities, but sound and word fears are rare and weak. Prolongations and complete blockages of the airway rarely appear. What may appear is a lengthening of the volley of repetitions and an increase in the tempo, pitch, and intensity, which is only terminated by complete exhalation. These appear to be emergent involuntary activities that are experienced as frightening losses of control. At such points situational fears and ultimately sound or word fears may appear, but avoidance behaviors rarely appear to be as organized or as powerful as they are in the advanced stutterers of the Track I type. Spontaneous remissions appear to be rare with this form of stuttering and in Van Riper's opinion, when the advanced form of this disorder is treated effectively, the speech is likely to show an increasing prominence of the cluttering features.

The two remaining developmental tracks or subgroups are less common. In both subgroups stuttering begins suddenly and usually later in development after the child has experienced several years of fluent connected speech.

The distinctive developmental features of Track III include an onset in which the child is suddenly unable to speak. Sometimes this can be linked to a specific frightening experience. As the total inability to speak passes it is evident that the child is having difficulty in starting an utterance. The articulatory position of the initial sound is assumed but the airway is closed, usually at the level of the larynx, so that phonation does not occur. A variety of struggle behaviors, including unusual patterns of breathing and speaking on residual air or on inhalation may appear in a very short time. It is apparent to observers that the child is aware that something extraordinary has happened and shows intense frustration or fear. It is common to observe a marked reduction in speech attempts, social withdrawal, and regressive behavior in self-care habits.

As the individual struggles to break through a closure, intermittent noises and sounds (vocal fry) may be apparent, increasing in audibility until an explosive release of the sound or syllable being attempted occurs. Once difficulty in the initiation of an utterance is overcome the rest of the utterance is likely to be delivered fluently. It is only after a pause and the initation of a new utterance that difficulty will be evident.

Van Riper (1971) asserts that this form of stuttering can disappear suddenly in most cases to reappear soon in a similar form. In most individuals fitting the present pattern, dysfluencies go through a phase in which audible prolongations appear seeming to supplant some of the blockages, and in some individuals a form of slow repetition may be evident that appears to be a testing or retrial procedure rather than the automatic volleys of repetitions encountered in the modal developmental pattern of stuttering.

To this point in development, stuttering has consisted of starting difficulties and if fluency is to occur spontaneously it will most likely occur here. Beyond this point stuttering spreads beyond the initiation of utterances. Sound and word fears develop rapidly as do situational fears. The avoidance behaviors of individuals in this group may be more complex, more extensive, and distort to a greater extent the social and emotional development of the individual than is the case with stutterers of other subgroups.

Individuals belonging to the last subgroup (Track IV) may show considerable variety in the initial form of their abnormal dysfluencies, which usually begin in later childhood, adolescence, or even in adult life according to Van Riper. The form of dysfluency, however, will be dramatic and have a spurious or contrived character. They may, for example, involve many volleys or word or phrase repetitions or consist of lengthy pauses during which tongue protrusions or other elaborate behaviors largely unrelated to understandable speech attempts to occur. Beyond this variety of dysfluency form, there are several consistent features. The initial form of dysfluency does not change but remains monosymptomatic and stereotyped as long as the condition persists. Another common feature is that the affected individual appears relatively unemotional in the midst of dysfluencies and his behavior actually suggests that the auditor is being tested, controlled, or punished by these excessive delays and struggles in the communicative process. A final feature is that it is usually apparent that at certain times or in certain relationships the individual is capable of extremely fluent speech.

Research is needed to confirm the accuracy and exhaustiveness of the categories or subgroups of stuttering just considered. To those clinicians and researchers who are persuaded that treatment should issue from a diagnostic process, the four tracks of Van Riper are particularly appealing in their detail, developmental perspective, and their congruence with clinically acquired impressions. While this approach strikes many as sound and promising, others argue that to consider stuttering as other than unitary errs in needlessly seeking complexity or at best is scientifically premature. Clinicians who are primarily concerned with the treatment of older individuals are particularly likely to hold such views. It is possible to argue, for example, that as adults those individuals supposedly from the first three tracks of Van Riper will have much in common. All are likely to have acquired sound and situational fears and to have developed a variety of behaviors they believe operate to avoid or minimize the appearance of elemental stuttering behaviors. All are likely to have developed complex social avoidance patterns. They may arrive at this status by different developmental routes, but having arrived they appear to present similar problems for treatment.

To appreciate this issue further it is important to consider what is known of treatment practices and their effectiveness, and it is necessary to consider the problems of preventative management in young children.

The Treatment of Adult Stutterers

Several writers (e.g., Van Riper, 1973; Wingate, 1976) have provided accounts of treatment procedures that have been used in the remote and in the less-distant past, indicating their continuity with some present scientifically respectable methods. The use of externally provided rhythm as a pacing procedure for speech is one such example. The reader is reminded that stuttering is largely confined to that speech that is executed in the habitual manner of the individual. A great many variations from the circumstances of habitual speech are attended with fluency or a marked reduction in the frequency of stuttering. Speaking in unison, shadowing the words of another speaker, and speaking in time with a metronome are some of the many conditions associated with these effects. A great deal of speculation has been engaged in to account for these phenomena. Experimental work (Fransella, 1967) has permitted us to rule out distraction of attention from speech since not all distracting activities have these effects, but an adequate explanation still appears remote. Despite our lack of understanding why these factors operate as they do, it has occurred to many that they represent a way of demonstrating to the stutterer that fluency is a possibility and suggesting that if these controlling procedures were used, in a modified form that reduced their social saliency, a high degree of fluency could be attained in everyday social life. Thus the slow rhythms of arm swinging in the speech school that permitted fluent concurrent speech might become the faster but regular rhythms of the imagined bouncing ball permitting a certain sort of fluency, a sometimes unnerving one to auditors in everyday conversation. Innumerable variations of this sort have provided the bases for many therapeutic programs, which also included the belief that if enough time were spent in this new fluent behavior the individual would overcome the suspected source of stuttering, such as lack of self-confidence. Apparently, for some individuals these programs helped but for most they proved to be a failure in various ways. Some stutterers, having learned the general nature of treatment, rejected it or failed to comply with the requirements. Some tried to comply but found that the advocated controlling devices did not work for them. These, then, were some of the primary treatment failures.

Some individuals encountered failure because they were unable to carry on the behavior controlling stuttering or mediating fluency in everyday life because of embarrassment or because of other pressures. In other cases, although the controlling behavior was exercised faithfully, it failed to have the efficiency encountered in school or clinic. These are some of the failures to transfer clinic-won gains to the ordinary life situations of the stutterer.

In still other cases, the device worked well for some time but was abandoned when the individual was confronted by some new life crisis following which stuttering returned. Or in other instances, perhaps because the new way of speaking became highly automatic or habitual, the device lost its ability to control stuttering. These are some of the failures of longer term maintenance of gains.

The treatment program sketched above is one that attempts to teach a new method of speaking. It is an old approach to the treatment of stuttering, but enhanced with the terminology of learning theory and some of the technology of behavior modification, it is also one of the contemporary approaches to stuttering. For want of a better term these contemporary treatment schemes might be termed "educational" since they focus on the assets of the individual and seek to teach a set of new skills that will supplant older, less desirable ones. Such programs have acquired considerable sophistication and efficiency over earlier programs. There is, for example, evidence of greater success in acquiring patient compliance in treatment and perhaps greater efficiency in reducing stuttering to low levels of frequency in the treatment period. Still, however, there is considerable failure in transferring treatment fluency to everyday life and very considerable failure in having clients maintain fluency gains over a longer period of time. These two areas of generalization and maintenance still remain the priority problem areas in the management of stuttering. A simple contemporary treatment program might profitably be considered at this point.

Ingham and Andrews (1973) describe a program designed to produce a stutter-free way of speaking with special attention to those factors working against generalization. Treatment occurred in small groups of four who were hospitalized for a three-week period. Those administering the program exercised control over the food and luxuries that the patients received and these were paid for by tokens earned by attaining the various objectives set out for them. In the first week, a pseudooperant procedure was employed, in which the response observed and reinforced was percentage of spoken syllables stuttered, with the amount of reinforcement determined by the extent of departure from the baseline level. This was evaluated and reinforced in special assessment sessions occurring nine times each day. In the second week, the patients were taught to speak with continuous voicing, in which vowel sounds were to be prolonged and smooth transitions emphasized. Assistance in this type of speaking was provided by having the patients speak under conditions of delayed auditory feedback. Both normal speakers and stutterers are prone to prolong vowels when their normal auditory feedback is restricted and what is received has been delayed about 250 msec. Over the period of a week, the delay of feedback was reduced progressively by 50 msec. as goals for speech form were reached and the increasing rates of speech required in assessment periods was attained. During this phase, token control extended beyond the formal assessment periods to all staff–patient interaction. In the third week, subjects were required to carry out speaking assignments in person and by telephone in four situations that were perceived as constituting increasing threat or demand. These were recorded and evaluated for token reward.

In evaluating the results of this program with 39 stutterers, Andrews and Ingham (1972) report that 65% were found free of stuttering when formally assessed nine months after treatment. The authors add an interesting cautionary note to the effect that, when fifteen months after treatment the speech of patients was evaluated by a stranger not known to be related to the project, the evaluations showed that most subjects had shown some relapse. These investigators are among the few to report follow-up and clearly indicate the need for the use of external auditing procedures in treatment research (Ingham, 1975).

There are, of course, many procedures for instating fluency through teaching a new way of speaking that is then gradually shaped to a less abnormal one that might pass for normal speech. This is particularly the case when treatment is massed or when there has been a total transformation of the individual's living conditions. It is possible to set up various forms of control in various sequences and contexts in order to carry out comparisons that appear scientifically sound, since the treatment packages appear replicable and are largely independent of the sort of stutterers being treated (e.g., Ingham, Andrews, & Winkler, 1972). This study, however, compared treatments on the basis of fluency instatement in treatment and it is possible that some procedures less impressive in the treatment phase might have important transfer advantages. Brady's (1971) use of a hearing-aid-style metronome might be an example of a technique with such special advantages. There are certain difficulties that many treatment approaches stressing fluency face.

Use of the new form of speaking avoids most stuttering but not all. What is the individual to do if there is a greater or lesser resurgence of stuttering? Since these approaches to treatment do not offer instructions as to how to deal with stuttering, the only solution is that of returning to treatment with the hope that on this new cycle more effective habits of fluent speech will be acquired. To many this does not appear to be either an economic or reasonable way to deal with relapse.

Another difficulty is that the fluency attained by the end of treatment is a peculiar one that, while offering a new facility in communication, may also offend both speaker and auditor by its lack of authenticity. Undoubtedly our notions of fluency and its attainment are still at a primitive level. In this connection, it is of interest to note that Webster (1974, 1977), in a treatment termed "precision fluency shaping," holds that faulty voice onsets are the major shortcomings of stutterers and has developed procedures for systematically shaping vocal onset behavior so that it will fall within established norms. This is assisted by an electronic device that continuously monitors vocal behavior and provides information for self-correction.

Developments which identify specific behavioral targets and target sequences offer considerable encouragement to those who would like to pursue fluency instatement by a purer form of operant conditioning. Hegde (1971, 1978) has commented on the immaturity of much of the work in fluency-oriented treatment research and has outlined some of the theoretical and research needs of what appears to be a promising area of development.

Turning now to those therapeutic approaches that emphasize the management of dysfluency, it is more difficult to characterize briefly the procedures and programs because of the varying positions within this general viewpoint and because of the complexity of procedures sometimes advocated. Perhaps, however, the viewpoint that stuttering behaviors are operants that can be brought under control with behavior modification principles may be accorded only scant attention. While a number of investigators have shown that the dysfluencies of normal speakers can be brought under operant control (Siegal & Martin, 1966, 1968), stuttering behaviors appear to be more complex. A number of investigators have found that some stuttering behaviors are unaffected or increase with negative reinforcing contingencies (Oelschlaeger & Brutten, 1976; Siegal, 1970) and it has also been shown that some stuttering behaviors decrease with positively reinforcing contingencies (Patty & Quarrington, 1974). Accordingly, a more tenable position would appear to be that set out by Brutten and Shoemaker (1967), which holds that two sorts of learning are involved in the acquisition and modification of stuttering behaviors. First of all, there are the repetitions and prolongations together with other behavioral components (e.g., fluttering eyeblink, sudden inhalation) that are aspects of intense negative emotional responses and that become conditioned to sounds, words, and other stimuli, and second, there are the varied instrumental escape and avoidance responses that are maintained by anxiety reduction. Word avoidance, word substitution, and all those idiosyncratic motor behaviors that appear to terminate stuttering tremor or to release closures on the one hand are regarded as instrumental behaviors capable of operant control or conditioning. The basic stuttering behaviors of repetition and prolongation on the other hand are regarded as part of a configuration of emotional responses that may be reduced in their evoked range by desensitization procedures such as reciprocal inhibition therapy.

Prins and Lohr (1972) have studied the covariation of a large number of stuttering behaviors and suggest that their factor analysis of these relations lends support to the two-factor view of Brutten and Shoemaker (1967). Only those behaviors belonging to the class of avoidance responses have been shown to be modifiable by contingent reinforcement (Bastijens, Brutten, & Stes, 1978; Quist & Martin, 1967).

This two-factor view of stuttering has not given rise to packaged programs of treatment but has undoubtedly become incorporated into the practices of many speech clinicians, particularly those trained in the mainstream views of stuttering and stuttering therapy in America. Perhaps the best embodiment of such views, which Wingate (1976) has termed "reconstructive," is to be found in the writings of Van Riper (1971, 1973). The ideas expressed there have had a very extensive influence on the practice of stuttering therapy in the United States and throughout the world. A sketch of Van Riper's approach is offered here, but the reader is cautioned that a brief personal summation of a complex therapeutic approach may be misleading.

The first phase of treatment advocated by Van Riper assumes that the advanced stutterer has acquired many rigid habits of speech and speech-related behavior of which he is largely unaware. The stutterer has avoided reading, thinking, and talking about stuttering, so that the first phase is designed to be a cognitive awakening to personal and general knowledge of stuttering. In a variety of ways the stutterer is brought to a readiness to observe and to experiment with his speech and speech-related behavior. An important outcome of this phase is a detailed analysis of the various sorts of behaviors and feelings involved in stuttering, together with some appreciation that these are lawfully connected with one another and with external factors.

The next phase of therapy is one in which the patient learns to cancel stuttering, and learning from the cancellation of stuttering is of central importance. Briefly, cancellation refers to a series of steps to be taken immediately following stuttering behavior or a chain of stuttering behaviors. These steps are designed to acquire the information believed needed to inform the stutterer what his actions were, to consider what other actions might have been taken, and to determine what actions would be required to produce an integrated delivery of the troubling word. These are complex cognitive tasks, at first undertaken with considerable external assistance and personal verbal rehearsal and regulation, but becoming more automatic with practice.

Van Riper (1973) points out that for some stutterers once the techniques of observation and systematic analysis and manipulation have been acquired no further help is required and the individual becomes a self-directing therapist. For those unable or unwilling to assume such responsibilities, the further phases of therapy will next attempt to bring forward the new skills of observation and control to bear on the ongoing stuttering behavior. This is the technique of "pull-out," which is not anticipation or prevention of stuttering, but is a controlled manifestation of stuttering which minimizes the experienced tension and reduces the observable abnormality of a stuttering occasion. A final phase of therapy has the objective of training the stutterer to enhance the use of proprioceptive feedback and to reduce reliance on auditory feedback in monitoring speech. This is based on a conviction Van Riper shares with some other workers that accords faults in the auditory feedback system or abnormal usage of auditory feedback an important role in the etiology of some types of stuttering.

An important point to be made is that Van Riper (1973) and other practitioners of similar persuasion are not committed to particular technical procedures but advocate flexible usage of operant, desensitizing, or other technologies as the problems of the particular individual in attaining treatment goals of the various phases are encountered. This means that it is not possible to specify a treatment package that will be delivered in a uniform fashion to clients, which creates obvious problems in executing rigorous studies of outcome for treatments of this kind. It must be mentioned that while fluency is the ultimate goal of these practitioners, the general view is that fluency, if it will be realized, will evolve slowly over a period of several years. The early attainment of fluency during the treatment period is not considered a proper or appropriate goal. This difference in early objectives between treatment approaches rules out easy short-run comparisons of treatment effectiveness. Unfortunately, substantial information about long-range treatment outcome for reconstructive approaches of all sorts is scanty. Some of the best known practitioners (Bloodstein, 1975; Van Riper, 1973) in assessing treatment outcome, largely in terms of clinical impression, are rather pessimistic particularly with regard to the difficulties of maintaining gains over time.

In a brief review of treatment approaches, only passing reference may be made to a wide variety of psychotherapeutic approaches to stuttering. One subclass of these approaches regards stuttering as a neurotic symptom that will dissipate with some reorganization of intrapsychic forces (e.g., Glauber, 1958). Sheehan (1970) has reviewed the literature considering the evidence for stuttering as symptom formation and concludes that the evidence is rather better that stuttering can, in some cases, give rise to neurosis. There is certainly no evidence of the general effectiveness of psychoanalytic approaches. The other subclass of psychotherapeutic approaches regards stuttering as having given rise to a number of rigid cognitive-affective states or habits that have distorted and restrained personal growth and that have encapsulated stuttering behaviors. Therapy is seen as necessary to permit the individual new perceptions of self and others and to foster attitudes that will ultimately result in new acceptable forms of speaking or fluency (Hejna, 1960; Sheehan, 1970). Again substantial evidence of treatment effectiveness is lacking. One exception perhaps is that of Fransella (1972), who has treated stutterers with a cognitive psychotherapeutic system developed by Kelly (1955). The evidence offered is that for some of the individuals treated the gains in speech and attitudes attained in clinic group appeared to transfer readily to the everyday world. In general, however, it is not at all clear that gains in self-acceptance, reductions in anxiety, or other commonly entertained objectives in psychotherapy will result in significant reductions in stuttering in the short or in the long run. In this connection, however, it is interesting to consider studies of individuals' recovery from stuttering, which has persisted for many years, without professional assistance. Such cases are rare. One investigation of a series of such cases (Quarrington, 1977) indicated that the majority had undergone experiences that had significantly altered their views of self prior to undertaking rather simple ways of modifying their speech that were applied with persistence over a period of several years. It is not known whether such cases are very unusual and to what extent they have features common to adult stutterers seeking treatment, but such findings do suggest that for some stutterers recovery has been initiated and sustained primarily by new conceptions of responsibility, competence, and self-control that are commonly psychotherapeutic goals.

Treatment Research

One rather intriguing aspect of treatment evaluation involves the perceived quality of fluency attained with treatment. Reference has already been made to the clinical observation that some stutterers abandon fluency-instating programs because they feel their fluent speech lacks authenticity. This matter has concerned some investigators who have tried to determine if naive auditors could recognize abnormality in the fluent speech of treated individuals and if the specific features of such recognition could be determined. Results of research to date indicate that auditors can differentiate the fluent speech of treated stutterers from the speech of nonstutterers, but the source of this distinction has not been identified (Ingham & Packman, 1978). The stutterers investigated in this study had been treated by the method described by Ingham and Andrews (1973) outlined earlier here. It will be recalled that this method importantly involves prolonged speech and accordingly one might suspect remnants of this particular sort of controlling behavior are being identified. The results of another investigation call this into question (Runyan & Adams, 1978). This study employed speech samples of successfully treated stutterers from six different approaches to fluency instatement including precision fluency shaping, as well as from the reconstructive approach of Van Riper. In addition, a group of partially treated stutterers was included and the fluent speech of all these subjects compared to the speech of a comparable number of nonstutterers. Auditors were able to differentiate stutterers from nonstutterers. Differences between the treatment groups were not apparent but differences in pretreatment severity measures of stuttering were related to the accuracy of auditors' discriminations. These results suggest that while various treatment approaches may be reasonably successful in eliminating the obvious features of stuttering they may either introduce abnormal features or, what appears more likely, fail to eliminate less obvious abnormalities.

If present treatment research is unable to identify clearly the treatment of choice for stuttering, is it known which individuals are most likely to profit from therapy? Several studies have sought answers to this question without evidence that personality as assessed by psychological tests (Gregory, 1969; Lanyon, 1966), or selected features of speech, or stuttering (Lanyon, 1965; Prins, 1968) are significantly related to treatment outcome. An exception here might be that of pretreatment severity of stuttering, which Gregory (1969) found negatively related to measures of stuttering severity at the conclusion of treatment, but far less potently related to measures of stuttering severity nine months after treatment. Indeed there is some evidence that greater reductions in stuttering frequency in the treatment period are related to the likelihood of subsequent relapse (Prins, 1970).

One of the few studies to yield more substantial findings was reported by Guitar (1976). All subjects had been treated by the fluency instatement approach described by Ingham and Andrews (1973). The treatment outcome measures were derived from five-minute recordings of conversations that expatients had with a covert evaluator 12 to 18 months after completion of treatment. The characteristics related to fluency measures from this source included three questionnaire scales assessing behavior and attitudes indicative of strong emotional response to stuttering such as avoidance tendencies, and specific fears, questionnaire measures of extroversion and neuroticism, and several pretreatment measures of stuttering severity. These pretreatment measures, when intercorrelated, showed a clustering of the avoidance and emotional reactivity measures, which were significantly correlated with neuroticism but essentially unrelated to the stuttering severity measures. This unrelatedness may surprise some, but has been commented on by clinicians and recently Lanyon, Goldsworthy, and Lanyon (1978) have, with factor analytic procedures applied to questionnaire data, demonstrated the independence of a severity dimension of avoidance behaviors and negative emotional attitudes. It seems clear that stuttering severity, whether assessed by the frequency of stuttering or by the extremity of qualitative features, is not directly related to the severity of psychological responses to stuttering as manifested in avoidance behaviors, experienced anxiety, etc.

Returning to Guitar's (1976) study it was found that those scales measuring psychological response to stuttering were most potently related to posttreatment measures of stuttering severity. Since all stutterers discharged from this treatment have the near perfect fluency of the controlled speech taught, one can interpret the results as indicating that the likelihood of regression in speech increases with the extremity of pretreatment psychological response to stuttering. It is of particular interest that the relationships under discussion suggest better than chance predictability of outcome with this particular treatment form, and this degree of predictability has held up surprisingly well when applied to a new sample of patients. The power of this predictability lies primarily in the measures of psychological response to stuttering and the measure of neuroticism.

There are a number of ways to assimilate the results of the foregoing study. One can conclude that stutterers with strong psychological responses to stuttering will be less likely to profit from any treatment program. More positively, one can conclude that the standard program needs to be modified to meet better the needs of those with strong psychological reactions. Such results are of great interest to those with a diagnostic orientation, for they suggest that efficacy and economy may lie in the recognition of treatment needs differing importantly among those seeking treatment, and the revealed differences may suggest what different treatment forms may be most appropriate. In some instances some stutterers, whose stuttering may be mild or moderately severe, present themselves primarily for help with their anxiety or paralyzing fears, and claim to be willing to accept their present form and severity of stuttering.

It is of some interest that Guitar's (1976) findings appear to relate to Van

Riper's (1971) description of Track III stutterers. This is the subgroup with nonvocalized dysfluencies appearing at onset. Van Riper emphasizes the severity of speech and situational avoidances and other aspects of strong emotionality in members of this group. These views are in some accord with other workers who attempted to differentiate stutterers on the basis of the predominant forms of dysfluency shown as adults stutterers (Douglass & Quarrington, 1952). Those stutterers showing a clear predominance of inaudible (nonvocalized) stuttering, most of whom might qualify as Track III members, were found to have more extreme avoidance behaviors and emotional reactions than stutterers with a predominance of vocalized stuttering. A specific concern for the inaudibility of their dysfluencies was demonstrated in nonvocalized stutterers (Quarrington & Douglass, 1960). An examination of therapeutic processes in a reconstructive approach to therapy indicated specific difficulties that could be anticipated if the differences, psychological capabilities, and emotional reactivity of vocalized and nonvocalized stutterers were ignored.

In most of the approaches to treatment considered here the major hazards have appeared in the phase of transferring clinic-won gains to everyday life and in the long-term maintenance of these skills and attitudes over a period of time. There are many other disorders in which treatment shows similar hazards. Most of the disorders of consummatory behaviors such as smoking, obesity, and alcoholism can be shown to come under immediate control with a radical transformation of the life situation of the afflicted individual and the introduction of new directive principles or controls. Successful transfer to the community of these new patterns usually requires the individual to initiate changes in old social relationships, to persist aggressively in these changes until they are generally accepted and to continue a pattern of lengthy or lifelong vigilance and control. A recognition of these and other needs in a variety of behavior disorders has prompted many behaviorally oriented workers to view the client not simply as a reactive organism, but to consider what conditions might be necessary to encourage the development of a cognitively active agent capable of self-direction, self-management, or self-control (Cautela, 1969; Thoresen & Mahoney, 1974; Watson & Tharp, 1972). Self-control techniques in other disorders have been demonstrated to have a variety of advantages over treatment approaches that do not require the client to assume responsibility for monitoring and evaluating ongoing behavior. There is evidence that these approaches work with children and indeed some of the procedures may be peculiarly appropriate to the levels of cognitive development encountered in younger children (Meichenbaum & Goodman, 1971). Space does not permit an examination of these developments, but an account of self-control principles and procedures as they might be incorporated into stuttering therapy has been offered by Hanna and Owen (1977). The terminology may in part appear unfamiliar but the reader knowledgeable with reconstructive approaches to treatment will find interesting counterparts and relationships with more familiar terms and procedures.

The Management of Stuttering in Children

It is unfortunate that so much attention has been given to the treatment of adults when there is every reason to believe that the management of stuttering in children is more urgent and more promising. One reason for this emphasis is attributable to the influence of Johnson's diagnostic theory of stuttering development, which suggested that the critical factors that could be readily manipulated were the responses of parents and others to dysfluencies. As noted earlier this view has not been supported by research findings and accordingly more direct approaches to management have been advanced. Luper and Mulder (1964) have presented a rather comprehensive position that assumes the validity of Johnson's views and regards stuttering that persists to follow the developmental scheme set out by Bloodstein. With beginning stutterers then, management consists of parent education and counseling. By implication, stuttering that persists and shows the changes of the form and locus of stuttering represents a failure in this preventative program. Subsequent management is seen as consisting of a combination of direct and indirect procedures to retard or prevent stuttering from proceeding to the anticipated features of the next phase.

In contrast to this conservative position there are therapists (e.g., Wingate, 1976) who insist that older views of the development and developmental dynamics of stuttering are incorrect and not appropriate in formulating management procedures. His general approach, which is educational in orientation, features the instatement of fluency by emphasizing vocalization and the prosody of speech. It is his view that as long as the child is able to follow the instructions involved in remedial exercises that treatment may be given without concern. While this is a refreshing shift from older approaches, which unduly emphasized the fragility of the stuttering child and left all concerned puzzled and guilt-ridden when stuttering advanced, it may be too incautious. While one might dismiss some views of the role of parents in stuttering development, it is possible that interaction with parents or others contributes in ways not yet clearly seen. It is quite possible that the same parental behavior has quite different functional implications depending on the type of stuttering involved.

Evidence was cited earlier indicating that the parents of children who have been stuttering for some time are different from control parents in being more punitive and perhaps less consistent in their discipline. Kasprisen-Burrelli, Egolf, and Shames (1972) have demonstrated differences in parents of older stutterers from controls in that they show more of a variety of negative interactional behaviors. Egolf, Shames, Johnson, and Kasprisin-Burrelli (1972) have shaped therapeutic procedures on the premise that what can be observed as negative and salient in parent–child interaction probably maintains stuttering. Therapists, in dealing with the child, strive to avoid engaging in the particular sorts of negative behavior prominent in the repertoire of particular parents in the hope of reducing stuttering. The parent is gradually included so that the therapist can model a healthier interactional role and fluency gains will generalize to parent–child interactions. The nonspecific nature of the negative behaviors demonstrated by Kasprisin-Burrelli *et al.* (1972) to be more frequent in the parents of older stuttering children are entirely consistent with the interpretation offered earlier that without assistance parents can become disturbed in themselves and in their relations with others as stuttering persists in their child. It is my impression that these developments can be largely prevented if the parents of beginning stutterers receive accurate information about stuttering and if they know that access to further consultation is available at any time requested. In those parents who have not received such assistance and have become unstable, depressed, guilt-ridden, or disturbed in other ways, a direct approach to their discomfort is usually effective. It is very doubtful that in most cases the approach suggested by Egolf *et al.* (1972) would be very economical or effective. Their approach suggests that they entertain an implicit belief that these negative parental behaviors in interaction have caused stuttering as well as operate to maintain it, which is an unwarranted assumption.

It is my impression that research has offered little to support the notion that stuttering is a unitary condition. Almost certainly it is an aggregate of several disorders, which have yet to be clearly recognized, but which are probably best glimpsed at the present time by the four subgroups or developmental tracks of Van Riper (1971). If this is accepted, then the management of stuttering children should begin with an attempt to improve diagnostic procedures and the classificatory system available. If the classifying scheme today does not lead to descriptions of preferred treatment then it can at least alert us to particular concerns or features that deserve clinical attention. For example, stuttering occurring in the context of language retardation and cluttering symptoms (Track II) might prompt one to assign priority or management to the encouragement of language development and to the provision of unpressured but corrective language assistance at school and at home. Stuttering on the other hand would be accorded a low priority in direct therapy but would be included in parental instruction and guidance.

Some other suggestions for management derived from Van Riper's classification might be mentioned. Most clinicians will have encountered one or more young stutterers who fitted the description for Track IV. They may, however, have difficulty in recalling these cases because their contacts are usually broken early by the parents, due in part to our failure to perceive the role that stuttering is playing in an interparental struggle for the child's dependency and affection. This at least would be my speculation as to a common familial dynamic in this type of stutterer. Andrews and Harris (1964) suggest that no child should receive direct treatment until stuttering has persisted for two years. If the child fits the features of Track I this advice appears appropriate. This is the group in which children show the greatest oscillation in stuttering frequency and great variation in dysfluency form with the frequent complete dissipation of stuttering. This same advice appears most inappropriate for the child with the sudden blockages of the Track III onset. In these children there appears little justification for withholding direct treatment and one might be understandably willing to introduce fluency instatement procedures as suggested by Wingate (1976) or as proposed by others.

Admittedly, clinicians will differ considerably in their interpretations of what seems most appropriate when the classificatory system is in the early stages of development, but probably there would be greater agreement among clinicians than obtains at the present time with the assumption of the unitary nature of stuttering that is simply not in accord with so much of clinical experience.

The Causes of Stuttering

A good deal of past research and theoretical construction has been concerned with those factors that may account for the occurrence of elemental stuttering or the high frequency of these sorts of dysfluencies in certain children. Earlier, for example, reference was made to the tendency for stuttering to occur frequently in some families and to the sex ratio of stuttering, both of which might suggest the importance of genetic factors in stuttering. Of particular interest here are studies of identical and fraternal twins in which stuttering occurs in at least one twin. Two studies that appear to have dealt adequately with the hazards of twin research, such as accuracy of zygoticity identification, have found the concordance rates in identical twins to be 60% or higher while the concordance rate for same-sex fraternal twins was less than 20% (Howie, 1976; Nelson, Hunter, & Walter, 1945). Such discrepancies strongly suggest that hereditary factors play a role of importance in stuttering, at least as it is a presenting phenomena in children and adults. Sheehan and Costley (1977), in a review of several lines of research, concluded that the facts are entirely consistent with a view that genetic factors play a role in the disorder of stuttering. Kidd, Kidd, and Records (1978), studying familial patterns of stuttering, have suggested that a sex-modified inheritance model in which females, having a higher threshold for the emergence of stuttering, would be expected to show higher levels of familial predisposing factors. Their data demonstrate that female stutterers show higher frequencies of stutterers in near family members than do male stutterers. While it is likely that hereditary factors play an important role in the onset or persistence of stuttering in some individuals, the biological mechanisms mediating genetic influences are unknown. Past and present searches for the biological substrate of stuttering behaviors have been extensive and here only a few of a larger number of possible mechanisms can be considered.

A good deal of this research has attempted to infer the causes of stuttering from the specific conditions in which stuttering occurs. In adults stuttering is not randomly distributed in speech, but is much more likely to occur on the initial syllable or sounds of words and on words in the initial or early positions of an utterance or phase. The stuttering loci of young incipient stutterers who show fewer secondary features is of particular interest in such research. It has been found to show similar trends in reproduced or experimentally elicited speech (Williams, Silverman, & Kools, 1969), as well as in spontaneous conversation (Bloodstein & Gantwerk, 1967).

Studies of adults stutterers' experimentally elicited speech have also indicated a higher likelihood that longer words, content words, and words of higher information value are more likely to be stuttered (Brown, 1945; Soderberg, 1967). Williams *et al.* (1969) found that in their sample of older children stuttering was associated with word length and with content words also. It is of interest, however, that Bloodstein and Gantwerk (1967) found that function words such as conjunctions and pronouns were more likely to be stuttered in the spontaneous speech of young stutterers, while nouns and interjections were less likely than chance to be stuttered upon. Bloodstein (1974) in a subsequent study of spontaneous speech of young stutterers was prompted to emphasize the importance of the initiation of a syntactic unit as the probable locus of stuttering over other word-related characteristics. Such findings suggest that problems or uncertainties in the choice of syntactic structures may play an important role in the stuttering of some children.

Early research on the sounds most frequently associated with stuttering indicated that initial consonants were more frequently stuttered on than initial vowels and that the difficult consonants were variable from one stutterer to another (Johnson & Brown, 1935). Various writers have pointed out that, while the initial consonant is repeated or is the locus of the prolonged posture, one must consider this to represent a difficulty of transition from consonant to vowel. Work by

Agnello (1974) indicated that young stutterers show unusual delays in voice onset when required to reproduce experimental consonant–vowel combinations /pa, ba/. This suggests generally faulty phonatory processes in consonant–vowel transitions may underlie stuttering.

Wingate (1976, 1977b) believes stuttering involves defective phonatory functioning and points out that it is the transition from consonant to stressed vowel that requires the production of a stress prominence that is most likely to be associated with stuttering. Furthermore, it is argued that the words most likely to be stuttered are those that are associated with stress peaks within an utterance. At such points of linguistic stress, the greatest demands are made for energizing the processes involved in phonation and it is at these points that deficiencies in phonatory function will be most apparent. This view appears to synthesize many of the findings considered here although some relationships still beg explanation. The specific sound difficulties that are characteristic of individual stutterers suggest the importance of specific anxiety-laden expectations that are unrelated to a general factor of linguistic stress. It would also appear unlikely that the relationship of word frequency or the information load of words to stuttering can be neatly subsumed by the concept of linguistic stress. Most importantly, Bloodstein's (1974) findings of stuttering occurring at syntactic junctures in the spontaneous speech of young stutterers strongly suggests that language-encoding difficulties may be of central importance.

Wingate's thesis does, however, show considerable potency and has reinstated interest in older views such as Kenyon's (1953) which stressed the central significance of phonation in stuttering over other factors such as articulation or emotional evaluation.

Currently, there is a good deal of research interest in specifying the phonatory behavior and in detailed study of the laryngeal functioning of stutterers. Both the duration of vowel and consonant production have been found to be shorter in the stuttered speech than in the nonstuttered speech of stutterers (Di Simoni, 1974). Brayton and Conture (1978) have shown that several conditions such as noise and rhythmic stimulation, known to reduce stuttering frequency, are associated with increases in vowel duration. Freeman and Tatsujiro (1978) have directly observed laryngeal muscles resulting in failures of the initiation of phonation and interruptions or breaks in phonatory transitions.

Faults in phonatory processes appear to be likely proximal causes of elemental stuttering, but the sources of these may prove to be complex and elusive. The system of speech processes is complex. To appreciate the phonatory processes requires that one consider not only the immediate production aspects of phonation, but also the various control or integrative aspects such as the motor and auditory feedback processes. Phonatory functions cannot be divorced from breathing and articulatory activities, which must be coordinated if fluent speech production is to occur (Adams, 1978; Perkins, Rudas, Johnson, & Michael, 1976). Finally, the speech system cannot be divorced from the cognitive and affective inputs to the system, which are capable of influencing phonatory functioning in a variety of ways.

Some investigators have been pursuing notions that auditory feedback mechanisms may be the primary sources of faulty phonatory functioning. Stromsta (1972) and Timmons and Boudreau (1972) have argued that the phase-angle discrepancy between air- and bone-conducted feedback differs in stutterers from that in nonstutterers in a way that presents stutterers with a condition analogous to delayed auditory feedback. Others (Butler & Stanley, 1966; Webster, 1974; Webster & Lubker, 1968) subscribe to the concept of critical feedback fault resident in anomalous middle-ear muscle functioning, which immediately prior to the initiation of speech introduces faulty control information. These views have stimulated research, some of which appears to be crucial experimentation (Altrows & Bryden, 1977) and which fails to support the auditory interference theory of Webster (1974).

It is likely that research into the proximal and more distal causes of stuttering will increase in volume and rigor in the near future. It is also likely that such research will identify a variety of components in the speech system relatable to stuttering and, as suggested by Van Riper (1971), will also find that stutterers differ in the extent to which these components appear to play a critical role in the precipitation of stuttering.

Future Directions

While the search for causes of elemental stuttering has an understandable appeal, it should be evident that a complete account of stuttering will be complex. Such an account may begin with an elucidation of the faulty mechanisms of the speech system that are inherited by some individuals, but from this early point the story must leave the strictly biological level of explanation and shift to a complex psychosocial account of the developing individual. There is still much research needed to enlarge our appreciation of the social and psychological processes that are of critical importance in the evolution of stuttering pathology and those processes entailed in the dissipation of stuttering.

Furthermore, increased knowledge of the proximal causes of stuttering may not lead directly to new powerful methods of clinical management. It is likely that stuttering therapy and related research will develop for some years without direct guidance from research findings bearing on the causes of stuttering.

With respect to therapy research, several trends of an apparently conflicting nature are evident. One must be struck by the vigor of those advocating a focus on fluency instatement. Undoubtedly the reader will have been impressed by the amount and by the value of the research that has issued from the development of the treatment program described by Ingham and Andrews (1973). This work has given us sound information regarding treatment outcome, has developed outstanding methods of evaluation in follow-up, has pointed up issues in treatment evaluation, and has demonstrated important factors in the prediction of treatment outcome. Clearly, we need more programmatic research that answers questions of an applied nature of great interest to clinicians. This research is not importantly derived from theoretical considerations or the theoretical underpinnings are minimal. Treatment packages that are administered in a relatively uniform fashion to all subjects, and that are of a sufficiently simple nature so that they can be described and replicated appear to be critical needs that must be met if this research is to flourish.

To many clinicians the simplification and standardization of treatment is anathema. They are far happier to learn that workers in other areas of disorder share their problems and may have suggestions for new levels of clinical sophistication in dealing with the unique problems of the individual stutterer. The cognitive developments of behavior therapy are likely to be welcomed by many therapists of stutterers.

There are some workers who feel that those in the area of stuttering have been caught up in what might be called the special theories of stuttering to an excessive degree. It can be argued that such theories emphasizing the unique aspects of stuttering and the unique technical procedures isolate workers from theoretical and technical developments in other disciplines. If that has been the case then present developments seem to be reversing these trends and must be judged healthy.

There is, however, a lesson that workers in the area of stuttering could learn from fairly recent history of stuttering theory and research. The lesson is that in our search for theoretical generality and elegance, we have blundered by glossing over differences and assuming uniformity when it was not warranted. Several decades ago we permitted ourselves to see stuttering behaviors as continuous with a loose collection of dysfluencies found in the speech of normal speakers. Such a gloss permitted the development of the diagnosogenic theory that was not warranted because it should have been evident from the supporting data offered that true differences in dysfluencies existed between children called stutterers and nonstuttering children. Subsequently, we permitted ourselves the assumption that stuttering behaviors of a very diverse history and nature were really functionally equivalent. This permitted us to think about "moments of stuttering" and to engage in a variety of research enquiries of doubtful value until massive evidence convinced us of what we should have intuited long before. Today we know that there are at least those stuttering behaviors classically conditioned and those that are free operants. We are probably prepared now for further distinctions. For example, is it not likely that passive avoidance behaviors emerging from approach–avoidance conflict and active avoidance behaviors acquired through escape may require different learning conditions for their modification?

Perhaps the assumption of the unitary character of the stuttering syndrome is another instance where we are committing a similar error. If we ignore the gross developmental differences that stutterers present, it appears to facilitate the conduct of treatment research. Indeed, most research activities are apparently eased. In fact, we may be looking for relationships through a glass darkly. We can, of course, be very busy and confident that even though the view in our glass may be obscure, that science, in even this handicapped form, will ultimately make us realize our error and prompt us do what must be done to see more clearly. To some this is cold comfort.

References

Abbot, T. B. *A study of observable mother–child relationships in stuttering and nonstuttering groups.* Unpublished doctoral dissertation, University of Florida, 1957.

Adams, M. R. A clinical strategy for differentiating the normally nonfluent child and the incipient stutterer. *Journal of Fluency Disorders,* 1977, *2,* 141–148.

Adams, M. R. Further analysis of stuttering as a phonetic transition defect. *Journal of Fluency Disorders,* 1978, *3,* 265–271.

Agnello, J. G. Voice onset time and voice termination features of stutterers. In L. M. Webster and L. C. Furst (Eds.), *Vocal tract dynamics and dysfluency.* New York: Speech and Hearing Institute, 1974.

Altrows, I. F., & Bryden, M. P. Temporal factors in the effects of masking noise on fluency of stutterers. *Journal of Communication Disorders,* 1977, *10,* 315–329.

Andrews, G., & Harris, M. *The syndrome of stuttering.* London: Heineman, 1964.

Andrews, G., & Ingham, R. J. An evaluation of follow-up procedures for syllable-timed speech/token system therapy. *Journal of Communication Disorders,* 1972, *5,* 307–319.

Aron, M. L. *An investigation of the nature and incidence of stuttering among a Bantu group of school children.* (Doctoral dissertation, University of Witwatersrand, South Africa, 1958.)

Aron, M. L. The nature and incidence of stuttering among a Bantu group of school-going children. *Journal of Speech and Hearing Disorders,* 1962, *27,* 116–128.

Bastijens, P., Brutten, G. J., & Stes, R. The effect of punishment and reinforcement procedures on a stutterers Factor II avoidance responses. *Journal of Fluency Disorders,* 1978, *3,* 77–85.

Bleumel, C. S. Primary and secondary stuttering. *Quarterly Journal of Speech,* 1932, *18,* 187–200.

Bloodstein, O. Conditions under which stuttering is reduced or absent: A review of the literature. *Journal of Speech and Hearing Disorders*, 1949, *14*, 295–302.

Bloodstein, O. Stuttering as an anticipatory struggle reaction. In J. Eisenson (Ed.), *Stuttering: A symposium*. New York: Harper and Row, 1958.

Bloodstein, O. The development of stuttering: I. Changes in nine basic features. *Journal of Speech and Hearing Disorders*, 1960, *25*, 219–237. (a)

Bloodstein, O. The development of stuttering: II. Developmental phases. *Journal of Speech and Hearing Disorders*, 1960, *25*, 366–376. (b)

Bloodstein, O. The development of stuttering: III. Theoretical and clinical implications. *Journal of Speech and Hearing Disorders*, 1961, *26*, 67–82.

Bloodstein, O. The rules of early stuttering. *Journal of Speech and Hearing Disorders*, 1974, *39*, 379–394.

Bloodstein, O. *A handbook on stuttering*. Chicago: National Easter Seal Society for Crippled Children and Adults, 1975.

Bloodstein, O. Stuttering. *Journal of Speech and Hearing Disorders*, 1977, *42*, 148–151.

Bloodstein, O., & Gantwerk, B. G. Grammatical function in relation to stuttering in young children. *Journal of Speech and Hearing Research*, 1967, *10*, 787–789.

Bonfanti, B. H., & Culatta, R. An analysis of the fluency patterns of institutionalized retarded adults. *Journal of Fluency Disorders*, 1977, *2*, 117–128.

Brady, J. P. Metronome-conditioned speech retraining for stuttering. *Behavior Therapy*, 1971, *2*, 129–150.

Brain, R. *Speech disorders: Aphasia, apraxia and agnosia*. Washington, D.C.: Butterworths, 1961.

Brayton, E. R., & Conture, E. G. Effects of noise and rhythmic stimulation on the speech of stutterers. *Journal of Speech and Hearing Research*, 1978, *21*, 285–294.

Brown, S. F. The loci of stuttering in the speech sequence. *Journal of Speech and Hearing Disorders*, 1945, *10*, 181–192.

Brutten, E. J., & Shoemaker, D. J. *The modification of stuttering*. Englewood Cliffs, N.J.: Prentice-Hall, 1967.

Butler, B. R., & Stanley, P. E. The stuttering problem considered from an automatic control point of view. *Folia Phoniatrica*, 1966, *18*, 33–44.

Cabanas, R. Some findings in speech and voice therapy among mentally deficient children. *Folia Phoniatrica*, 1954, *6*, 34–37.

Cautela, J. R. Behavior therapy and self-control: Techniques and implications. In C. M. Franks (Ed.), *Behavior therapy: Appraisal and status*. Toronto: McGraw-Hill, 1969.

Cooper, E. B. Recovery from stuttering in a junior and senior high school population. *Journal of Speech and Hearing Disorders*, 1972, *15*, 632–638.

Cooper, E. B., Parris, R., & Wells, M. T. Prevalence and recovery from speech disorders in a group of freshmen at the University of Alabama. *American Speech and Hearing Association Journal*, 1974, *16*, 359–360.

Critchley, M. *Aphasiology*. London: Arnold, 1970.

Dickson, S. Incipient stuttering symptoms and spontaneous remission of the nonstuttered speech. *Journal of the American Speech and Hearing Association*, 1965, *7*, 371 (abstract).

Dickson, S. Incipient stuttering and spontaneous remission of stuttered speech. *Journal of Communication Disorders*, 1971, *4*, 99–110.

Di Simoni, F. Preliminary study of certain timing relationships in the speech of stutterers. *Journal of the Acoustical Society of America*, 1974, *56*, 695–696.

Douglass, E., & Quarrington, B. The differentiation of interiorized and exteriorized secondary stuttering. *Journal of Speech and Hearing Disorders*, 1952, *17*, 377–385.

Egolf, D., Shames, G., Johnson, P., & Kasprisin-Burrelli, A. The use of parent-child interaction patterns in an experimental therapy program for young stutterers. *Journal of Speech and Hearing Disorders*, 1972, *37*, 222–232.

Farmer, A. Stuttering repetitions in aphasic and nonaphasic brain damaged adults. *Cortex*, 1975, *11*, 391–396.

Floyd, S., & Perkins, W. H. Early syllable dysfluency in stutterers and nonstutterers: A preliminary report. *Journal of Communication Disorders*, 1974, *7*, 279–282.

Fransella, F. Rhythm as a distraction in the modification of stuttering. *Behavior Research and Therapy*, 1967, *5*, 253–255.

Fransella, F. *Personal change and reconstruction*. London: Academic Press, 1972.

Freeman, F. J., & Tatsujiro, U. Laryngeal muscle activity during stuttering. *Journal of Speech and Hearing Research*, 1978, *21*, 538–561.

Freund, H. Inneres Stottern und Einstellbewegung. *Zietschrift fur Neurologie und Psychiatrie*, 1932, *6*, 1243–1245.

Freund, H. Studies in the interrelationships between stuttering and cluttering. *Folia Phoniatrica*, 1952, *4*, 146–168.

Froeschels, E. Pathology and therapy of stuttering. *Nervous Child*, 143, 2, 148–161.

Glasner, P. J. Personality characterisitcs and emotional problems in stutterers under the age of five. *Journal of Speech and Hearing Disorders*, 1949, *14*, 135–138.

Glasner, J. P., & Rosenthal, D. Parental diagnosis of stuttering in young children. *Journal of Speech and Hearing Disorders*, 1957, *22*, 288–295.

Glauber, I. P. The psychoanalysis of stuttering. In J. Eisenson (Ed.), *Stuttering: A symposium*. New York: Harper, 1958.

Goodstein, L. D., & Dahlstrom, W. G. MMPI differences between parents of stuttering and nonstuttering children. *Journal of Consulting Psychology*, 1956, *20*, 365–370.

Goss, B., Thompson, M., & Olds, S. Behavioral support for systematic desensitization for communication apprehension. *Human Communication Research*, 1978, *4*, 158–163.

Gregory, H. *An assessment of the results of stuttering therapy*. Final Report, Research and Demonstration Project 1725-S HEW, 1969. Washington, D.C.: Department of Health, Education and Welfare, 1969.

Guitar, B. Pretreatment factors associated with the outcome of stuttering therapy. *Journal of Speech and Hearing Disorders*, 1976, *19*, 590–600.

Hall, P. The occurrence of disfluencies in language-disordered school-age children. *Journal of Speech and Hearing Disorders*, 1977, *42*, 364–369.

Hanna, R., & Owen, N. Facilitating transfer and maintenance of fluency in stuttering therapy. *Journal of Speech and Hearing Disorders*, 1977, *42*, 65–76.

Head, H. *Aphasia and kindred disorders of speech*. London: Macmillan, 1926.

Hegde, M. N., & Brutten, G. J. Reinforcing fluency in stutterers: An experimental study. *Journal of Fluency Disorders*, 1977, *2*, 315–328.

Hegde, M. N. Fluency and fluency disorders: Their definition, measurement, and modification. *Journal of Fluency Disorders*, 1978, *3*, 51–71.

Hejna, R. F. *Speech disorders and nondirective therapy*. New York: Ronald Press, 1960.

Holliday, A. R. *An empirical investigation of the personality characteristics and attitudes of the parents of children who stutter*. Unpublished doctoral dissertation, University of Washington, 1957.

Howie, P. M. *A twin investigation of the etiology of stuttering*. Paper presented at the annual convention of the American Speech and Hearing Association, Houston, Texas, November, 1976.

Ingham, R. J. A comparison of covert and overt assessment procedures in stuttering therapy outcome evaluation. *Journal of Speech and Hearing Research*, 1975, *18*, 346–353.

Ingham, R. J. A reassessment of findings from the Andrews and Harris study. *Journal of Speech and Hearing Research*, 1976, *41*, 280–281.

Ingham, R. J., & Andrews, G. An analysis of a token economy in stuttering therapy. *Journal of Applied Behavioral Analysis*, 1973, *6*, 219–229.

Ingham, R. J., & Packman, A. C. Perceptual assessment of normalcy of speech following stuttering therapy. *Journal of Speech and Hearing Research*, 1978, *21*, 63–73.

Ingham, R. J., Andrews, G., & Winkler, R. Stuttering: A comparative evaluation of the shortterm effectiveness of four treatment techniques. *Journal of Communication Disorders*, 1972, *5*, 91–117.

Jaffe, J., Anderson, S. W., & Rieber, R. W. Research and clinical approaches to disorders of speech rate. *Journal of Communication Disorders*, 1973, *6*, 225–246.

Johnson, W. A study of the onset and development of stuttering. In W. Johnson (Ed.), *Stuttering in children and adults*. Minneapolis: University of Minnesota Press, 1955.

Johnson, W., & Brown, S. F. Stuttering in relation to various speech sounds. *Quarterly Journal of Speech*, 1935, *21*, 481–496.

Johnson, W., and Associates. *The onset of stuttering*. Minneapolis: University of Minnesota Press, 1959.

Kasprisin-Burrelli, A., Egolf, D. B., & Shames, G. H. A comparison of parental behavior with stuttering and nonstuttering children. *Journal of Communication Disorders*, 1972, *5*, 335–346.

Kelley, G. A. *The psychology of personal constructs*. New York: Norton, 1955.

Kenyon, E. L. The etiology of stammering: The psychophysiological facts which concern the production of speech sounds and stammering. *Journal of Speech and Hearing Disorders*, 1953, *8*, 337–348.

Kidd, K. K., Kidd, J. R., & Records, M. A. The possible causes of the sex ratio in stuttering and its implications. *Journal of Fluency Disorders*, 1978, *3*, 13–23.

Kussmaul, A. *Speech disorders*. New York: William Wood, 1891.

La Follette, A. C. Parental environment of stuttering children. *Journal of Speech and Hearing Disorders*, 1956, *21*, 201–207.

Langova, J., & Moravek, M. Some results of experimental examinations among stutterers and clutterers. *Folia Phoniatrica*, 1964, *16*, 290–296.

Lankford, S. D., & Cooper, E. B. Recovery from stuttering as viewed by parents of self-diagnosed recovered stutterers. *Journal of Communication Disorders*, 1974, *7*, 171–180.

Lanyon, R. I. The relationship of adaptation and consistency to improvement in stuttering therapy. *Journal of Speech and Hearing Research*, 1965, *8*, 263–269.

Lanyon, R. I. The MMPI and prognosis in stuttering therapy. *Journal of Speech and Hearing Disorders*, 1966, *31*, 186–191.

Lanyon, R. I., Goldsworthy, R. J., & Lanyon, B. P. Dimensions of stuttering and relationship to psychopathology. *Journal of Fluency Disorders*, 1978, *3*, 103–113.

Luchsinger, R., & Arnold, G. E. *Voice–Speech–Language*. Belmont, Calif.: Wadsworth, 1965.

Luchsinger, R., & Landholt, H. Electroencephalographische Untersuchungen bei Stottern mit und ohne Polterkomponente. *Folia Phoniatrica*, 1951, *3*, 135–151.

Luper, H., & Mulder, R. *Stuttering: Therapy for children*. Englewood Cliffs, N.J.: Prentice-Hall, 1964.

McCroskey, J. C. Oral communication apprehension: A summary of recent theory and research. *Human Communication Research*, 1977, *4*, 78–96.

McDearmon, J. C. Primary stuttering: A reexamination of data. *Journal of Speech and Hearing Research*, 1968, *11*, 631–637.

Meichenbaum, D. H., & Goodman, J. Training impulsive children to talk to themselves: A means for developing self-control. *Journal of Abnormal Psychology*, 1971, *77*, 115–126.

Metraux, R. W. Speech profiles of the pre-school child 18 to 54 months. *Journal of Speech and Hearing Disorders*, 1950, *15*, 37–53.

Moncur, J. P. Parental domination in stuttering. *Journal of Speech and Hearing Disorders*, 1952, *17*, 155–165.

Morgenstern, J. J. Socio-economic factors in stuttering. *Journal of Speech and Hearing Disorders*, 1956, *21*, 25–33.

Morley, M. *The development and disorders of speech in childhood*. Edinburgh: Livingstone, 1957.

Mysack, E. D. *Pathologies of speech systems*. Baltimore: Williams and Wilkins, 1976.

Nelson, S., Hunter, N., & Walter, M. Stuttering in twin types. *Journal of Speech and Hearing Disorders*, 1945, *10*, 335–343.

Oelschlaeger, M. L., & Brutten, G. J. The effect of instructional stimulation on the frequency of repetitions, interjections, and words spoken during the spontaneous speech of four stutterers. *Behavior Therapy*, 1976, *7*, 37–46.

Patty, J., & Quarrington, B. The effects of reward on types of stuttering. *Journal of Communication Disorders*, 1974, *7*, 65–77.

Paul, G., & Shannon, D. Treatment of anxiety through systematic desensitization in therapy groups. *Journal of Abnormal Psychology*, 1966, *71*, 124–135.

Perkins, W., Rudas, J., Johnson, L., & Michael, W. Discoordination of phonation with articulation and respiration., *Journal of Speech and Hearing Research*, 1976, *19*, 509–522.

Phillips, G. M. Reticence: Pathology of the normal speaker. *Speech Monographs*, 1968, *35*, 39–49.

Prins, D. Pre-therapy adaptation of stuttering and its relation to speech measures of therapy progress. *Journal of Speech and Hearing Research*, 1968, *11*, 740–746.

Prins, D. Improvement and regression in stutterers following short-term intensive therapy. *Journal of Speech and Hearing Disorders*, 1970, *35*, 123–135.

Prins, D., & Lohr, F. Behavioral dimensions of stuttered speech. *Journal of Speech and Hearing Research*, 1972, *15*, 61–71.

Quarrington, B. Cyclical variations in stuttering frequency and some related forms of variation. *Canadian Journal of Psychology*, 1956, *10*, 179–184.

Quarrington, B. The parents of stuttering children: The literature reviewed. *Canadian Psychiatric Association Journal*, 1974, *19*, 103–110.

Quarrington, B. How do the various theories of stuttering facilitate our therapeutic approach? In R. W. Rieber (Ed.), *The problem of stuttering*. New York: Elsevier North-Holland, 1977.

Quarrington, B., & Douglass, E. Audibility avoidance in nonvocalized stutterers. *Journal of Speech and Hearing Disorders*, 1960, *25*, 358–365.

Quist, R. W., & Martin, R. R. The effect of response contingent verbal punishment on stuttering. *Journal of Speech and Hearing Research*, 1967, *10*, 795–800.

Rieber, R. W., Smith, N., & Harris, B. Neuropsychological aspects of stuttering and cluttering. In R. W. Rieber (Ed.), *The neuropsychology of language*. New York: Plenum Press, 1977.

Runyan, C. M., & Adams, M. R. Perceptual study of the speech of "successfully therapeutized" stutterers. *Journal of Fluency Disorders*, 1978, *3*, 25–39.

Sheehan, J. G. *Stuttering: Research and therapy*. New York: Harper and Row, 1970.

Sheehan, J. G., & Costley, M. S. A reexamination of the role of heredity in stuttering. *Journal of Speech and Hearing Disorders*, 1977, *42*, 47–59.

Sheehan, J. G., & Martyn, M. M. Stuttering and its disappearance. *Journal of Speech and Hearing Research*, 1970, *13*, 279–289.

Siegal, G. M. Punishment, stuttering and disfluency. *Journal of Speech and Hearing Research*, 1970, *13*, 677–714.

Siegal, G. M., & Martin, R. R. Punishment of disfluencies in normal speakers. *Journal of Speech and Hearing Research*, 1966, *9*, 208–218.

Siegal, G. M., & Martin, R. R. The effects of verbal stimuli on disfluencies during spontaneous speech. *Journal of Speech and Hearing Research*, 1968, *11*, 358–364.

Soderberg, F. A. Linguistic factors in stuttering. *Journal of Speech and Hearing Research*, 1967, *10*, 801–810.

St. Onge, K. The stuttering syndrome. *Journal of Speech and Hearing Research*, 1963, *6*, 195–197.

Stromsta, C. A spectrographic study of disfluencies labeled as stuttering by parents. *De Therapia Vocis et Loquellae*, 1965, *1*, 317–320.

Stromsta, C. Interaural phase disparity of stutterers and nonstutterers. *Journal of Speech and Hearing Research,* 1972, *15,* 771–780.

Thoresen, C. E., & Mahoney, M. J. *Behavioral self-control*. Toronto: Holt, Rinehart and Winston, 1974.

Timmons, B. A., & Boudreau, J. P. Auditory feedback as a major factor in stuttering. *Journal of Speech and Hearing Disorders,* 1972, *37,* 476–484.

Van Riper, C. *The nature of stuttering*. Englewood Cliffs, N.J.: Prentice-Hall, 1971.

Van Riper, C. *The treatment of stuttering*. Englewood Cliffs, N.J.: Prentice-Hall, 1973.

Watson, D., & Tharp, R. *Self-directed behavior*. Monterey, Calif.: Brooks-Cole, 1972.

Webster, R. L. A behavioral analysis of stuttering: Treatment and theory. In K. S. Calhoun, H. E. Adams, & K. E. Mitchell (Eds.), *Innovative treatment methods in psychopathology*. New York: Wiley, 1974.

Webster, R. L. A few observations on the manipulation of speech response characteristics in stutterers. In R. W. Rieber (Ed.), *The problem of stuttering: Theory and therapy*. New York: Elsevier, 1977.

Webster, R. L., & Lubker, B. B. Interrelationships among fluency producing variables in stuttered speech. *Journal of Speech and Hearing Research,* 1968, *11,* 754–766.

Weiss, D. A. *Cluttering*. Englewood Cliffs, N.J.: Prentice-Hall, 1964.

Williams, D. E., Silverman, F. H., & Kools, J. A. Dysfluency behavior of elementary school stutterers and nonstutterers: Loci of instances of dysfluency. *Journal of Speech and Hearing Research,* 1969, *12,* 308–318.

Wingate, M. E. *Stuttering: Theory and treatment*. New York: Irvington, 1976.

Wingate, M. E. The relationship of theory to therapy in stuttering. In R. W. Rieber (Ed.), *The problem of stuttering: Theory and therapy*. New York: Elsevier, 1977a.

Wingate, M. E. The immediate source of stuttering: An integration of evidence. In R. W. Rieber (Ed.), *The problem of stuttering: Theory and therapy*. New York: Elsevier, 1977b.

Wyatt, G. L. *Language learning and communication disorders in children*. New York: Free Press, 1969.

Zaliouk, D., & Zaliouk, A. Stuttering: A differential approach in diagnosis and therapy. *De Therapia Vocis et Loquellae,* 1965, *1,* 437–441.

Zisk, P. K., & Bialer, I. Speech and language problems in mongolism: A review of the literature. *Journal of Speech and Hearing Disorders,* 1967, *32,* 228–241.

AUTHOR INDEX

Aaronson, A., 37
Abbots, T.B., 316
Abrons, I., 83
Adams, M.R., 302, 309
Adler, S., 268
Agnello, J.G., 337
Alajouanine, T., 151, 153, 169, 171
Albert, M., 168
Alberti, R., 68
Alexander, L.W., 88
Alford, B., 70, 86
Allen, D., 188
Allison, G., 217
Altman, F., 78
Altrows, I.F., 339
American Speech & Hearing Association, 61
Anderson, H., 68–70
Andreason, N., 201
Andrews, G., 309, 311–313, 326, 336
Aran, J.M., 128
Arenberg, I., 86
Arndt, W.B., 6–7
Arnold, G.E., 34, 36–37, 262–263, 271
Aron, M.L., 311–313
Aronson, A.E., 263
Aungst, L., 217, 236
Azzi, A., 286, 292

Bailey, H.A., Jr., 277
Baker, L., 198
Baldwin, A., 158
Ballantyne, J., 277–281
Bankson, W.W., 6, 233
Bangs, J.L., 266–267
Bangs, T., 112
Barelli, P.A., 279
Barr, B., 81, 124, 280, 282
Barry, H., 198
Barton, R.T., 267
Bartosituk, A.K., 127
Bastijens, P., 328
Bateson, M., 185
Battin, R., 106, 127
Bay, E., 153
Beasley, D., 131
Beedle, R., 68
Bekesy, G., 65–66, 80
Belal, A., 58
Bellucii, R., 78
Benfant, B.H., 310
Benton, A., 196
Berg, J., 261
Berg, K., 282
Bergman, M., 89
Bergstrom, L., 82–83
Berlin, C.I., 130
Berry, M.F., 193, 195
Bess, F., 125
Beyn, E.S., 172
Bienenstock, H., 50
Black, J.W., 285
Blank, M., 189
Bloch, P., 37
Bloodstein, O., 302, 308, 320–321, 329, 334, 337–338
Bloom, L., 186–187
Bloomer, H.H., 47, 268
Bluestone, C., 125, 126
Blumel, C.S., 302
Blumstein, S., 169
Bocca, E., 74, 89, 130
Bohne, B.A., 84
Bollen, F., 158
Boone, D.R., 38–40
Bordley, J.E., 81
Boyd, W., 86
Bradford, L., 127
Brain, R., 309
Braine, M., 187
Brayton, E.R., 338
Brazleton, T., 181–185
Brockman, S.J., 278, 288
Bronditz, F., 33–35, 37, 38, 47, 49–50, 260, 270
Bronditz, F.R., 267
Brooks, D., 68–69, 120
Brown, R., 188, 195
Brown, S.F., 338
Brunt, M., 31
Brutten, E.J., 327–328
Burkowski, J.G., 157
Burr, B., 81
Butler, B.R., 339
Butler, R.A., 73
Bzoch, R., 42

Cabanas, R., 310
Calero, C., 74
Callearo, C., 131, 283
Calnan, J., 47, 268
Cansey, G.P., 285
Cantell, D., 199
Caramazza, A., 151, 159–160, 161
Carhart, R., 61, 63, 65, 66, 78, 86, 275, 279, 282, 283, 286
Carlson, L.C., 5, 6, 12
Carpenter, M.D., 201
Carrow, M.A., 195
Catlin, F., 81
Cautela, J.R., 333
Cavagna, C.A., 24
Celano, K.M., 215
Chaika, J.B., 20, 281–283, 289, 291
Chase, R.A., 285
Cherry, C.E., 203
Chomsky, C., 188
Chomsky, N., 195, 212–213, 226, 236, 249
Clarke, B.J., 181–186, 212
Cleeland, C.E., 262
Cody, D., 129, 290
Cole, R.A., 250
Coletti, E.A., 5, 6
Compton, A.J., 213, 242, 244, 246
Condon, N., 187
Conway, N., 83
Cooper, E.B., 217
Cooper, J.C., Jr., 69, 125
Cooper, M., 38
Cooper, W.A., Jr., 285
Cooper, W.E., 249
Corso, J.F., 58
Costello, J., 211, 212, 213, 230, 231, 253, 254
Critchlen, M., 309
Culton, G.L., 164–165
Cushing, H.W., 86

Damaste, P.H., 49
Daneloff, R.G., 229, 257
Darley, F.L., 10, 11, 31, 32, 36, 47, 164, 261
Davis, H., 59, 63, 72, 74, 78, 85, 128, 129, 194, 290
Davis, J., 277, 280, 292
Deal, I.L., 13, 14
Dedo, H., 37, 38
DeHirsch, K., 201
Dempsy, C., 116, 132
Derbyshore, N., 290
DeRenzi, E., 157, 158
DiCarlo, 194
Dickeson, S., 12, 313, 317
Diedrich, W.M., 39–41
Diehl, C.F., 259, 262
Disomoni, F., 338
Dodd, B., 14, 15
Doefler, L.G., 281, 282, 287, 288
Doehring, D., 195
Dorsey, H.A., 216
Douglas, B.L., 267
Douglass, E., 301, 303
Dirks, D., 62
Dix, M. R., 123
Dixon, R.F., 279, 281, 283
Djupesland, G., 68
Downs, M., 117, 127
Dubois, E.M., 252
Dworkin, J., 4

Eagan, J.P., 61, 63
Eagles, E.L., 76
Edwards, M.L., 241
Efran, R., 196
Egolf, D., 335
Eilers, R.E., 105, 241
Eimas, P.D., 104, 240
Eimind, K., 79
Eisenberg, R., 127
Eisenson, J., 173
Elbert, M., 211, 216
Elbrond, O., 68
Emas, P., 84
Erikson, E., 124
Everberg, G., 81

Fairbanks, G., 6, 27, 34, 217, 261, 269
Faircloth, M.A., 251
Farley, J., 276, 278
Feldman, A., 124–126
Fitzpatrick, P., 4
Fletcher, H., 61, 63, 283
Fletcher, S., 5–6, 44–47
Flloyd, S., 314
Flores, P.M., 10
Ford, F.R., 31
Foster, C., 195
Fournier, J.E., 278, 282, 284
Fowler, E.P., 67, 78, 276–278, 280
Framer, A., 309
Frank, S., 185, 188, 199, 201
Fransella, F., 324, 330
Freeman, F.J., 338
Freud, E., 26, 31
Froeschels, E., 37, 260, 270, 271, 307
Fujimura, O., 25
Furth, H., 196

Gacek, R.R., 88
Gaeth, J.H., 88, 113
Gaetzinger, C.P., 61, 286
Galamtos, R., 284
Gardner, R., 189
Gateley, G., 39
Gaymore, E.B., 279
Gazzaniga, M.S., 170–171
Geisler, C.D., 71, 290
Gengel, R., 136
Gerber, A., 211, 218
Geschwind, N., 104, 150, 153, 161, 162, 168
Gibbons, E.W., 285, 291
Glasner, J.P., 304, 312, 314–317
Glasner, P.J., 315
Glassock, M., iii, 82, 87
Glattke, T., 128

Glauber, I.P., 330
Glorig, A., 88, 283–290
Glorig, A., Jr., 84, 85
Goates, W., 48
Godfrey, C.M., 164
Goldenberg, R., 129
Goldman, R., 217
Goldstein, D.P., 62
Goldstein, K., 169
Goldstein, L.D., 316
Goldstein, M.A., 117
Goldstein, R., 71, 74, 118, 129, 194, 275, 289, 290
Goodglass, H., 150–153, 156–157, 159, 160–162
Goodhill, V., 178
Gossib, B., 309
Grabb, W.C., 41
Grare, T.G., 289
Gray, B.B., 212
Green, D., 66
Green, E., 159
Greene, M., 35
Greenwood, D., 61
Gregg, J.D., 82
Gregory, H., 331
Greisen, O., 68
Griffing, T.S., 119
Grinker, R.R., 87
Guess, D., 218
Guilford, F., 173
Guitar, B., 331, 332
Gunderson, T., 77

Hagan, C., 165
Hahn, E., 44
Hall, M., 217
Hall, P., 304
Hallberg, D., 84
Hallpike, C.S., 86
Hamrick, K.S., 218
Hanley, T., 46
Hanna, R., 334
Hansen, G., 217
Hardy, J.B., 123
Hardy, N.J., 291
Harper, H., 68
Harris, D.A., 282–288
Harris, I., 77
Harrison, W., 79
Hart, C., 82
Hattler, K.W., 287–288
Haughton, D., 125
Haynes, D., 127
Head, H., 150, 151, 152, 309
Heaver, L., 37, 270
Hecox, K., 129
Hedge, M.N., 327
Heighton, R.S., 6
Heilman, K., 75
Hejina, R., 330
Heller, M.F., 280, 282, 284
Helm, N., 168
Hemenway, W.G., 81
High, W.G., 64
Hiler, J.A., 80
Hilgard, E., 214
Hirose, H., 49
Hirsch, J.J., 63
Hiruno, M., 23, 33
Hiscock, M., 215
Hixon, T.J., 30, 31, 32
Hodyson, W.R., 123
Holbrook, A., 35, 44
Holdgrafer, G., 213
Holland, A.L., 151, 154, 156, 158, 163, 218
Holliday, A.R., 316
Hollien, H., 24, 26–28, 262
Holm, V.A., 126
Honse, H., 86
Hood, J.D., 61, 62, 66
Hood, W.H., 133, 288
Hopkinson, N.T., 277, 279, 282, 283, 284, 287, 291
Howell, W.H., 127
Howes, D., 160
Hubert, F., 280, 284, 291
Hughson, W., 61
Hulks, J., 76

Ingham, R.J., 312, 325, 326, 331, 340
Ingram, D., 189
Irwin, O., 184, 185
Irwin, O.C., 9, 10
Isshiki, N., 24, 37

Jackson, C., 33, 76
Jacobson, R., 4, 150, 184, 185
Jaffe, F., 306
Jahn, S. K., 76
Jarvis, J.F., 82
Jerger, J., 58, 62, 64, 65, 66, 68, 69, 73, 75, 76, 80, 86, 87, 287, 288
Jerger, J.F., 65, 66, 73, 74, 89, 283
Jerger, S., 125, 126, 131
Jewitt, D., 71
Juers, H.L., 283

Kacker, S.K., 282
Kallenborn, H.F., 268
Kanner, L., 198
Kaplan, E.N., 46
Kaplan, G., 126
Kapp, H.G., 15, 216
Karja, J., 66
Karlin, I.W., 158
Karlovich, R., 285
Kasprisin-Burrelli, A., 334–335
Katz, J., 75, 131, 289
Keith, R.W., 69
Kelly, G.H., 330
Kenin, M., 165–166
Kent, R., 10
Kenyon, E.L., 338
Kerschensteiner, M., 152
Kertesz, H., 167, 171–172
Kiang, N., 129
Kidd, K.K., 337
Kieffner, R., 198
Kien, R.E., 127
Kimura, D., 188, 189, 215
Kimurd, R., 84
Kinsbourne, M., 170, 215
Klien, J., 117
Klotz, R.E., 88, 277

Kobruk, H.G., 366
Kodman, F., Jr., 281, 285
Kohonen, A., 83
Konkee, D., 131
Kopetsky, S.J., 81
Kramer, E., 259
Krashen, S., 189, 190
Kriendler, A., 165, 173
Kronvall, E., 217
Kryter, K., 73, 85
Kupperman, G.L., 72
Kurdziel, S. H., 131
Kussmaul, A., 305

Lafallete, A.C., 316
Lamb, L.E., 285
Langova, J., 305–306
Lankford, S.D., 309
Lanyon, R.I., 331, 332
Lashley, K., 196
Lauder, E., 39, 40
Lawrence, Van L., 268
Lee, L., 195
Lempert, J., 77, 79
Lenneberg, E., 107, 184–187, 193
Lenneberg, G., 215
Leshie, G.H., 281
Levy, J., 170
Lewis, F.C., Jr., 213
Lewis, N., 106, 120, 127
Lewis, R.S., 76
Liden, G., 68, 83, 123
Lieberman, P., 24
Lightfoot, C., 62
Lilly, D., 68
Lim, D., 81, 82
Lindsay, J.R., 78, 81
Ling, D., 117
Linke, C., 28
Lloyd, L.L., 123
Locke, J., 218
Lofgren, R., 50
Lowell, S.E., 88
Luberman, A., 80
Luchsinger, R., 34–35, 37, 48, 49, 305, 320
Ludlow, C.L., 158, 159, 161, 163, 165
Lurid, A.R., 150, 151–153, 164, 169, 171–172
Lynn, G.E., 62, 74–75, 131

Madell, J., 136
Mahrer, M., 75, 199
Manser, R.B., 271
Marcus, L., 190
Marcus, R.E., 84
Markides, H., 15
Martin, F., 275, 279, 291
Martin, F.N., 61, 64
Martin, N.A., 279–280, 291
Maryszemski, M., 164
Mason, R.M., 516
Mattox, D.E., 83–84
Mawson, R.S., 78, 82
McCall, G., 38
McCandless, G., 110, 290
McCarthy, J., 197
McConnell, 7, 110, 135
McCroskey, J.C., 308
McDearmon, J.C., 304
McDonald, E.T., 216
McFarlan, D., 117
McGee, T.M., 83
McGinnis, M., 194, 198
McGlone, R., 28
McGranahan, L., 285
McKenzie, W., 79
McLean, J.E., 213
McNutt, J.C., 12
McReynolds, L.V., 213, 217, 230, 231, 236–237, 238
Meichenbaum, D.H., 333
Mencher, G., 113, 118
Mencher, G.T., 118, 119
Menzel, O.J., 282–283, 286–287
Metraux, R.W., 303
Metz, O., 68, 285
Meurman, O.M., 83
Meyer, D.R., 73
Michael, L.A., 62
Milisen, R.A., 209, 229
Miller, C., 105
Miller, G., 51
Miller, G.H., 283
Miller, M.A., 158
Miller, R., 181, 194
Minyuk, P., 195–196
Miyasaki, T., 44–45, 251
Mokatoff, B., 129
Moll, K.L., 7
Molrk, G.L., 186
Moncur, J.P., 186
Monnin, L.M., 217, 218, 230
Monsee, E., 194, 196, 198
Monsen, R.R., 14, 15
Monyuk, P., 241
Moore, G.P., 23, 33, 35, 38–39, 49, 262
Moore, J., 123
Moore, W.H., 210
Morales, G.C., 75
Morehead, D., 195
Morgenstern, J.P., 312–313
Morley, M., 303
Morris, H.C., 7
Morrissett, L.E., 279
Morse, P.H., 105
Moscovitch, M., 170, 171
Moser, H.M., 262
Moses, P.J., 37
Moskowitz, A.I., 244
Mowrer, D.E., 210
Mullendore, J.M., 31
Murphy, A., 22, 261, 266
Murray, T., 34
Myersen, R., 162–163
Myklebust, H., 193–194
Mysak, E.D., 21, 23, 26–29, 30–32, 35–37, 42–44, 46, 48, 310

Nager, G., 87
Nell, H.B., 69
Nelson, K., 86
Nelson, S., 336
Netsell, R., 23, 31
Neuhaus, E., 195
Newby, H., 79, 286
Nichols, A.C., 4, 215

Nielson, J.M., 169, 170
Niemeyer, W., 70
Nixon, J.C., 88
Nober, E.H., 65, 79, 89, 282, 289
Noffsinger, D., 75
North, F.A., 117
Northern, J.L., 77, 106, 112, 117, 119, 125

O'Conner, N., 94
Oelschluaeger, M.L., 327
Ogara, J.H., 317
Okada, T., 50
Oller, D.K., 213
Olsen, W., 64, 75
Orchuk, D., 120, 125
Osborne, G.S., 47
Ozaki, T., 82–83

Palva, T., 61, 66
Pang-Ching, C., 287
Paparella, M., 76
Paradise, J., 69, 125
Patter, P.K., 15
Patty, J., 327
Paul, G., 309
Payne, E.E., 76
Peck, J.E., 286
Penfield, W., 104, 169
Perkins, W.H., 22, 30, 31, 261, 271, 339
Peterson, J.L., 287
Philips, B.J.W., 7, 144
Phillips, G.M., 308
Pickett, J.M., 250
Pisinoni, D.B., 218
Pitman, L.K., 284
Poale, G., 214, 227
Pollack, G., 213
Porch, B.E., 156–158
Prather, E.N., 214
Prins, D., 236, 328, 331
Procter, D.F., 82
Prosek, R., 35
Ptacek, P., 25, 27–28
Pulec, J.L., 85, 86

Quarrington, B., 305, 316, 330, 333
Quick, C.A., 82
Quist, R.W., 305

Rahn, H., 72
Rainville, M.J., 62
Ramshell, D.A., 278
Randall, R., 47
Rapin, I., 1, 195, 285
Rathbone, J.S., 6
Rees, M., 4, 262
Rees, N., 196
Resnick, D.M., 287
Rieber, R.W., 66, 305
Riev, G., 217
Riesen, A.H., 106
Rigrodsky, G., 39
Ringel, R.L., 12
Rintlemann, W., 278, 283, 287–288

Sabirana, A., 171
Saffran, E.M., 153, 164
Saltzman, M., 280, 284
Sanchez-Longo, 74
Sander, E.K., 214
Sanders, D.Y., 83
Sanders, J.W., 66
Sands, I., 164
Sarno, M.T., 154, 168
Satwoff, J., 88
Saunders, W.H., 82
Schein, J.D., 81
Scherrl, V., 113
Schmanuel, V.A., 252
Schuell, H., 150–153, 156, 164, 173
Schuknecht, H., 73, 76, 80, 82, 87, 88
Schulman, C., 127
Schulman-Galambas, 129
Schwartz, D., 137
Seiderman, M., 126
Sellars, I., 22
Selters, W.H., 73
Semenou, H., 282
Shambaugh, D.E., 71–81
Shapiro, T., 190, 199
Sharp, H.S., 76
Sheehan, J.G., 311, 316–336
Sheehy, J., 84, 86
Shelton, R.L., 8, 45, 233
Shelton, R.L., Jr., 8
Shepherd, D.C., 282
Sherman, D., 217
Shimizu, H., 73
Shipley, E., 187
Shipp, T., 4, 24, 25
Shprintzen, R.J., 25, 44–45
Shurin, P.H., 69
Sieyel, G.M., 327
Silverman, G., 49
Simmons, F.B., 118
Skelly, M., 530
Skinner, P., 128
Skolnick, M., 44
Sloan, R.F., 215
Slobin, D.I., 244
Slorock, N., 215
Smith, B.B., 74
Smith, F., 195
Smith-Frable, M., 49
Snidecor, J.C., 6, 39, 41
Snow, C.E., 186, 227
Snow, K., 6, 210–211, 237
Soderberg, F.A., 337
Sommers, R.K., 215, 217
Sparks, R., 75, 168
Speaks, C., 132
Spika, B., 285
Spoor, A., 88
Spreen, D., 156–157, 159
Sprusterbach, D.C., 42, 231, 236
Stansell, B., 6
Stark, J., 195, 287
Stein, L., 287
Stern, D., 185
Stewart, J.P., 81–82
Strayer, L.C., 215
St. Onge, K., 320
Street, B., 158
Strome, M., 80–81
Stromsta, C., 30
Strudebaker, D., 61
Subtenny, J.B., 5, 8

Surapathan, L., 82
Swisher, L.P., 158

Tallal, P., 196
Tapp, S., 116
Tato, J.M., 77
Taub, S., 41
Taylor, D., 103
Taylor, G., 284, 286
Taylor, G.B., 133
Teixeira, L.A., 13
Templin, M.C., 214–217, 227, 241
Terkildsen, K., 68
Terrm, M., 158
Thomsen, K.A., 285
Thoresen, C.E., 333
Thorne, B., 280
Tiffany, W.R., 285
Tillman, T.W., 62, 65, 283
Timmons, B.A., 339
Tonndorf, J., 61, 85
Torok, N., 86
Travis, L., 217
Truex, E.H., Jr., 279
Tucker, G., 199
Tyack, D., 196

Unger, T.K.J., 216

Van Disholck, H.A., 80
Van Hattum, R.J., 212
Van Riper, C., 35, 225, 232, 300, 302, 304, 306, 308, 311, 312, 314, 315, 320–323, 324, 328, 329, 331, 333, 335, 339
Van Wagoner, R.S., 70
Varpella, E., 83
Verbruggle, R.R., 251
Veutry, I.M., 282, 283, 287
von Leden, H., 33–34
Vuori, M., 81

Waldrop, W.F., 39
Walsh, T.E., 78
Walski, W., 267
Walton, D., 267
Walton, W.K., 69
Ward, M.M., 5
Ward, W.D., 61, 62
Warren, R.M., 249
Warrington, E.K., 153
Watson, D., 333
Watson, J.E., 286, 288
Watson, L.A., 286
Weaver, M., 87
Webster, A., 68
Webster, R.L., 327, 339
Wedenberg, E., 284
Weigel, E., 172
Weir, R., 186
Weir, R.H., 228
Weisenberg, T.H., 150–151
Weiss, D.A., 306, 321
Weiss, H., 280
Wellenford, J., 116
Wellman, B., 216
Wendler, J., 49
Wepman, J.M., 150, 151, 152, 153, 171, 173, 194, 217
Werner, H., 186, 188
West, R.W., 32, 47, 266, 268–269
Westlake, H., 44
Wever, E.G., 77
Whitehead, R., 49
Whitehouse, P., 159
Wier, J.A., 82–83
Wilbar, L., 68
Williams, D.E., 337
Williams, G.C., 218
Williams, G.S., 238, 240
Williamson, A.B., 263
Williamson, D., 276
Wilmont, T., 80
Wilson, D.K., 35
Wingate, M.E., 300, 302, 324, 328, 334, 336, 338
Winitz, H., 3, 218, 228, 229, 231, 232, 236, 238, 241, 242, 248, 250, 252, 253, 254
Wood, N.E., 193, 194
Wright, V., 231, 252
Wyatt, G.L., 314

Yoss, K.A., 11, 12

Zaidel, E., 170
Zalionk, A., 261
Zalionk, D., 320
Zaner, A., 124, 136
Zangwell, O., 198
Zariff, E.B., 152, 160
Zemlia, W., 23
Zisk, P.K., 310
Zwislocki, J., 68
Zwitman, P., 41

SUBJECT INDEX

Abnormal resonance, 268–271
  assimilation nasality, 269
  hyperrhinolalia aperta, 268
  hyperrhinolalia clausa, 269
  hyporhinolalia, 269
  supraglottal cavities, 268
  therapy, 269, 270
    for hyperfunction, 270
    for hypofunction, 271
Adult apraxia, 11–13
  dichotic listening, 13
  production, 13
Anatomical articulation disorders, 1–8
  clusters, 4
    terminal, 4
  consonants, 4
    terminal, 4
  emphysema, 4
  endocrine function, 48–50
  esophageal speech, 39–41
    and speakers, 4
  female, deficits, 27
  infraglottal origin, 30–32
  laryngeal valve, 23, 24
  larynx, 1
  lungs, 1
  male, deficits, 25–27
  palatal anomalies, 7–8
  subglottal pressure, 41–48
  teeth and jaws, 6–7
  tongue, 4–6
  velopharyngeal valve, 23–25
  vowels, 4
Aphasia, adult, 149–167, 169–173
  anomia, 152
Aphasia, adult (*cont.*)
  Broca's, 151
  conduction, 153
  neurolinguistic studies, 158–163
  oral production, 165
  reception, auditory, 165
  recovery, 163–167
  syndromes, 150
  tests for, 154–158
  theories, 169–173
  transcortical sensory, 153
  Wernicke's, 151
Apraxia, 3, 11–13
  childhood, 11
  dichotic listening, 13
  gross motor abilities, 12
  left hemisphere, 12
  production, 13
  sensory-motor deficits, 12
Articulation, 1–11, 13–15, 209–233
  and adult apraxia, 13–14
  anatomical, 1–8
  and childhood apraxia, 13–14
  and deaf and hard of hearing, 14–15
  learning, 209–233
  neurological, 8–11
Audiology, 57–75
  acoustic impedance, 68–69
  audiometer calibration, 59
  audiometry, 63, 65, 281–291
  and auditory knowledge, 236–247
    threshold, 60–66
  central auditory impairment, 55
    tests, 73–75
  conductive mechanism, 58

Audiology (*cont.*)
disorders of, 75–79
deafness, 59
and articulation, 14
hereditary, 79–80
sudden, 80
diagnosis, 57–58
dysacusis, 59
etiology of, 80–82
hearing loss, 58
ISO standard, 60
peripheral disorders, 58
speech audiometry, 63–65
sensorineural, 58
tests, 62, 66, 67, 75, 131
tumors, 72
eighth nerve, 86
Audiometry, 63, 65, 281, 282, 288–290
air conduction, 282
ascending vs. descending, 282
audiogram, 282
and audiology, 63, 65
bone conduction, 282
conditioned play, 124
electrodermal, 289
electrophysiologic, 288
looked response, 290
and pseudoneural hypoacusis, 281–282, 288–290
otological techniques, 281
pure tone audiometry, 282
speech reception, 282
Auditory disorders and articulation, 236–239
central processing tests, 73–75
and discrimination, 244–245
knowledge, 236, 245
and articulation control, 247
normal development, 240–241
pretraining, 240

Childhood apraxia, 11–12
gross motor abilities, 12
left hemisphere, 12
sensory motor, 12
Childhood deafness and hearing, 101–139
age of onset, 107
amplification, 136
assessment methods, 122–133
Childhood deafness and hearing (*cont.*)
audiometry, conditioned play, 124
electrophysiologic, 125
auditory association, 130
central processing, 137
classification, 133
educational, 133, 135
etiology, 113
identification, 117
early management, 138
medical, 133, 134
moderate deficits, 113
normal auditory development, 103–106
psychological, 116, 132, 133
screening, 117–119
and middle ear function, 120
sensory deficits, 106, 107, 110
severe deficits, 110–113
special schooling for, 102
tests, central function auditory, 130–131
tympanometry, 126
visual input supplements, 111
Conductive mechanism, disorders of, 75–79
atresia, 77
cholesteatoma, 77
mastoiditis, 77
otitis media, 75
etiology of, 76
otosclerosis, 77–79

Deafness and hearing loss, 42, 79–88
acoustic trauma, 84
aging loss (presbycusis), 88
childhood, 101–137
and eighth-nerve tumors, 86
etiology of, 42, 80, 81
hereditary, 79–80
Meniere's disease, 86
noise-induced hearing loss, 84, 85
ototoxic agents, 82, 83
presbycusis, 88
sudden onset, 80
vascular disorders, 83
Dysarthria, 9–11
acquired, 10–11
and articulation, 10

Dysarthria (*cont.*)
congenital, 9–10
x-ray techniques with athetoid speakers, 10
Knowledge, auditory, 233–235, 247, 253
categorical perceptions, 234
conversational speech, 248
duration, 249
errors, number of, 250
and motor theory, 235
phonation, context, 252
semantic factors, 249
syntactic factors, 249
redundancy, 251
stimulus-imitation method, 248
tests, with machinery, 233, 234

Language disorders of childhood and adolescence, 179–201
atypical psychosis, 192
autism, 198–201
developmental aphasia, 193
manic-depressive, 192
mother tongue, 186–188
normal development, 180–189
receptive, expressive, 194
retarded, 194
schizophrenia, 179, 192

Learning and articulation, 209–225
auditory perception (discrimination) and articulation, 217–219
dichotic listening, 215
early diagnosis of articulation deficit, 217
experimental designs, 217
interpersonal therapy, 212
intrapersonal therapy, 212
maturation and articulation, 214
operant conditioning, 212
and phonology disorders, 213
emotional, 219
lack of tests, 215
and self-esteem, 220
training, 218
two-factor theory, 210

Neurological disorders of articulation, 8–14

Operant conditioning, 212
Otological techniques, 281

Phonology, 259–272
abnormal resonance, 268–279
evaluation, 260
hyperfunction, 262–265
hyperkinetic, 260
hypofunction, 265–267
hypokinetic, 260
knowledge, 249, 252
and learning, 213, 215, 219
loudness, 261
nosology, 260
pitch, 261
quality, 261
therapy, 269–271
Pseudoneural hypoacusis and malingering, 278–292
audiological techniques, 281
audiometry, 281–282, 288–290
auropalpebral reflex, 284
delayed auditory feedback, 285
discrimination tests, 283
Doerfler–Stewart test, 286
Hood's masking test, 287
identification, 279
inconsistency, 280
lateralization, 285
Lombard voice reflex, 284
malingering, 279, 280
masking, 286
otological techniques, 181
psychological techniques, 280
sensorineural acuity level, 288
Stenger test, 286
story tests, 283
tone-in-noise test, 287

Schizophrenia, 179, 192
Stuttering, 299–341
adult treatment, 324–326
and aphasia, 309
in children, treatment, 334–336
chlorpromazine, 306
cluttering, 305
communication apprehension, 308
definition, 300
developmental, 302
theories about, 318, 319

Stuttering (*cont.*)
and Down's syndrome, 310
dysfluencies of childhood, 303, 304
early diagnosis of, 303–304
etiology, 336–339
family history, 313
incidence, 312
neurological deficit, 306
onset, 314
operant conditioning, 326–327
parental communication, 316–317
psychology, 307
psychotherapeutic, 330
research, 330
retardation, 310
sex ratio, 313

Tests
acoustic impedance, 68
air conduction, 60
alternate binaural loudness balance, 67
aphasia, 154–158
audiometry, 63, 65, 281, 282, 288–290
auditory knowledge and machine tests, 233–236
bone conduction, 61
central auditory processing, 73
central function auditory, 130–131
competing message, 75
dichotic nonsense syllable, 75
Doefler–Stewart test, 286
dye contact, 87
Hood's masking test, 286
lack of tests for phonology, 215
lesion, 64
other tests, 62

Tests (*cont.*)
Rainville, 62
short increment sensitivity index, 67
staggered spondaic word, 75, 131
Stenger test, 287
synthetic sentence identification, 75
threshold and suprathreshold, 63
tone decay, 66
tone-in-noise test, 287
vestibular, 87
x-ray polytomography, 87
Treatment, 35–38, 44–48, 212, 213, 217, 218, 225, 226–230, 231–233, 269, 270, 271, 272, 324–327, 334–336
abnormal resonance, 269, 270
articulation, errors, 232, 233
training, 218
auditory, 240
input, 228
carry-over, 232
conversational speech, 213
and articulation errors, 232, 233
experimental designs, 217
explicit training, 225
glottal, management, 35, 36, 38
therapy, 40
causal therapy, 44
symptom therapy, 44–48
hyperfunction, 270
hypofuncton, 271
interpersonal, 212
intrapersonal, 212
isolation, 229
movement, 231
operant conditioning, 212, 326–327
phonatory disorders, 269–272
phonetic feature training, 230
stimulation, 229